ON INTELLIGENCE

On Intelligence

A Bioecological Treatise on Intellectual Development

Expanded Edition

STEPHEN J. CECI

Harvard University Press

Cambridge, Massachusetts
London, England
1996

First Harvard University Press paperback edition, 1996

Library of Congress Cataloging-in-Publication Data
Ceci, Stephen J.
  On intelligence: a bioecological treatise on intellectual development
    / Stephen J. Ceci. — Expanded ed.
      p.   cm.
    Expanded ed. of: On intelligence—more or less, © 1990.
    Includes bibliographical references and index.
    ISBN 0-674-63456-X
    1. Intellect.   2. Intelligence levels.   3. Nature and nurture.
I. Ceci, Stephen J.   On intelligence—more or less.   II. Title.
BF431.C36   1996   96-23821
153.9—dc20   CIP

This book is dedicated to my daughter:

Nicole Genevieve Rossiter Ceci

Contents

CHAPTER 3

Mismatches Between Intelligent Peformance and IQ

29

CHAPTER 4

A Social-Organizational Analysis of Intellectual Development

45

CHAPTER 5

The Impact of Schooling on Intelligence

69

PART TWO

THE BIOECOLOGICAL FRAMEWORK   91

CHAPTER 6

The Role of Context in Shaping
Multiple Intelligences   93

CHAPTER 10

Taking Stock of the Options   193

Epilogue   221

Endnotes   253

References   263

Subject Index   285

Preface, 1996

The species *Homo sapiens* has a marvelous, if not unique, capacity to be creative, or at least many of its members do: some see novelty where others see only the hackneyed; some detect patterns where others see only randomness; and some bear witness to chaos where others unearth predictability. Such subjective constructions are the essence of art, but do they also occur in science? Do scientists see vastly different things when they look at the same constellation of data? Or do the scientific canons of evidence constrain the use of data, ineluctably leading to a single interpretation?

Scientists may eventually come to a consensual interpretation about a pattern of data, but until and unless they do there can be vast differences of opinion. Nowhere is this more evident than in the evaluation of the data that have served as the cornerstone of both this book and a recent spate of books written by a group of social scientists my colleague Urie Bronfenbrenner and I have called the "New Interpreters." This group of scholars (Richard Herrnstein and Charles Murray; Seymour Itzkoff; Philippe Rushton; Daniel Seligman) has peered into many of the same journals and analyzed the same data that formed the foundation of the argument put forward in this book, and yet they have reached almost the opposite interpretation from the one proffered in these pages.

Herrnstein and Murray's book *The Bell Curve* is the latest, albeit most prodigious, restatement of the genetic-meritocratic syllogism I attacked in the first edition of *On Intelligence*. Like other New Interpreters, their central thesis is that individuals and groups differ from each other mainly because of their genes for intelligence, and that these intellectual differences ultimately determine the hierarchical social status of individuals and groups in the larger society. This proposition was first presented more than a century ago by Francis Galton in a series of publications between 1874 and 1892. What is intriguing about its reaffirmation by the New Interpreters, however, is that they draw support for their thesis from some of the very data that I have used to reach the opposite conclusion in this book. What is going on here? Are the data so ambiguous that they can be promiscuously recruited in the service of any thesis? Have massive new data analyses become available since I wrote the first edition of this book, data that refute its claims? I do not think so, and thus I have appended an Epilogue to this edition in which I will explain why the genetic-meritocratic syllogism was wrong prior to the current spate of books by New Interpreters, and why it is still wrong.

Preface to the Original Edition

This book has been a long time coming. It began over five years ago as a short empirical paper that was to focus on several experiments dealing with the influence of context on thinking and reasoning. In these studies, my colleagues, graduate students, and I consistently found rather substantial changes in the way people solved problems as a function of their familiarity with the type of problem, its concreteness, and their level of motivation to solve it. It wasn't only that problem-solving performance could be elevated or lowered, depending on the context in which the task was performed (though that is an interesting finding in its own right), but sometimes individuals in these studies would behave in an exceedingly complex manner that was totally unpredictable from what we knew about them, including their IQ scores. Insofar as the mental complexity these subjects exhibited in solving these problems depended on aspects of the context as much if not more so than it did on their IQs, I began to question the interpretation of the latter as an indicant of disembedded, acontextual problem-solving aptitude.

Some follow-up work by my colleagues and me suggested that the nature of context was quite broad and included not just the physical, historical, and social addresses at which cognition unfolded, but also the relevant background knowledge that individuals bring with them to the testing situation and the attainment value they attach to successful performance. I thought it might be of interest to write about some of these demonstrations because of a pervasive disdain among some of my colleagues in cognitive psychology for "contextualism." To be sure, many psychologists do acknowledge the importance of context, especially those researchers in social, personality, and developmental psychology, but even here the meaning of contextualism is rather narrow and refers merely to environmental elicitors of behavioral responses, and not to the fundamentally symbiotic nature of context and cognition. (See Chapter 6 for a distinction between contextualism, situationalism, and interactionism). But even this watered-down version of contextualism has never been fully appreciated by those of us who conduct cognitive, bahavior-genetic, and psychometric research. All too often we adminster batteries of standardized tests or well-fashioned laboratory tasks to all sorts of individuals, from all sorts of backgrounds, and slavishly try to expunge all aspects of context from playing a role in their performance, unless they are so pervasive and well understood that they can be statistically manipulated. Pointing out "context effects" on task performance to research- ers from this tradition is akin to commenting on the emperor's sparse ward- robe.

In the search for cognitive universals, the emphasis has usually not been on contextual influences that may, collectively, throw a spanner into the outcome by altering the factor structure of the sequence of processing steps. This is because most of us were trained to view context-specific test variance as "noise," to be statisically controlled or, better yet, eliminated completely, rather than built into our models of the way people think. Susan Whitely, one of the most astute thinkers in the area of cognitive testing, referred to theories that guide test construction in the following way: "A good theory should contain variables that are general across tasks. Thus, if across-task generalizability is not shown for individual differences measurements on the theoretical variables, the quality of the theory can be questioned" (Whitely, 1983, p. 195). But such a research strategy takes no account of the difference between sources of random error (noise) and sources of systematic differenti- ation in the manifestation and meaning of cognitive performances in differ- ent contexts. A performance is assumed to have the same meaning across contexts. The very idea that individual differences are moderated in a funda- mental manner by aspects of the context seems inimical to much of what modern psychology has striven for.

William Marks (1982), in his historical analysis of the concept of intelli- gence, commented that early psychology tended to ignore individual differ- ences in thinking until James McKeen Cattell bemoaned this fact around the turn of the century and spurred on research into individual differences in earnest. But even the "individual differences" approach itself has come to mean a standardized view of performance, assessed in a sanitized context. The possibility that there exists a more restless relationship between intelli- gence and context, in which thinking changes both its nature and course as one moves from one situation to another, is enough to cause shudders in some research quarters. It represents a move toward a psychology of situa- tions, and we cognitive psychologists still know relatively little about that! In fact, were psychologists to admit that something as fundamental as one's intelligence was so labile as to vary dramatically depending on when and where it was used in the past and the present, this would be the first step down the slippery slope to admitting that our current psychometric arma- mentarium has captured only a snapshot of the organism, not its full range of capabilities. We like to think that our models of human behavior are dynamic, not composites of a few glimpses, filmed in a few contexts. (In a later chapter I shall argue that, contrary to George Herbert Meade's assertion, the problem is not the result of our use of quantitative models *per se*, but rather our reluctance to insert ecological complexity into these models.)

Many researchers of individual differences are aware of this danger, I think. They do not go around saying that their statistical models have captured all or even most of what is important in predicting academic performance, occupational success, or life satisfaction. And they have taken a beating in recent years by the "press," sometimes unfairly. But it would be a mistake to think that the public and the press are completely wrong about the misuses of IQ tests and those who promulgate them. As I shall show, the promulgators of IQ tests and other indicants of so-called general intelligence really *do* believe that these measures tell us something important about the cognitive complexity of those who have been assessed with them and, more importantly, about the neuroanatomical substrate that is presumed to underpin this cognitive complexity. That is, a large number of researchers believe that their statistical models go beyond simple prediction; they believe they index an underlying construct called "intelligence," often thought to be biologically fixed in its unfolding, and to be the basis for success (as opposed to simply being a correlate of it) in a wide range of real-world endeavors. Depending on the theorist and the statistical approach taken, the documentation of context effects on thinking and reasoning performance provokes either a dismissal of the finding as unimportant for the purposes for which tests of intelligence were designed (prediction of academic outcomes), or unimportant for theories of intelligence because they fall outside some of the assumptions that underlie traditional psychometrics.

Concerning the first objection, it may be that the insertion of context into the testing format will be shown to add to the predictiveness of traditional tests, at least for some types of criteria. It is an empirical question, and I shall consider it later in the book. Concerning the latter objection, some of the assumptions that underpin psychometric theories of intelligence have been held by researchers for such a long time that they have begun to resemble articles of faith, to be challenged only by the heretical. Some of these assumptions have given rise to some of the most stubborn tautologies in the field of intelligence research today. For instance, if a new test of intelligence is to gain acceptance in many quarters, it must conform to traditional assumptions about a normal or Gaussian distribution of scores. (This is the perfectly symmetrical "bell-shaped" curve that begins to taper off dramatically from its center peak as it moves gradually toward its tails.) The assumption behind this Gaussian curve is that whenever a large number of independent factors are involved in producing an effect, their joint influence frequently yields such a distribution, with many products at the center, and then progressively fewer as you move away from the center toward the tails. Geneticists have documented such (polygenic) effects for many physical traits, such as height, with most individuals falling around the center or average of the height distribution and progressively fewer persons located at the extremes of tallness or shortness. Because many of the early researchers in the field of intelligence were hereditarians, believing intelligence to be largely a genetically determined and transmitted trait, it was only logical that a test of intelligence should yield an approximately bell-shaped distribution. Because of this assumption, tests that did not do so were systematically refashioned, their difficulty revised, and their content reconfigured until they did yield a bell-shaped distribution. And these tests were put to one other test: They were expected to correlate highly with previous tests that also had yielded a bell-shaped distribution of scores! So *if* the insertion of context into our notions of intelligence were to result in a non-Gaussian distribution of IQ scores, then some would take this as evidence of the weakness of such notions of intelligence rather than as a sign that the assumption itself deserved scrutiny.

There are many such assumptions associated with traditional theories of intelligence that one bucks up against whenever something is proposed by someone who, like myself, is outside the psychometric tradition. For example, in Chapter 8 the reader will encounter in some detail the notion of a *positive manifold*, that is, the assumption that performances on all intellectual tasks should be moderately intercorrelated. If a battery of tests does not fit this pattern, some see this not as evidence that the positive manifold assumption should be challenged, but as grounds for dismissal of the new battery of tests. So in tackling the problem of the nature of intelligence, I soon realized that, because of the rich history of research in this area, I was not free to speculate at will. Many before me have given thought to such questions, and the current IQ-general intelligence edifice is built upon their findings and assumptions. It would be necessary for me, if I were to make headway in this venture, to make explicit the assumptions of these earlier researchers while at the same time trying to unravel them.

Once I began to discuss with colleagues how context effects forced me to rethink some traditional assumptions concerning the nature of intelligence, it became clear that something more was warranted than a short empirical article that described the experiments my colleagues and I had done. Each discussion I had with intelligence researchers raised new issues: "Aren't you simply showing that some people have rather isolated pockets of talent, and if you catch them in a highly specific context, they will perform well; otherwise they are quite dull? If so, then this is not to be confused with being intelligent, because these people lack the generality that term implies!" Or "Do you mean to suggest that those persons who perform well in some contexts can do so in all contexts, given similar types of problems to solve, provided they receive training? Isn't this wishful egalitarian thinking in the guise of science?" And "What theory could account for results such as yours without being suspiciously *post hoc*?" And finally, "These data fly in the face of research demonstrating intellectual consistency across tasks. You need to explain your *anomalous* findings." So it went, until finally I decided to try to answer some of these questions in a middle-length paper, say, 70–100 pages long. After a year or so, I had finished drafting such a monograph and circulated it among a group of nationally visible intelligence researchers. I quickly discovered it raised almost as many questions as it answered, convincing me that a book-length treatment of the topic was necessary.

I confess that I fought this realization for a while. I have this negative feeling about book-length treatments of scientific topics: Very few people read an entire book, except the afficionados in an area, and I wanted to reach a broader audience than just the IQ afficionados because I was proposing that the nature of intelligence cannot be determined solely from psychological research on IQ-like constructs. Hence, I wanted a more diverse "jury" to judge the merits of my arguments!

Once I decided to expand the size of the project, the writing quickly got out of control. The typescript soon had expanded from 100 pages to over 600 pages. And it could have easily doubled again, as I drafted additional sections to be inserted. I had lost sight of my original purpose and at some point apparently opted to cover, in addition to my own bioecological framework, many aspects of *intelligence* that might be of interest to someone teaching an advanced course on that topic. The result was a mishmash, with parts of the book written at the level of a research monograph, parts written at the level of an advanced textbook for a course on intellectual development, and parts written for an intelligent "lay" audience. In the last analysis, it fell to friends to point out sections that could be deleted without sacrificing clarity or derailing the description of my own bioecological framework. The main sacrifice has been that this monograph is not a comprehensive treatment of all aspects of intelligence that might be covered in a textbook for a course. But it was never my purpose to write a survey text of the field of intelligence research. Rather, this monograph is aimed at problems I have identified with existing theories of intelligence. It is a critique of what I have come to regard as the conceptual myopia that has influenced the way researchers think about intelligence and its development. In this critique I attempt to provide a new framework and evaluate it by criss-crossing empirical findings with logical analyses. Readers unacquainted with developmental psychology, cognitive psychology, behavior genetics, and psychometrics may occasionally want to skip over some of the technical arguments because I usually have not provided the requisite background information to understand them fully. This is another sacrifice of writing a monograph instead of a survey text.

In an effort to draw on the nonpsychological literature to inductively develop a framework for thinking about individual differences in intellectual development, I have been forced to exercise some degree of exclusiveness, for the most part neglecting the nonempirically based analyses of anthropologists, historians, and philosophers, though, as will be seen in Chapters 3 and

4, many exceptions seem to have penetrated this protective mesh. As I said, I wanted to keep this monograph short for reasons having to do with cost and a belief that something short is more likely to be read than something long. There are many instances throughout these pages where I have stopped short of providing a complete description of empirical findings or of providing a complete listing of relevant citations, or of not extending an argument to its ultimate conclusion because of an assumption that the reader can do this unaided. In addition to these shortcomings, there is undoubtedly another cost in adhering to this plan: The force or direction of particular arguments may occasionally suffer because of the abbreviated nature of the rationale or description of the methodology. The reader will have to judge how serious a problem this is.

Acknowledgments

No project like this is complete without some acknowledgment of the many sources of assistance given to the author. I received help from many quarters, and I want to thank as many of these persons as I can remember for their generosity. Two of my former graduate students were especially helpful to me. Jacquelyn G. Baker and Narina N. Nightingale provided numerous critiques of my writing, word choice, and thinking. Although their comments were often brutal, the end product is better because of them.

Many colleagues have influenced my thinking through letters, journal reviews and commentaries, and colloquia repartee. Even an abbreviated list of their names and suggestions would unduly stretch the length of this section. But several were especially important. Their comprehensive letters and phone calls beame a sort of seminar for me to sharpen my ideas. Bob Sternberg, Sylvia Scribner, Doug Detterman, Robin Barr, Pat Worden, Howard Gardner, Barbara Rogoff, Robbie Case, Sandy Scarr, Maggie Bruck, James Flynn, Doug Herrman, Glen Elder, Elizabeth Loftus, and Jonathan Baron were, at one time or another, unwitting participants in this "seminar." On occasion, all of them provided detailed feedback on a draft of one or more chapters of this book. Bob Sternberg has been particularly generous and unsparing in his criticism of my ideas. I have tried to reciprocate with my treatment of his own "triarchic theory" in Chapter 10. In addition, Maggie Bruck and Mike Howe provided detailed comments on various sections of the manuscript. Their help was extremely important to me. Besides detecting conceptual inconsistencies and sloppy language, they made many important suggestions about the book's organization and pointed out numerous instances of arguments in need of repair. Finally, the monograph was completed while I was a visiting professor at the University of North Carolina at Chapel Hill's Frank Porter Graham Research Center. Associates there, particularly Craig Ramey, Glen Elder, Bob Cairns, Earl Schafer, Peter Ornstein, and its director, Sharon Landesman Ramey, provided numerous suggestions that I am indebted for.

I work at a rather magical institution, Cornell. It has a sense of itself that is uncharacteristic of other institutions I know. Cornell has been a home to contextualism from the very beginnings of psychology in America, from Frank Angell to Urie Bronfenbrenner—with Freeman, Neisser, and the Gibsons in between. During my tenure there, several colleagues have been especially influential in my thinking. John Doris, who first introduced me to this field of inquiry through his own writings, has been unstinting in his feedback and guidance throughout the project. The title of the book is only one of his many contributions to it. My friend and colleague Steve Cornelius has been a constant and reliable source of feedback who, over the years, has exposed me to new ideas and standards that have changed the way I think about some of the issues covered in this book. John Harding made me aware of Haddon's *Cambridge Anthropological Expedition to the Torres Straits* in 1898 to gather anthropometric data on its inhabitants. Dick Darlington first raised the possibility of assessing the abstractness of IQ items in the manner I report, and it was Dick who also brought to my attention the interesting results regarding Caucasians outside North American and Europe. Daryl Bem, Steve Robertson, Henry Ricciuti, and Frank Keil have from time to time passed on comments and suggestions that have found their way into the monograph. Finally, Ulric Neisser and Urie Bronfenbrenner, each in their own way, have helped shape and sharpen my ideas, and I want to take this opportunity to thank them for their constant advice and criticism. It was Urie Bronfenbrenner who educated me about the roots of Vygotsky's notion of sociohistorical development in W. H. R. Rivers' "psychological ethnology," and it was Ulric Neisser who challenged me not to construct mental models for phenomena that have no real-world counterpart. I feel especially privileged to call these two men not just my intellectual mentors and professional role models, but also my friends. On numerous occasions, *all* of my Cornell colleagues have given me lessons in psychology (and in humanity). My visions often have taken form in the light of their reflected wisdom. If there are instances in this book when I have been farsighted, it is because the horizon is more visible from atop the backs of giants. This book is for all of them.

It is customary to thank one's children whenever a project of this duration has been completed, to acknowledge their many hours of functioning without dad because he was in his study pecking at the word processor. I will not do that here because I do not believe my daughter, Nicole Genevieve, was made to sacrifice my presence due to my involvement with this project. In fact, she was (and, to the best of my knowledge, still is) unaware that I have have been writing a book for the past five years. I worked on it by day and, clandestinely, in the wee hours of the night and morning, when she was asleep. Had I wished it to be different, the book might have been finished years ago, by working more evenings and weekends. That I did not wish it to be different attests to her importance to me. I have dedicated this book to her.

ON INTELLIGENCE

PART ONE

SOCIAL CONTEXT AND INTELLECTUAL DEVELOPMENT

In the first part of the treatise I shall try to convince the reader that something is amiss in the way we view intelligence, particularly in the way we imagine it to develop. I shall review and critique a variety of findings that call into question static views of intelligence and suggest that an empirically adequate account of the development of intelligence must go beyond present psychometric, information processing, and genetic presentations. It must also go beyond (well beyond, in fact) current so-called cultural accounts. I address the importance of various social contexts (schooling, cultural values, and social organization) on intellectual development. Before that, however, I present some basic concepts and data about intellectual development that derive from research in cognitive science, as these will be picked up in Part II when the bioecological framework is described in more detail.

CHAPTER 1

Why a Treatise on Intelligence?

This is a treatise on *intelligence*. Specifically, it is an argument about its modular nature and its contextual determinants. In these pages, the reader will be taken on an expedition through the literature of diverse disciplines, decades, and continents. Some of this literature may be unknown even to those in the intelligence research community. This is because I have attempted to enlarge the traditional psychometric treatment of individual differences in intelligence by incorporating developmental, cognitive, and social psychological research into the equation. In addition, I have tried to incorporate the work of scholars from fields outside of psychology, such as education, anthropology, sociology, and genetics.

Throughout this expedition, an attempt will be made to identify and chart the development of the "wellsprings" of mental performance and suggest how they build upon relevant experiences. I shall critique the existing literature on intelligence, put forward a new view, called the bioecological framework for understanding individual differences in intellectual development, sketch its components, suggest how these components might interact, and assess the adequacy of this framework in terms of its usefulness in explaining findings that either are ignored or mishandled by existing theoretical frameworks. I have termed this enterprise *theory discovery* (as opposed to *theory confirmation*), because I regard it as a "first cut," lacking some of the formal specification needed for theoretical confirmation or disconfirmation. This expedition is squarely ensconced in the discovery mode rather than in the theory-confirmation mode. Its aim is not to test hypotheses, but to generate them and, in so doing, provide structure and direction for a new approach to biology—ecology interactionism. Later I describe this new approach in detail, but for now I shall say only a little about it. In a nutshell, it posits that intellectual development is driven by processes that encode, transform, manipulate, and store information. These processes have their efficiency established through a chain of reactions involving their biological bases shaping,

3

and being shaped by, aspects of the environment. For example, to understand how efficiently one can draw inferences, we need to know about the biologically based unfolding of the size of working memory where the information from which inferences are made is temporarily stored. We also need to know about the individual's level of prior experience with the process of inferring, as well as how the developing contexts for its use fostered or inhibited its growth. Contexts are normally thought of as physical or social addresses, but they can also be cognitive, as when we speak of the manner in which information needed for the inference is structured in one's long-term memory. Different structures of the same knowledge are more or less congenial to making an inference. From the very beginning, mental development is an indivisible whole in which biological dispositions interact with their proximal and distal contexts, to shape and be shaped by the environment. It is tempting to try to separate these biological and nonbiological sources, as many do (e.g., heritability analysis requires their separation), but I argue that this is not how nature operates.

Readers may recognize that the foregoing depiction is a type of "person × process × context" model. *Person* variables refer to biological bases for either cognitive or personality dispositions, while *processes* are mental operations, such as inferring, and *context* is broadly construed to refer to environmental potentiators as well as to the structure of one's knowledge. If the biological efficiency of a particular process is known, its actual efficiency would be a result of a series of reactions it had with contexts. All of this will become clearer later.

FIVE EASY FACTS

There are a few basic facts that are known to all members of the intelligence research community, and it is important to make these explicit before going any further. There are two reasons why I wish to make these facts explicit at the outset. First, doing so may provide readers who are unacquainted with the intricacies of this field with some psychometric findings that will not only aid them in evaluating existing theories, but also help them to evaluate their own hunches about the nature of intelligence. The second reason for making these facts explicit is to assure those who *are* familiar with them that I have not attempted to craft an account of intellectual development that is oblivious to them or in any way tries to dodge them. Their early explication provides an opportunity to say what I find lacking about the use of such facts as premises for existing theories. So I shall delay presenting the bioecological framework for intellectual development until after these facts and their implications for it have been articulated. Otherwise, some of the features of the bioecological framework may seem arbitrary or unneeded.

The first fact known to researchers in the field of intelligence is called the *positive manifold* of correlations among test scores. Simply put, if a group of persons is administered a battery of cognitive tests or tasks, the performances of the persons tend to be consistent across the tests or tasks. So if we were to rank these persons from highest to lowest on one task, a somewhat similar ranking might obtain on the other tasks, too. This positive manifold of correlations suggests to some that there is a common ability or intellectual resource that is responsible for some of the individual differences on all of the tests or tasks in the battery, thus "explaining" their intercorrelatedness. The notion of a positive manifold is central to many psychometrically derived theories of intellectual development, and Chapter 8 is devoted to the explication of these theories.

The second fact is that if the correlations among a battery of test scores are analyzed by factor analysis, an entity known as the "first principal component" can be extracted. The size of the first-principal component reflects the average correlation among the test scores. The greater the intercorrelatedness found among the test scores, the greater will be the size of the first-principal component. The typical size of the first-principal component is around .30, for example, although it is possible to alter this magnitude in many ways—by changing the assumptions we make about the nature of intelligence, by changing our statistical practices, by substituting different tasks, or by modifying the context in which the tasks are embedded. And it appears that the size of the first-principal component decreases as children get older, a reflection of greater differentiation of abilities (Sternberg and Powell, 1983a). None of this is controversial, although the implications *are*, as will be seen in Chapters 6 and 7.

A third fact is that the first-principal component is viewed by many researchers as a surrogate for general intelligence, or "*g*." General intelligence or *g* is thought of as a singular intellectual resource that underpins virtually all intelligent behavior. Note that this view does not deny that specific intellectual skills exist in addition to *g*, only that most intellectual tasks require some degree of this general resource for their successful performance. If a test or task is made more difficult (e.g., by requiring greater memory capacity), it tends to rely more heavily on *g* in statistical analyses. Thus, in the jargon of the trade, it may be said to be more or less *g*-loaded or *g*-saturated, depending on its complexity. This finding forms part of a powerful construct validation for what I take to be a fallacious premise of researchers of intellectual development.

The fourth fact is that a number of correlations exist between *g* and academic and social accomplishments. For example, *g* is strongly correlated with IQ—anywhere from .4 to .9, depending on the type of IQ test and the battery of tasks used to derive either the first-principal component or *g*. Moreover, as will be seen later, *g* is somewhat predictive of school grades, work efficiency, everyday problem solving, social attainment, criminality,

and a variety of other important life outcomes (mental health, marital dissolution rates, years of school completed, managerial level attained, etc.). In
some people's minds, such correlations validate g as a measure of the biological capacity of the organism to adapt in order to shape its environment:
"Persons with high g can retrain themselves to do many different tasks in one
lifetime and often at a highly creative level." (Itzkoff, 1989, p. 85)

The fifth and final fact to be made explicit at this time is that a number
of studies have reported that g is quite heritable, with the exact size of the
heritability estimate depending on a number of factors, such as the age of the
individuals being tested. Moreover, the g-loading of each particular test or
task is correlated with its heritability, the higher the g-loading the higher its
heritability. Since racial and ethnic groups are most dissimilar on tests that
have high g-loading *and* a high heritability, the inference has been made by
some that racial and ethnic differences in IQ are the result of genetic differences (Rushton, 1988). I am referring not simply to the old claims of Arthur
Jensen, but to a rash of newer studies by him and others that will be described
in later chapters. These studies claim that the hereditary mechanisms involved in the transmission of g from parent to offspring are also responsible
for the "apparent" influence of the environment on g; that is, heredity is
behind the apparent environmental influences on general intelligence
through genotype-environment correlations. Certain genes propel the individual to construct environments that are associated with later varieties of
intelligence and social outcomes. For example, genes for what Bronfenbrenner (1989) calls "instigative characteristics" (e.g., novelty seeking, reward
dependence, social engagement) may lead to the selection of environments
that are subsequently found to be correlated with later IQ. But some have
argued that a path model of such gene-environment correlations or interactions indicate that these seemingly environmental influences on later IQ are
in actuality genetic influences on IQ.

To recap these five facts, an individual's performance across the types
of cognitive tasks or tests that researchers administer tends to be consistent or
correlated, and this correlation is the basis for extracting the first-principal
component, or general intelligence g. General intelligence is predictive of
many important life outcomes and is thought to be fairly heritable, with
varying magnitudes depending on characteristics of the specific group being
studied. Figure 1.1 is a simplified path model that depicts what for some are
the causal links that connect these facts. A singular, underlying biological
resource pool (e.g., the "signal-to-noise ratio" in the transmission of information in the central nervous system) sets constraints on the efficiencies of
various microlevel processes that are involved in learning (e.g., encoding and
storage and retrieval of information). In turn, these microlevel processes
ultimately determine macrolevel outcomes such as IQ scores, school grades,
acquired knowledge, and job attainment. Although the model is highly deterministic, numerous proponents can be found for various versions of it. For

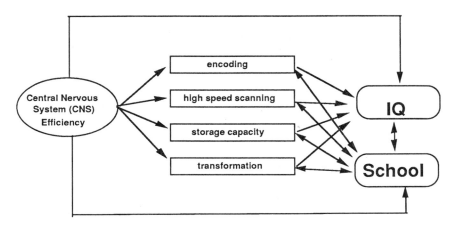

FIGURE 1.1 Path diagram depicting a causal model linking a single underlying biological resource pool (central nervous system efficiency) with microlevel cognitive processes such as encoding, which themselves are instrumental in macrolevel task performance (e.g., IQ and school achievement tests).

example, Hans Eysenk has concluded from the neuropsychological literature that

> .... individuals with neuronal circuitry that can maintain the encoded integrity of stimuli will form accessible memories faster, and will enable problems to be encoded and solved more quickly, than individuals whose circuitry is more "noisy."... for individuals of low neuronal integrity, the maintenance of long sequences of pulse trains . . . will be practically impossible; the total information can never be stored, . . . thus preventing accessible memories from being formed. Hence, one would observe differences within individuals' knowledge bases. (Eysenk, 1988, p. 97)

As this bioecological treatise unfolds, I shall mount an attack on the theoretical implications of the foregoing five facts, as well as on the interpretation of some of the facts themselves. In particular, I shall critique the cognitive and genetic implications of psychometrically defined general intelligence g and argue that g does not deserve to be thought of as general intelligence. Instead, I shall argue that *prediction* from correlations and *explanation* based on those correlations are at times fundamentally different enterprises and that several alternative models exist for the five facts. To present one example of how this difference plays out, consider the traditional wisdom that high IQ is a necessary but not sufficient condition for high levels of achievement in mathematics, science, and language (motivation and special-

ized talents also being necessary). Numerous writers espouse this view and
the frequent implication has been either (a) that it means that IQ is a most
important attribute (e.g., Economos, 1980; Jensen, 1980), or (b) that although
the restricted set of abilities measured by IQ tests *are* predictive of advanced
mathematical and scientific reasoning, they may not be predictive of many
everyday activities that are also important (e.g., Bane and Jencks, 1977; Flynn,
1988). I shall propose in these pages that in fact these correlations reflect no
primary causal role for IQ in higher level understanding. A case can be made,
and I will try to make it, that the relationship between IQ and advanced
academic understanding is epiphenomenal; that persons with low IQs may
have the cognitive architecture needed to appreciate advanced mathematics,
language, and science, but they lack the relevant background experience.
Were they provided such background experience, then both their IQs and
their linguistic, scientific, and mathematical understanding might be elevated. But IQ would be not the cause of the advanced understanding, but
simply a concomitant attribute which is susceptible to the same type of
experiential influences. It then becomes no more correct to assert that individuals are good at language, science, and mathematics because they are
intelligent (i.e., have high IQs) than it is to assert that individuals have high
IQs because they are good at mathematics, language, and science. The pathways that link these outcomes are complex, as will be shown.

I shall also put forward the position that intelligence is a multifaceted
set of abilities and that a specific facet might become more or less effective as
a result of the physical, social, cultural, and historical contexts in which it has
been crystallized and the contexts in which it is subsequently assessed. A
corollary of this view, as will be seen in Chapter 9, is that the generality of
intellectual functioning is more illusory than real, a phenomenon inextricably
linked to the paradigms, values, and assumptions made by those committed
to a deterministic viewpoint.

I have opted in this treatise to take on the entire structure of the biol-
ogy/intelligence/IQ/meritocracy argument depicted in Figure 1.1. That is, I
shall argue against the view that there exists a singular, biologically based
resource pool g that underpins macrolevel measures of intelligence (IQ) via
constraining the microlevel processes (e.g., encoding) that it is presumed to
be built upon. Further, I shall take issue with the claim that individual
differences in intelligence, as indexed by IQ scores, *explain* individual differences in real-world success.

From an expositional standpoint, it might have been better if I had
carved out a discrete piece of this argument and argued against it rather than
try to take on the entire argument in all of its ramifications. However, I
decided that attacking anything less than the entire argument was problematic, because doing so allows pockets of logically discreditable hypotheses to
be entertained by readers. I am certain that this decision has resulted in some
unevenness in my own argument by virtue of requiring a broader based, yet

interrelated, treatise. However, my decision was motivated by the belief that unevenness is a price worth paying if the alternative is a fragmentation of the argument that allows invalid explanations to be espoused.

OLD WINE IN NEW BOTTLES?

Much of the justification for the bioecological framework espoused here comes in the form of argumentation: Current theoretical explanations and their empirical bases are critiqued, and the intelligence quotient (IQ) is reexamined in light of a number of recent findings that call into question some important assumptions about the psychological mechanisms that underpin it. A new view of intelligence, based on this bioecological critique, will be put forward and compared with traditional views (both in this chapter and in later ones). So in a very real sense, the bioecological critique and framework are defined as much by the limitations of existing theoretical frameworks as by their own positive theoretical views. In theoretically uncertain areas like intelligence research, such a hedge would seem not only acceptable but desirable. The idea of making a theoretical contribution, even if one can claim no more than to be a step ahead of one's peers, is a sound one, I think. It is my own fallibilist view of science that progress often results from the accretion of imperfect formulations, each of which provides some new and valid insight even while falling short of the total picture. In this sense, the value of scientific theories is like the story of the two hikers who waded through thick mountainous brush to find themselves in an open clearing, face to face with a large grizzly bear. One hiker immediately threw off the backpack and donned running shoes, while the surprised colleague remarked, "Don't you realize that you can't outrun a grown grizzly bear?" To which the reply came: "I don't have to outrun a grown grizzly bear; I only have to outrun you!" Of course, this does not suggest that a theory should be viewed as helpful simply because of the weaknesses of alternative theories. A theory ought to handle all of the points of confirmation that a competitor theory handles and at the same time provide explanations that the competitor cannot. I have tried to do this wherever possible, though on occasion I have been notably unsuccessful.

As might be inferred from the name, a bioecological account acknowledges the joint role of biological and environmental sources of influence on intellectual development. Because of this, readers may be tempted to envision it as "old wine in a new bottle." After all, what serious theory of intelligence does not endorse the joint roles played by nature and nurture? As will be seen, however, this particular bioecological account contains much that is new, including much that is incompatible with traditional nature-nurture accounts. For one thing, it adopts a developmental perspective in order to understand how individuals could begin life possessing comparable intellectual potential but differ in the level and types of intelligence they ultimately manifest. It does this by positing an interplay between various biologically constrained cognitive potentials, such as the capacity to store, scan, and retrieve information, and the ecological contexts that are relevant for each of their unfoldings. At each point in development, an interaction between biology and ecology results in changes that may themselves produce other changes until a full cascading of effects is set in motion. And because the size of the first principal component gets progressively smaller as children get older, a developmental perspective allows inferences about the mechanisms that produce increasing differentiation of abilities, as well as the shift in the form that g takes, going from something akin to sensorimotor alertness in the first 20 months to later symbol manipulation.

Each biologically constrained cognitive potential is shaped and re-shaped by a chain of interactions with the social, cognitive (e.g., beliefs and the structure of knowledge), and physical environment. Because of the long train of processes and events that intervene between the initial inheritance of a biological boundary for some form of cognition and the extent of its eventual manifestation, it is exceedingly difficult to *predict*, let alone to *explain*, how biological differences between individual genotypes will affect any later cognitive differences between them. Adoption studies that have so ably charted the continuity of mental development are open to the criticism that what appears to be continuous may not be, despite significant correlations between offspring and parent IQs. This argument will be developed later in greater detail. Before doing so, however, more should be said about the need for a developmental perspective to inform this line of inquiry.

It will be argued that environmental influences are specific to each type of cognitive potential, both in the timing of onset and the rate of unfolding. Thus, the timing of events in one's personal and historical development may have profound implications for some facets of behavior, but little, if any, for others, as demonstrated in the lifecourse literature (e.g., Elder, 1986). Whether a disposition thought to be influenced by genes creates difficulties or is used to one's advantage depends on the personal and historical niche that the proband occupies, a case in point being the influence of exposure to economic adversity on intellectual and economic outcomes as a function of how old the child was at the time of exposure (e.g., Elder and Caspi, in press). Unlike psychometric theories of intelligence, the bioecological framework is inherently developmental. It posits the existence of periods of development when specific cognitive muscles may be differentially sensitive to the environment. Although biology and ecology are interwoven into an indivisible whole, their relationship is continually changing, and with each change a new set of possibilities is set in motion until soon even small changes produce cascading effects. Hence, developmental change is not always or even usually linear, but rather, is synergistic and nonadditive. A small environmental influence on a protein-fixing gene may initially result in only tiny changes, but over time the chain of events may produce a snowballing of effects on other processes. At any point in development, a system strives for homeostasis, and the intellectual system is no exception.

The usefulness of incorporating a developmental perspective can be seen in the following example. Suppose that we administer a battery of cognitive tests (spatial, verbal, perceptual-motor, concept formation, etc.) to a group of hypothetical seven-year-olds. Based on the extent to which the individual test scores for the same child are correlated with each other, a measure can be derived through a statistical procedure known as principal component analysis that, as already noted, is accepted by many in the research community as an indicant of general intelligence or g.[1]

Now suppose that we had previously measured aspects of the home environment of these same seven-year-olds at two points: when they were one and four years of age. We can then ask whether the timing of various aspects of their home environment, such as its degree of organization, is more or less influential for their later general intelligence. The developmental perspective being put forward in this monograph would posit that it is indeed: Some developmental periods may be more sensitive than others to the effects of the environment, and thus contribute more to shape certain cognitive muscles. Luckily, we needn't rely on hunches about such things: Plenty of good research from neurobiology and behavior genetics demonstrates that timing *is* quite important during the developmental cycles. For instance, some neuroarchitecture is extremely sensitive to environmental variation during specific periods of development, but not during others that are separated from the latter by only a few days or weeks (see reviews by Epstein, 1979; Goodwin, 1988). Craig Ramey and his colleagues have reviewed some of this literature and concluded that a substantial body of animal and human research, some dating back to Hebb's work on "critical periods" for intellectual development, indicates that the relative contributions of the environment (and genetics) to intellectual outcomes change with the developmental status of the organism (Ramey, Bryant, and Suarez, 1987; Yeates, MacPhee, Campbell, and Ramey, 1983). Recently, Bob Siegler (1989) at Carnegie-Mellon has reviewed the most current evidence for the specificity of the neural-cognitive links. I shall return to this point shortly.

An additional assumption of this monograph is that certain epochs in development can be thought of as sensitive periods during which a unique predisposition exists for a specific cognitive ability to crystallize in response to its interaction with the environment. During such periods, neurons within specific compartments rapidly "overarborize" (spreading their tentacle-like synaptic connections to other neurons). Even though some of the arboreal connections laid down during these periods of brain spurts will not be used at that time, they can be recruited to enable future behaviors to occur, provided they are not "pruned" because of atrophy. Siegler (1989) concluded that "the timing of the sensitive period seems to be a function of both when synaptic overproduction occurs and when the organism receives relevant experience" (p. 358). He suggests that while some neural processes are more fully under maturational control, others are responsive to experiences, and synapses are formed in response to learning that may vary widely among humans.

So the available neurophysiological evidence is compatible with the bioecological framework's developmental orientation: Various brain compartments, comprised of specific types of neurons, begin growing at different times, and at different rates. The rate of connective growth during the first six years of life appears to be astounding, averaging in the neighborhood of 2.5 trillion connections per year, or 4.7 million wires branching out from neurons (synapses) per minute (Goodwin, 1988). These are crude averages, of course, and individual compartments develop according to their own schedule as well as according to external environmental influences. The research indicates quite unequivocally that experience accounts for some of the specificity in nervous system development.[2] To give one fairly precise estimate, the synaptic density of the third layer of the middle frontal gyrus increases tenfold between birth and 12 months of age. The density keeps increasing until around the second birthday, after which time it begins decreasing. It decreases until around age seven, when it reaches the mature adult level. Thus, from about birth until seven, the synaptic density in this area is greater than it is among adults.

Going back to the earlier hypothetical example about testing seven-year-olds' general intelligence and relating it to their earlier home environments, Rice, Fulker, Defries, and Plomin (1988) have shown that although there is no significant correlation between a child's general intelligence (g) at age seven and aspects of the home environment at ages three or four, there is a significant correlation between their general intelligence at age seven and aspects of the home environment at ages one and two! So clearly, the timing of environmental events can be important for later intellectual development. Yet, as will be seen in Chapter 10, the only theories of intellectual development to have taken into account this changing genotype-phenotype relationship over ontogenesis have not been concerned with individual differences (e.g., Piaget, 1952).

The bioecological framework shares with traditional theories the view that there exists some construct that deserves to be called *intelligence*, despite failures among those in the research community to agree on its definition and measurement. It further shares with some of these theories the belief that not all individuals are equivalently intellectually endowed. Where it differs from these other theories most seriously, however, is in the role it assigns to the nature of the entity called *general intelligence* and to the meaning of the IQ score specifically, and to the psychometric findings more generally. As will be argued later, because of the rather limited usefulness of IQ scores for inferring intellectual aptitude, much of the so-called evidence for traditional ways of thinking about intelligence is ill wrought. For instance, evidence for the long-term decline in national intelligence and the correlated decline of SES (Galton, 1892, and Herrnstein, 1971, but see Flynn, 1987a, 1987b, 1988 for a counterargument), as well as "evidence" that each new generation's intelligence has been declining (e.g., Burt, 1946), has been based largely or wholly on trends in IQ scores—and even here there is room for contest, as will be described in a later chapter. When intelligence as a concept is broadened to include cognitive functioning outside the realm of IQs, a very different picture emerges.

Finally, even when the bioecological framework does cover old ground, it appears to me to do so by providing a finer grained analysis than is true of other theoretical approaches. In Chapter 10, the bioecological framework is contrasted with nine classes of theories and is shown to differ not only in the *scope* of its evidentiary base, but also in the *pattern* of its assumptions. All of this is my way of saying that these pages do contain new ideas, and perhaps even some convincing arguments against older ideas. But before delving into the various problems with existing theories of intelligence and intellectual development, it is useful to begin with a description of the barest essentials of each class of theories so the reader will be prepared for what is to come.

TRADITIONAL PSYCHOLOGICAL THEORIES OF INTELLIGENCE

As already noted, the majority of psychologists who study intelligence believe that there exists some unitary mental energy g that flows into virtually all cognitive endeavors. Psychometric researchers have quantified this entity by extracting the first-principal component independently of rotation in a factor analysis. The majority of these researchers, however, would argue that this singular form of mental energy is not identical to past learning. It would be rather empty if these theories argued that individual differences on mental tests largely reflect differences in the amount of prior learning, because if prior learning were all that was being measured, then there would be no basis for speculating about any innate capacity or about the nature of intelligence. Differences among children on intelligence tests would be chalked up to differences in their previous learning, which would leave unanswered the question of the origin of the differences in prior learning. Although some psychologists do interpret test scores in just this manner, they are in a decidedly small minority in the research community. In later chapters of this book, I shall lay out the various theoretical camps' assumptions and try to make clear why most of these camps have maintained that more than prior learning is being measured on IQ tests. For now, let me just say that the bioecological framework that will be elaborated here differs from these other theories on this as well as on many other points.

Two working assumptions of the bioecological framework, for which empirical evidence and logical analysis will be mustered later, are that (1) there exists not one underlying intelligence g, but multiple forms of intelligence, and (2) it is logically impossible to separate these intelligences from acquired knowledge (and the elaborateness with which this knowledge is organized), even though in principle such a distinction can (and in many theories must) be made.

Excellent reviews of psychometric, genetic, and information processing perspectives exist (e.g., Bouchard, 1984; Carroll, 1976; Eysenck, 1979; Horn, 1978; Horn and Donaldson, 1980; Hunt, 1980; Lewontin, et al., 1984; Keating, 1984; Scarr and Kidd, 1983; Sternberg, 1981, 1985). I shall not summarize any of them here, except by way of noting that each of the perspectives relies for its argument on the relationship between various constructs that will be described later and IQ (or a surrogate of IQ such as verbal ability, SATs, or GREs).[3] For example, Hunt's (1980, 1985) theory of intelligence is built on the well-established correlation between the earlier mentioned positive manifold (i.e., the association between linguistic and short-term memory functions) and IQ. The fact that measures such as *speed of encoding, verbal fluency,* and *retrieval efficiency* all correlate modestly with IQ (for the most part, with r's ranging around .3) is taken by some researchers as evidence for the existence of a unitary, underlying intellectual force that permeates a broad range of verbal activities. Since students who score in the top quartile of the verbal SAT tend to be faster at scanning and retrieving the contents of their memories than those who score in the bottom quartile, one can infer that the correlation of memory scanning speed with IQ is due to the causal role of the former in IQ performance, with high-verbal subjects having more efficient memory processes that lead them to glean more information from their environments than low-verbal subjects (Hunt, 1985).

Similarly, genetic and psychometric theories have been built upon the intercorrelations found among various measures of individual differences (e.g., Horn, 1978; Jensen, 1981, 1984; Sternberg, 1985). In general, persons who score high on one measure tend to score high on the others. A later criticism of these approaches that can be adumbrated at this time is that factor analytic solutions of individual test score variances as well as behavior-genetic and componential analyses of task parameters and IQ, are limited both by the nature of the test battery and by the limited contexts in which the tests are administered. It will later be shown that factor loadings, heritability indices, and cognitive performances shift, often dramatically, with changes in the composition of the cohort being tested, the battery being administered, and the context in which it is administered. Humphreys (1962) sounded this alarm with regard to factor analysis many years ago. He argued that the differential number (and specificity) of tests included by factor analysts makes the interpretation and rotation of factors difficult. If one area of testing is more developed (by researchers) than another, "the more likely one will find a hierarchy of tests of varying degrees of specificity" in that area (p. 478). Moreover, a factor structure from any particular sample can be reanalyzed by disaggregating the sample and reforming it into subgroups (e.g., based on gender, SES, hobbies, or preferences), and the result may be a dramatically different factor structure (Lorenz, 1987). Similar problems exist for the other approaches and these will be addressed later.

Furthermore, on logical grounds, all of these approaches must assume that knowledge and learning are not confounded with what is being called intelligence (see Gregory, 1981, for a logical analysis of this assumption). For example, Hunt, et al. (1975), attempt to explain individual differences in verbal ability not in terms of differences in learning or vocabulary, but in terms of differences in microlevel cognitive processes (e.g., the speed with which internal memory representations can be integrated or altered) that are not merely the result of differences in past learning or prior experiences. In fact, some have argued that the success of the IQ test is due to its success at indirectly tapping products of these lower order cognitive processes, such as memory scanning.

Yet, notwithstanding such an account of the knowledge-free functioning of microlevel cognitive processes, it is impossible to separate acquired knowledge from what is termed intelligence in these theories. It will be argued later that virtually all items used in test batteries and problem-solving tasks rely on a combination of past learning (what Gregory, 1981, terms potential intelligence) and the generation of novel insights (termed kinetic intelligence), with the former usually dominant. Although more will be said about these arguments later, I shall present findings from several disciplines that, taken together, are not easily explained or predicted by current psychological theories of intelligence.

ABOUT THE REST OF THIS BOOK

Before launching this expedition across "disciplines, decades, and continents," it may be helpful to spell out the book's organization, lest the reader imagine this to be a journey with many ports of call but no recognizable destination. I begin later in this chapter, and carry over into Chapter 2, by differentiating four different but related concepts that are useful to distinguish when thinking about intelligence. The first of these is *cognitive processes*. Cognitive processes are hypothetical mechanisms, such as memory and sensory detection, for gathering and interpreting information from the environment. The potential efficacy of cognitive processes is in part biologically constrained and in part constrained by the nature of one's knowledge.

Which brings us to the second concept. *Knowledge* refers to the information, rules, beliefs, attitudes, etc., that are gleaned from the environment by means of the aforementioned cognitive processes. Knowledge is organized into structures in long- term memory that vary in their degree of elaborateness and interrelatedness.

The third concept, *cognitive complexity*, refers to the degree to which one's processes are able to operate on one's knowledge structures in a complex, efficient, and flexible manner. Cognitive complexity is limited by the efficiency of cognitive processes, which in turn are involved in the acquisition of knowledge and its structure.

The final concept, *IQ*, refers to the score one achieves on a test that purports to measure general intelligence. In later chapters, I shall argue that the first three of these concepts interact to account for all intellectual products, including IQ. IQ is seen as "phenotypic" of only one concatenation of processes and knowledge and is not representative of other equally important concatenations. Further, to refer to such tests as measuring intelligence is to assume something about them that should be the basis of empirical and theoretical verification which, I shall claim, has not been accomplished in the past and does not exist at present. Finally, I shall argue that even the specific concatenation of processes and knowledge required for IQ test performance is no basis for inferring anything about the capacity of an individual to reason complexly with these same processes, because the efficiency of such processes is tied to specific knowledge contexts.

In Chapter 2, the relationship between the preceding constructs and cognitive performance is further spelled out. The roles of motivation and context are introduced as well.

In Chapter 3, I review studies from the fields of anthropology, experimental psychology and sociology, both to emphasize the limitations of many of the traditional views of intelligence and to highlight what I believe to be advantages of the bioecological framework, as the diversity of these results are problematic for many psychological theories to explain.

In Chapter 4, I describe a social-organizational view of intellectual development, review and reanalyze older studies, and conclude with the suggestion that intellectual development is a function, at least in large measure, of the individual's role in a larger social structure.

Taken together with the review of studies from education that I describe in Chapter 5, these chapters indicate that traditional ways of indexing intelligence have led to ungenerous views about the capabilities of much of the world's people, including many from non-Western-educated nations who perform poorly on IQ tests. Although no pretense of comprehensiveness is made in this literature review, the studies cited are illustrative of those reported in the respective fields and no attempt has been made to "slant" conclusions by deliberately ignoring studies. Any omissions are the result of the author's ignorance, not contrivance.

Chapter 5 deals with the specific influence of schooling on cognitive processes, in general, and on IQ, in particular. Schooling plays a central role in the bioecological theory because it provides a context in which biologically constrained cognitive processes instrumental in IQ performance are honed and crystallized to varying degrees of efficiency. Consequently, the processes associated with schooling are seen as an important mechanism underlying one of the major components of the bioecological theory, namely, the elaborateness of one's knowledge. As I researched this section on schooling, I became increasingly convinced of the unacknowledged potency (on several different levels) of completing grades. Prior psychological analyses of schooling, though important, have been unrevealing in this respect, I think. So this section of the book attempts to explicate the role of schooling in intellectual development and, specifically, in IQ test performance. Contrary to what many have alleged, I conclude that schooling conveys a direct and powerful influence on IQ and cognitive complexity that cannot easily be dismissed as the indirect result of genetically mediated variables.

In Chapter 6, the concept of heredity is discussed. Although the bioecological theory posits an important role for biology to play in mental development, it does so in a different manner than is customary in traditional psychological theories.

In Chapter 7, the bioecological framework is described in some detail and a preliminary justification is provided for each of its components.

In Chapter 8, the concept of a singular, "primordial" source of intellectual energy is discussed. The notion that general intelligence or "g" represents such a primordial source is contested. The relevant empirical studies are discussed in some detail. The concept of g and its presumed biological basis combine to form a type of construct validation for many current researchers.

In Chapter 9, the concept of "abstractness" is examined, as is the relation between it and generalization. It is argued there that abstractness is best viewed as something inseparable from knowledge and possibly no different from knowledge, unlike contemporary thinking on this issue.

Finally, Chapter 10 deals with the seminal findings that have served as the bases for many of the traditional theories of intelligence. I describe what I take to be the essence of each of these theories and discuss what I see as their advantages and shortcomings. The bioecological framework is evaluated vis-a-vis several traditional psychological theories (information processing, genetic, modularity, and psychometric), as well as several theories that, like the bioecological framework, go beyond the field of psychology for their justification, namely, the triarchic theory (Sternberg, 1985), the theory of multiple intelligences (Gardner, 1983), knowledge-based theories (Chi and Ceci, 1987; Keil, 1984; 1985), contextualist theories (e.g., Cole and Scribner 1974; Charlesworth, 1979), structural/developmental theories (e.g., Case, 1985; Piaget, 1951), and modular theories (Fodor, 1983).

Some of the findings I discuss throughout these pages, such as the interrelatedness of test performances and the heritability of IQ, could easily have formed the basis of entire books themselves. I have addressed them only insofar as they bear on my principal objectives. Throughout this book I have revealed my proclivity to mount attacks on prevailing ideas through the combined effort of trying to unearth contrary empirical findings and intermittently engaging in logical analysis. The result of this tendency may be inchoate at times for readers used to less eclectic forms of argumentation. I apologize in advance to anyone who gets lost amidst such eclecticism, though I suspect that readers who share this proclivity may actually enjoy such a form of argumentation.

CHAPTER 2

Toward an Inductive Theory of Intellectual Complexity

Theories of intelligence have a long history, predating modern forefathers such as Galton, Spencer, and Binet by at least 2,000 years. From the time that Plato (427–347 B.C.) first described the characteristics of intelligence in terms of the soul's entrapment of ideas, through medieval notions of intelligence in terms of trivium and quadrivium, to current conceptualizations based upon information processing constructs, there have been hundreds of theories of intelligence. Elsewhere (Doris and Ceci, 1988) my colleague John Doris and I have described some of the assumptions made by these theorists and I shall not go into them here.

While it would be highly inaccurate to suggest that the introduction of each new theory has been in direct response to a specific new set of empirical and/or political realities, unconstrained by the long and diverse legacy of prior researches, it is not misleading to state that modern *psychological* theories of intelligence have not been motivated by a desire to assimilate as much of the world literature concerning intellectual development as possible. Rather, the motivation of psychologists has usually been to address some fairly specific counterevidence to a single prevailing theory of intelligence; attempts to construct a theory on the basis of a more extensive corpus of the scientific literature have been scarce, with some notable exceptions (Gardner, 1983; Sternberg, 1985; Vernon, 1969). Thus, researchers with a hereditarian bent have spent considerable energy amassing evidence for the importance of genetics (e.g., Bouchard, 1984; 1986; Gordon, 1980; Horn, 1983; 1985; Jones, 1955; Loehlin and Nichols, 1976; Plomin, 1983; Rice, et al., 1986; Scarr, 1968; Scarr and Kidd, 1983; Scarr and McCartney, 1983), while those with an environmental bent have devoted their considerable talents trying to refute the hereditarian evidence and amassing evidence of their own (e.g., Hoffman, 1985; Lewontin, et al., 1984; Wachs and Gruen, 1982; Walker and Emory, 1985; Yeates, et al., 1983). To be sure, both groups have endorsed a degree of interaction between genetic and environmental components, differing mainly in their relative contributions. Seldom, however, have members of either group of researchers endeavored to extend their theoretical accounts beyond well-defined nature/nurture issues to include findings from such areas as information processing, sociology, anthropology, or education.

TOWARD A DEVELOPMENTAL FRAMEWORK

Within the information processing tradition, researchers have devoted considerable energy to carrying out componential analyses of cognitive tasks like IQ or searching for microlevel correlates of IQ, such as encoding, and formulating mental models that reflect these correlations. But information processing researchers have only recently sought to integrate their findings with those from either of the other two main approaches of psychological research on intelligence, the genetic and psychometric approaches, and the current level of integration is low (although see Vernon, 1986, for a promising beginning). The same can be said of the genetic and psychometric approaches' failure to integrate their findings with those from information processing and with each other. And to make matters worse, I think it is fair to say that *none* of these approaches has attempted to integrate its own findings with those from sociology, education, developmental psychology, and anthropology. In fact, to date, no psychological theory of intelligence of which I am familiar has tried to harvest the great variety of empirical findings and philosophical analyses provided by researchers working within as well as outside of psychology and to cast the resulting insights into a developmental framework. To presage one of my later arguments, it is only through an integration of findings across different disciplines, embedded in a developmental perspective, that one can hope to understand many of the phenomena that are difficult to explain on the basis of existing psychological theories. A corollary, of course, is that nothing poses difficulty if it remains unknown and certainly ignorance provides no stimulus to extend our beliefs: *Ex nihilo nihil fit!*[1]

So the distinction of the effort presented here lies in its *a priori* intent to go beyond the collection of evidence (especially psychological evidence) for or against a prevailing theory of intelligence and attempt to construct a developmental framework capable of assimilating diverse empirical findings, many of which may not be widely known by psychologists or others interested in intelligence. By aiming at such a goal, the framework-construction process has resulted in a different form of framework from that typically developed on the basis of refutation and experimentation (i.e., deductive or "functional" theories that consist of hypotheses suggested by prior research findings that are then tested in subsequent research and eventually may even be used to modify the theory). The present attempt was begun in an avowedly "inductive" manner (in the sense that Skinner's theory is inductive) and entails a set of descriptive statements that summarize a diverse set of data, initially with less inference involved than is typically found in deductive theories. The strengths and weaknesses of this type of approach are well known (Marx, 1976; Miller, 1983) and soon will become apparent. As a "first cut" at synthesizing the diversity of findings produced by researchers from different disciplines, such an approach seemed useful to me. Later in the book the framework-construction process shifts to a more deductive level— although I fall short of producing anything polished at this level. Once the bioecological framework is posed, an effort is made to describe the sorts of tests needed for its validation, and a few tests are provided in a later section, thus making the effort not entirely inductive. But, as noted in Chapter 1, the goal of the enterprise is not to test hypotheses but to generate them. The bioecological framework is in an exploratory mode rather than a confirmatory mode—a result of casting my net over such a wide range of phenomena.

Thus, the fruit of this book is the description of a bioecological framework for understanding individual differences in intellectual development, focusing on the implications of some recent as well as historical empirical findings from the fields of cultural anthropology, experimental psychology, sociology, genetics, developmental psychology, and education. As such, it shares more with so-called "contextualist" theories (e.g., Charlesworth, 1976; 1979; Cole, 1975; Gardner, 1983; Rogoff, 1981; Sternberg, 1985; Witkin and Berry, 1975) than it differs from them. Its chief difference is a breadth and specification of components not found in contextualist accounts, plus many new interpretations of older findings.

TERMINOLOGICAL CONSIDERATIONS

Over the years, cognitive psychologists have introduced terms that have been confusing and difficult to distinguish for noncognitivists. In this section a number of related terms will be introduced and some distinctions will be made among them. These terms play an important role in the theoretical discourse that appears in later chapters.

In 1921, the editors of the *Journal of Educational Psychology* convened a symposium (through the mail) in which 14 top researchers were asked to define intelligence and discuss its nature. Recently, Douglas Detterman and Robert J. Sternberg repeated this experiment with a group of contemporary researchers and reported the results of an analysis of the stability and change in the way researchers think about the nature of intelligence (Detterman and Sternberg, 1986). From their report, it is clear that although some things have remained constant and others have changed, the overriding impression is one of a diversity of views (Ceci, 1987). Nearly everyone today has a view on the nature of intelligence, as was also the case back in 1921. Some choose to emphasize the role of knowledge and past learning, while others choose to focus on context, process, or complexity. Some point to abstract reasoning, others to adaptation to novel environments. Seen narrowly, the only consensus is among small enclaves, though Sternberg and Powell (1983a) have suggested that on a broader level there do appear to be two main themes running through these divergent views: the capacity to learn from experience and environmental adaptation. What I propose to do here is introduce several related terms that will help frame what I take to be the nature of intelligence.

Elaborated Knowledge Structures versus Cognitive Processes

The term "intelligence" is often used synonymously with "IQ", "g," or "general intelligence", especially in some of the psychometric literature. In many of the studies that will be reported in later chapters, however, the ability to engage in cognitively complex behaviors will be shown to be independent of IQ, g, or general intelligence. As the reasons for this independence become clearer, *cognitive complexity* will be seen to be the more general of the two notions and the one most theoretically important to keep in mind when referring to intelligent behavior. Performance on an IQ test reflects but a small sample of one's intelligence (and even that sample is constrained by the context of the test questions) and says much less about the capacity for cognitively complex behavior outside the confines of the test than some recent researchers have alleged. For example, Schmidt and Hunter (1981), Jensen (1984), Gordon (1980), Gottfredson (1986), and many others have argued that g-loaded tests, of which IQ is the best known example, are important predictors of supervisor ratings, work samples, criminality, and other characteristics. It has even been estimated that the use of g-loaded tests for all job assignments in the United States economy would save this nation up to 178 billion dollars (in 1986 dollars) even if the predictive validity of such tests does not improve from its currently moderate level (Hunter and Schmidt, 1982). But prediction is not the same as explanation, and I shall try to show that the ability to succeed is not explained by the ability to engage in intelligent behavior.

A prerequisite for cognitively complex behavior in a given realm is the possession of a well differentiated yet integrated knowledge base that gets operated on by efficient cognitive processes. (More about this later.) The knowledge and beliefs we possess in a specific domain, whether it be the domain of sports, cuisine, electronics, or whatever, provide the raw materials for the operation of various cognitive processes during moments of problem solving—including even such mundane forms of problem solving as perceiving and remembering events. The construct "elaborated knowledge domain" refers to a knowledge structure that is organized by a relatively large number of dimensions. These dimensions are integrated in such a fashion that fine-grained classifications can be made, as in the case of someone who can differentiate a variety of foods (dairy, meat, dessert, etc.) along different dimensions, and the dimensions themselves can be integrated to produce various combinations (full course dinners, Indian foods, etc.). Elaborate knowledge domains are required for conceptualizing individual differences in thinking and reasoning that are domain specific and uncorrelated with IQ in the normal and above normal range. Various researchers have conceived of elaborated knowledge structures differently, including the number of ideas one possesses about an object or the number of independent attributes onto which an object can be projected in a multidemensional semantic space. A complexly represented object is one that is assigned to many attributes or dimensions which themselves can be temporarily associated (i.e., integrated) to make sense of novel data or to distinguish "shades of gray."

The foremost researcher in this area is Siegfried Streufert who, together with his colleagues, has provided extensive reviews and analyses of the meaning of cognitive complexity and its relationship to a variety of everyday achievements in the world of work (Streufert and Streufert, 1978; Streufert and Nogami, 1989; Streufert and Swezey, 1986). According to Streufert, complexity theorists are concerned with the dimensionality of human cognition, that is, the number of dimensions or attributes that are used to conceptualize, organize, and understand incoming data. Integration is dependent on a finely differentiated dimensional structure. A variety of tests are used to assess the number of dimensions and their integratability, and the interested reader is referred to Streufert and Nogami (1989) for a review. Research has documented that, although cognitive complexity is largely unrelated to IQ, it is highly relevant for performance in situations that require continual readaptation and nonentrenchment of thinking, namely the conditions of a manager or a planner in a complex operation. Importantly, Streufert has shown that knowledge per se is not the defining characteristic of the cognitively complex person, but rather how that knowledge is structured:

> A wealth of data indicate the value of cognitive complexity in managerial tasks involving uncertainty and change . . . . cognitively complex managers interacted at a faster pace, were better at cue utilization, and finished their task in half the time. Several researchers have reported that integrative capacity is an aid in decision making . . . . A series of simulation-based efforts . . . clearly establish the greater effectiveness of cognitively complex individuals in strategic and planning activities . . . . Since correlations with ability tend to be absent, the less complex person is equally likely to be knowledgeable and generally competent. (Streufert and Nogami, 1989, p. 119)

Particular patterns of cognitive outcomes can be predicted on the basis of (1) the degree of integratability and dimensionality of the resident information in long-term memory that governs a subject's representation of knowledge in a given domain of experience, (i.e., elaborated knowledge), and (2) the efficacy with which an individual deploys a relevant cognitive process such as "encoding," "comparing," "high speed scanning," etc. in a given domain of experience.

"Elaborated knowledge domains" will be used to refer to a set of attributes (dimensions) in terms of which events are understood (encoded, retrieved, inferred, etc.) within a particular domain. And a knowledge domain is defined by the concepts contained in it and the dimensions along which these concepts are appraised. The term "cognitive processes" will be used to refer to mechanisms involved in the translation and interpretation of sensory information into mental representations (e.g., encoding or detecting visual events from the environment) and in bringing previously translated information back into consciousness (e.g., scanning the contents of long-term memory). "Processes" are a common currency in the field of cognitive psychology, and thinking about them abounds, as for instance when one talks of memory or attentional capacity or efficiency. It is possible, in principle, to assess the efficiency of cognitive processes like attention, memory encoding, and inferential reasoning in isolation. In practice, however, it is difficult to do so because their efficiency depends to a large extent on the elaborateness of the knowledge they access, i.e., elaborated knowledge domains.

Not all domains of knowledge are equivalently structured; some are more elaborately interconnected by dimensions than others. Cognitive processes and elaborated knowledge domains are best viewed symbiotically: Efficient cognitive processes help add structure and complexity to an existing knowledge domain, and in turn, this structure may enhance the efficacy of the cognitive processes that operate on it. Thus, it is easy to imagine someone effectively deploying a particular cognitive process in one domain but not in another. When this happens, it would suggest that the cognitive process itself is "intact," but the dimensional structures of the different domains of knowledge are not equivalently congenial to its expression. Another possibility that is suggested by such an outcome is that there exist separate, duplicative processes in each domain, a proposal I review in Chapter 10 under the heading of "radical modularity." If none of this makes sense, bear with me because later I shall give several examples of constraints imposed on the efficiency of a cognitive process by a subject's undifferentiated knowledge structure.

The notions of a cognitive process and an elaborated knowledge domain are similar to constructs used by researchers across various branches of psychology. Experimental psychologists frequently have examined cognitive processes, though they often do so in the absence of considering the structure of the knowledge domain that such processes must access. Similarly, the concept of elaborated knowledge has been described by researchers both in the area of semantic memory and in the area of personality (e.g., Bieri, 1966;

Ceci, et al., 1981; Noble, 1952; Zajonc, 1968). In these fields it is generally assumed that concepts, including phenomenal objects, are represented by either geometric or algebraic descriptions of the dimensions by which they are classified. Concepts that are represented by many richly interwoven dimensions are characteristic of complex domains of knowledge, especially when the dimensions are well differentiated (Gabennesch, 1972). For example, attitude change is, to some extent, a function of the degree of dimensionality in a domain of knowledge (Streufert and Swezey, 1986).

In many studies it has been shown that memory development is also partly constrained by the level of age-related knowledge elaboration (e.g., Chi, 1978; Chi, et al., 1989; Schneider, et al., 1989), and in several studies it has been shown that long-term memory performance is a function of the number of dimensions by which events are classified and the degree of differentiation among these attributes (e.g., Coltheart and Walsh, 1988). To give an example that is most familiar to me, my colleagues and I once presented 7- and 10-year-old children with information about popular television characters that was either congruent or incongruent with the dimensional structure of the children's knowledge about these characters. We discovered that for those children with a less differentiated knowledge structure, there was a tendency to incorrectly reconstruct this information in a simplistic fashion. For instance, if the information residing in the domain of personal knowledge (knowledge about the way one construes attributes of oneself and others) was structured in such a way that dimensions like "strength," "attractiveness," "cleverness," etc., were highly undifferentiated, children had difficulty thinking about a character who was described as "ugly but clever," or "weak but handsome," as these combinations of attributes were considered incongruous by the children (Ceci, et al., 1981). Older children usually possessed more fully differentiated knowledge structures (statistically, their dimensions could be separated even though they could be integrated with each other when necessary in order to make distinctions between larger groupings of characters) and were easily able to imagine characters who were high on one positive dimension but low on another positive one. The point of this example is simply that the way knowledge is structured, and in particular its degree of complexity, influences the way we interpret, reason, and recall events.

Similar demonstrations have been published for expert bird watchers' recollections (Coltheart and Walsh, 1988) and expert business executives' decisions (Streufert, 1984). In the latter case, Streufert has demonstrated that highly successful executives possess elaborate knowledge structures that are differentiated but also integrated. Such complexity has been shown to influence their processing efficiency in a manner that is somewhat unrelated to their IQ in the range of 95–160. And in the field of personnel psychology, it has been shown repeatedly that individuals who score high on some measure of cognitive complexity exhibit a greater degree of differentiation among organizational characteristics and are better able to distinguish organizational attributes from task attributes, work group attributes, or superior attributes (Jones and Butler, 1980). But this complexity is unrelated to sheer amount of knowledge or to general measure of ability such as IQ. Finally, in the previously mentioned research by Schneider, et al. (1989), children with lower IQs who were highly knowledgeable about soccer outperformed their higher IQ peers on reading comprehension, inferential, and memory tasks that involved learning new information in the domain of soccer. There is a growing body of evidence illustrating that complex performance on reading, memory, and inferential reasoning tasks is predicted by individual differences in knowledge at least as much, if not more so, than it is by differences in IQ scores (Bjorklund and Muir, 1989; Means and Voss, 1985; Walker, 1987).

Knowledge versus Intelligence

The term "knowledge" differs from the term "cognitive process," as well as from the construct of an elaborated knowledge domain. Unlike the latter, it makes no reference either to the degree of differentiation in its structure or to the operation of the cognitive processes by which it was acquired. I use the term "knowledge" to denote a fairly static entity, that is, discrete units of information accumulated through experience, including procedural information (e.g., rules, short-cuts, and beliefs about strategy efficacy), that make up one's data base. Certainly, this is a caricature of knowledge, as existing knowledge can be used to generate new knowledge so that it cannot be said to be truly static. But it is important to distinguish between the contents or units of one's data base (knowledge), its dimensional structure (elaborated knowledge domain), and the efficiency with which cognitive processes (encoding, retrieval, inference, etc.) operate on it. As Bjorklund and Muir (1989) have noted, it is necessary to make such distinctions if one is to account for differences in memory and comprehension that are due, not to whether knowledge is present, but to how it is represented by experts and novices in a given domain. Thus, it is possible to be knowledgeable without being intelligent, for instance, if one's reservoir of knowledge in a given domain is not elaborately structured, rendering ineffective efforts to access it to solve novel problems. Or, alternatively, knowledge can be well structured, but the relevant cognitive processes that access it in the course of solving a problem may be inefficient. Cognitive complexity results from efficient processes operating in elaborated knowledge domains. It is what the lay person most likely means when marvelling at someone's skill or intelligence in a specific area.

It is generally assumed that cognitive complexity is domain specific, or at least that it can be so (see Scott, et al., 1979, p. 55; Streufert and Streufert, 1978, 1986). Processes, rules, algorithms, or strategies that may become general or "trans-domainal" in adulthood (i.e., applying with equal efficiency across many domains of knowledge) often have their child-hood origins in a particular domain and cannot be deployed effectively outside that domain of knowledge until ample experience or insight has been acquired. An example of this development is the case of the child who initially can add or divide only certain items, but who eventually extends arithmetic rules to other domains of knowledge (Chi and Ceci, 1987). One of Piaget's earliest observations was of this sort: A four-year-old, when asked how many pieces she would have if an apple or pear were cut in half, replied, "Two"; however, when asked how many pieces there would be if a water-melon was cut in half, she replied, "It depends on how big the melon is."

What passes for a cognitive process, algorithm, or strategy is, in reality, often a form of discrete domain-specific knowledge. For example, a child may first acquire the specific knowledge that an apple cut in half results in two pieces. This knowledge is acquired in much the same way that knowl-edge is acquired about the apple's color, size, taste, etc. It is only with experience that this domain-specific knowledge becomes an algorithm, and then only in a domain-specific manner. For example, the child may progress from knowing that an apple cut in half yields two pieces to an understanding that any concrete object cut into halves will yield two pieces. At such a point, it would seem that the knowledge has emerged as a transdomainal algo-rithm. But in fact, the algorithm may still be tied to a domain of concrete objects. The child does not immediately know that *all* objects in *all* domains e.g., horsepower, volume, intensity, light, may be divided with the same result as when concrete objects are divided. Only later, if at all, will such knowledge become truly transdomainal procedural rules.

Because learning in one form or another through the effective applica-tion of cognitive processes to a knowledge base is the ordinary mechanism for acquiring knowledge (Horn, 1978; Horn and Donaldson, 1980), and be-cause individuals' learning histories are presumably somewhat .varied, it follows that although a particular set of concepts may comprise an "identifi-able domain for one person, for someone else the same objects may. . . (1) be scattered over several domains, (2) be included as a subset of some larger domain, or (3) not enter into a cognitive domain at all. Furthermore, any given object may be considered as belonging to several domains, depending on the basis of classification employed" (Scott, et al., 1979, p. 56). These observations will later be exploited to demonstrate the nature and develop-mental course of intelligence and show how it differs from the way that intelligence as a concept is embodied in an IQ score. The point for now, however, is to recognize that intelligence is a function of cognitive complex-ity, which in turn is dependent upon the operation of cognitive processes on a specifiable knowledge structure and, conversely, cognitive processes are dependent upon the sheer quantity of knowledge a person possesses as well as the organization of this knowledge. And, finally, cognitive complexity is tied to specific domains of knowledge or information (Scott, 1979; Streufert and Nogami, 1989). People have not been found to respond with similar levels of complexity in diverse domains of experience: "while some level of complexity will be common *within* a domain, other levels of cognitive complexity may exist in unrelated domains of the same individual" (Streufert and Nogami, 1989, p. 110).

At this point it is desirable to introduce two more constructs, motivation and context, to set up the bioecological framework of individual differences in intelligence that will be described later. Attempts by psychologists to conceptualize cognitive complexity in terms of psychometric, genetic, or information-processing theories alone cannot account for the full range of findings that will subsequently be reviewed. A fully adequate account of the findings appears to require a specification of the joint roles of motivational and contextual influences at two points in development: in the initial crystallization of various biologically constrained cognitive potentials, and in the subsequent process of eliciting fully developed cognitive potentials at the time of their measurement. In a later chapter, I shall discuss some ways in which motivational and contextual variables affect cognitive efficiency. But let us move on now, because to say more about the joint roles of context and motivation on intellectual performance at this point would be to put the cart before the horse.

CHAPTER 3

Mismatches Between Intelligent Performance and IQ

Research by anthropologists, experimental psychologists, and sociologists provides a coherent picture of the limitation of IQ test performance as an index of intelligence. My purpose in this chapter is to describe some of the main lines of work in each of these fields. Although not exhaustive, the following studies and others like them should be considered by anyone proposing to account for individual differences in intelligence by relying on psychometric or behavior-genetic findings in general, and on IQ test scores in particular, as an index of intelligence. This chapter and the two that follow will provide the backdrop for my critique of the psychometric and behavior-genetic research.

CULTURAL ANTHROPOLOGY

Anthropologists and ethnographers have provided fascinating glimpses into the intellectual functioning of nonliterate and semiliterate peoples for a long time. A popular thesis in the anthropological research has been that different environmental demands associated with different cultures lead to the development of different patterns of ability across cultures. Perhaps the first anthropologist to make this statement explicitly, and one who influenced important psychologists like Vygotsky and Luria, was W. H. R. Rivers (1926). Since Rivers's time, many others have made similar claims (e.g., Burnett, 1952; Edgerton, 1981; Lancy and Strathern, 1981). Rivers was a member of the Cambridge Anthropological Expedition to the Torres Straits in 1898, and he carried out detailed analyses of islander dwellers living there. It was Rivers whom Vygotsky credits with influencing his notion of sociohistorical development, which in turn influenced A. R. Luria's ideas.[1] And it was Rivers's careful field notes that formed the basis for refuting many of the dominant beliefs of his day regarding the superiority of European intelligence. Responding to the popularly held belief that the inhabitants of "lowly cultures" were cognitively impaired due to their lack of concentration, Rivers wrote,

> I had hoped that this belief in the difficulty of mental concentration among peoples of lowly culture had long been exploded. Apart from its incompatibility with all that we know of the strenuous conditions to which early man must have been exposed if he were to keep himself and his young alive, the statement is definitely contradicted by all careful observation of existing lowly peoples. The belief rests on the uncritical observations of travellers who found that the attention of members of native savage and barbarous races rapidly flags when they are questioned about their customs. Such observations rest partly on the wish of those questioned to avoid such inquiries, partly on the fact that the inquirers have failed to arouse the interest of those questioned, or have even diverted interest by their obvious failure to understand the elements of the matter they were curious about. (Rivers, 1926, p. 51)

Recently, cognitive psychologists like Sternberg (1985, 1986), developmental psychologists like Charlesworth (1976, 1979), and social psychologists like Berry and Dasen (1973) have sounded a similar theme of "intelligence as adaptation." For example, Sternberg has defined practical intelligence as "intelligence that operates on real-world contexts through efforts to achieve adaptation to, shaping of, and selection of environments" (Sternberg, 1986, p. 301). It is only recently, however, that anthropological insights have begun to influence the thinking of experimental and developmental psychologists interested in intelligence (Ceci and Liker, 1986a, 1986b).

There are many reasons why psychologists have ignored the insights of anthropologists, but perhaps no single reason is more important than the clash of disciplinary values. Simply stated, psychologists and anthropologists have operated under differing assumptions about the nature of knowledge and scientific explanation. As a result, the rules of evidence can be quite different in the two disciplines. One suspects that many psychologists, while viewing anthropological studies as interesting, tend to regard them as scientifically unsound (in the Weberian sense) or at best as merely suggestive of causal relationships (Wakai, 1985). Moreover, my impression is that many psychologists who study individual differences in intellectual development, especially those who do so within the psychometric tradition, feel at odds with what they perceive to be the modal anthropological tenet, namely, that "everyone has everything, but different cultures bring to fruition different intellectual traits." Such egalitarianism runs counter to the bulk of their research, particularly the heritability research that will be reviewed in Chapters 5, 6, and 7.

Recently, it has become difficult to maintain a separation from the anthropological literature, as many anthropologists have begun to publish results that would appear to satisfy psychological standards of evidence. For example, Lancy and Strathern's (1981) cultural-linguistic analysis of memory differences between two neighboring tribes in the Melpo Islands is replete with traditional psychological constructs such as clustering and free recall. Lancy and Strathern initially were confronted with substantial differences between two neighboring villages in the amount of items they could recall. Ever since Margaret Mead studied the Melpo islanders, it had been thought that the lowlanders, who earned their living mainly by the sea, were smarter than the highlanders, who were agricultural. (In fact, it is one of Mead's unfortunate legacies that she wrote quite unfavorably about the limited cognitive abilities of highlanders, whose children now have had the opportunity to read her accounts of their parents "deficits.") Lancy and Strathern discovered that the highlanders' memory "deficits" disappeared when care was taken to select test stimuli that were based on their language's inherent polarity. It is a nice demonstration of the role of cultural artifacts and ecology in producing *apparent* group differences in intellectual functioning.

Since Lancy and Strathern's study, numerous others have been documented. For example, Jeyifous's (1985) account of the Yoruba of Western Nigeria demonstrates that their metaphorical skills are under the influence of their ecology. (This will be described in Chapter 5.) Similarly, studies have shown that visual-spatial abilities and memory skills as well as the understanding of metaphor, depend on the developing organism's ecology. For example, aboriginal children of Australia outperform white children in that country on visual-spatial memory tests and this advantage has been linked to their use of a visual strategy (Kearins, 1981), whereas white children outperform aboriginal children (and even some aboriginal adults) on Piagetian types of logical operations—*unless the latter live in urban communities* (Dasen, 1973). Mexican-Americans are more field-dependent than Anglo-Americans, yet by their third generation in the U.S. they are more field-independent than their ancestors (Saracho, 1983). And within India, urban dwellers are more field-independent than rural dwellers (Chatterjea and Paul, 1981). Finally, the type of job one does in business influences their field independence—accountants are less field dependent than students still in business school (Pincus, 1985). So, the ecology one occupies appears to be quite influential in a variety of cognitive attainments. None of this, of course, is meant to deny the role of biological factors.

Let us consider in greater detail a few recent anthropological studies that have included mental test scores as part of their investigation of cognitive functioning. Each of these studies raises questions about the adequacy of standardized assessments of cognitive potential.

Grocery Shopping

During the past decade, Jean Lave and her colleagues at the University of California at Irvine have been studying everyday cognitions of Americans, for example, planning for retirement and counting calories on diets. In one of their studies, a student of Lave's examined the consumer purchases of California grocery shoppers. Murtaugh (1985) accompanied shoppers on their tour of a California supermarket and attempted to determine the cognitive basis for their purchases. The supermarkets did not post unit prices that would have informed customers of the cost of each product per unit of its volume. Often, shoppers were confronted with comparisons between same-brand items of unequal sizes and prices (e.g., a brand of rice might sell for .47/lb, .99/32 oz, and $1.39/48 oz). Occasionally, shoppers prefer to pay more for the convenience of a smaller container. However, Murtaugh found that when their goal was to optimize volume for price, the large majority of shoppers made the correct unit price decision. Lest one think that the correct unit price decision can be made simply by selecting the largest size, it is worth noting that this relationship between size and cost did not hold for approximately one-third of the selections and therefore could not have been the basis for shoppers' decisions.

Shoppers in this study usually achieved high levels of performance and did so without the aid of pocket calculators or paper-and-pencil calculations. A dialectic existed between the problem itself and its potential solution, so that shoppers often modified the goal as a function of what was needed to arrive at an answer. For example, if the first "run-through" of the solution process revealed that the shopper would not have room in the refrigerator for the quantity at issue, the goal was modified to allow for only a few days' supply rather than a week's. Hence, problem formation is tied to problem solution—a qualitatively different situation than one might encounter on a standardized test or in school. Those who were able to describe the thought processes leading to their choices reported that they performed the arithmetical calculations mentally, using addition, subtraction, multiplication, and division. But the shoppers reported departing from purely algorithmic decisions, using qualitative manipulations instead, if these manipulations provided additional information of value. For example, if the quantities were close to a two-to-one ratio, then doubling followed by a comparison would be used. If the quantities were close in size, for instance, 20 ounces @ $1.79 vs. 24 ounces @ $1.89, a comparison might take the form of "you get four more ounces for ten additional cents." Notice how much richer this routine is than a purely algorithmic one ($.08095/ounce vs. $.07875/ounce), even though both procedures yield the same "best buy." It also makes calculations easier, even compared to a hand calculator.

Of course, one can inquire about Murtaugh's same shoppers when they attempt to calculate sums in other domains, e.g., their tax returns. However, even if they were shown to resort to algorithmic solutions in these other domains, this does not diminish the importance of the previous findings, nor does it establish the superiority of algorithmic approaches in many real-world exercises. Consider the experience of a colleague of this writer with a cashier at a local hamburger restaurant. He handed the cashier a five-dollar bill for several burgers, drinks, and french fries, only to receive over three dollars in change. Clearly, the change was excessive, but the cashier refused to discuss the matter, pointing to the computerized amount the register indicated was owed to him.

An interesting postscript to Murtaugh's study occurred when the shoppers were given the M.I.T. mental arithmetic test, which assesses the same mental arithmetic operations shoppers purported to have used in making their supermarket decisions. No relationship was found to exist between performance on the test and subjects' shopping accuracy (Lave, et al., 1984; Murtaugh, 1985). In effect, the shoppers' behavior was unpredictable from their standardized test scores. I shall return to this and similar examples of context-specific cognition later.

Dairy Workers

During the early 1980s, Sylvia Scribner (a psychologist) and a group of psychologists and anthropologists studied industrial literacy in America. In their examination of a Baltimore dairy factory these researchers discovered that experienced assemblers, the people responsible for filling cases of pre-specified sizes (gallons, half gallons, quarts, pints, and half pints) of whole milk, two-percent milk, skim milk, buttermilk, and chocolate milk, made rapid assessments of nearby cases that were partially filled in order to avoid beginning an order with an empty case. Beginning an order with an empty case requires a good deal of back bending to fill the order, and the workers appeared adept at selecting partially filled cases that allowed orders to be filled with the least amount of bending. For example, if an order called for 6 pints of whole milk, 12 pints of two-percent milk, and 3 pints each of skim milk and buttermilk, an experienced assembler might select a case for 24 pints that was already half-filled with two-percent milk and one-third filled with whole milk, rather than try to prepare the order from scratch with an empty case. Using the half filled case would enable the assembler to fill the order by removing 2 pints of whole milk and adding 3 pints each of skim milk and buttermilk, for a total of only three bends. Moreover, when the orders were not evenly divisible into cases, the assemblers were able to shift between different representations of the order, a feat equivalent to shifting between different-base systems of numbers:

> On all occasions the mode of order-filling. . . was exactly that procedure which satisfied the order in the fewest moves—that is, of all alternatives, the solution the assembler selected required the transfer of a minimum number of units from one case to another. Assemblers calculated these least-physical effort solutions even when the "saving" in moves amounted to only *one* unit (in orders that might total 500 units). Least

effort solutions required the assembler to switch from one base number system to another. The mental effort involved in the problem transformations was increased by the fact that assemblers typically went for a group of orders at a time, thus having to keep in mind quantities expressed in different base number systems. (Scribner, in press, p. 10)

In a follow-up experiment to confirm these observations, Scribner (1984) demonstrated the assemblers' skill at making rapid assessments of the possibilities of partially filled cases for making up orders. As a group, assemblers were the least educated workers in the factory, yet they were more successful at making rapid assessments of partially filled cases than were more educated white collar workers who occasionally substituted on the assembly line when an assembler was absent. Interestingly, there was no relationship between this skill and various high school test scores (including IQ, arithmetic test scores, and school grades).

!Kung San Hunters

Numerous ethnographies have documented the complex thinking and reasoning skills of peoples of the world who are known to be unremarkable in their performance on Western-style cognitive tests (e.g., Gladwin's, (1971) discussion of the elegant navigational skills of Pacific islanders who fail to solve many of the problems we expect Western children to solve, and Lewis' (1976) description of the route-finding skills of Australian aboriginees over a 500,000 square mile area of the Western Desert region, noting subtle land depressions and other features of the landscape). Perhaps the best chronicled ethnography for present purposes is that of the !Kung San people of the Kalahari Desert, a hunting and gathering society that ranges over portions of Western Africa. In the past decade, these people have been extensively studied by several cultural anthropologists and cross-cultural psychologists (e.g., Reuning, 1988). Super and his colleagues at Harvard have provided elaborate descriptions of the !Kung San men's behavior during hunts (Super, 1980). To a reader of these accounts, it appears that the men behave in a cognitively complex manner during hunts, gleaning clues from their environment, differentially weighting the import of these clues for the hunt's success, and arriving at some decision. The following excerpt captures some of this complexity:

We catch a glimpse of the kinds of reasoning !Kung San men use: defining the problem of whether it is worthwhile tracking a giraffe that is known to be wounded but may be capable of evading them for days while leading them away from their home territory; searching for available clues such as the pattern of crushed grasses, the color of droppings, and whether the blood on a twig fell there before or after it was bent; evaluat-

ing the possible interpretations and their import for the hunt's final outcome; and planning a course of action. (Super, 1980, p. 61)

!Kung San hunters evaluate clues such as whether a twig was splattered
with a giraffe's blood before or after it was bent because if it was before, the
giraffe was still standing tall while feeding on the branch and thus could be
expected to have the resources to evade the hunters for several more days,
whereas if the branch was already bent when the giraffe's blood marked it,
this might suggest that the animal was craning low to the ground during
feeding as a result of its injury. Similarly, !Kung San hunters evaluate the
pattern of crushed grasses and animal droppings to yield further information
about the giraffe's behavior. Not all of the clues are equally important in the
decision to continue or abandon the hunt.

It is difficult not to be impressed by the !Kung San hunters' adroit
problem-solving abilities. Evidence for their cognitive complexity abounds in
these ethnographies. Yet the !Kung San men perform only on par with
Western children when they are administered IQ tests, Piagetian tests, and
some information processing tasks (see Reuning, 1988), even when these
tasks are adapted to take account of the local culture (e.g., by substituting
familiar vegetables for unfamiliar fruits on the IQ test). How can one reconcile these men's obvious cognitive complexity in meeting the most important
environmental challenges in their lives with their poor performance on
(Western) micro- and macro-level aspects of traditional intelligence? And lest
one think that the example under discussion is a rarity in cross-cultural
research, similar observations have been made about unschooled peoples by
Scribner (1976, 1986), Cole (1975), Rogoff (1981), Gladwin (1971), Wakai
(1985), and many others (see chapters in Irvine and Berry, 1988).

Is the answer to the foregoing question that these forms of everyday
problem solving are simply hypertrophied skills, honed through prolonged
practice with specific experiences, but lacking in the flexibility, abstractness,
and generality that conform to traditional definitions of intelligence? (See, for
example, Sternberg, et al., 1981.) Truly intelligent persons, it might be
thought, are able to function at a more abstract level, enabling them to use
their school-based knowledge flexibly in and out of school contexts. And
even though there may be instances in which they transcend their school-
based knowledge when operating in the nonschool world, they should never
fall below it. After all, it is better to find examples of people superseding their
school-based algorithmic knowledge in some real-world contexts than to find
evidence for the reverse, that is, instances in which real-world performance
was worse than school-based performance. Imagine what an indictment of
American schools it would be if someone who passed high school calculus
could not figure their bowling average! Or is the answer to the above question that the traditional ways of assessing intelligence, and the theories that
underpin them, are insufficient in capturing the richness and the contextually

driven nature of complex cognitions outside of traditional verbal analysis? In Chapter 9, I shall tackle the issue of the presumed greater "abstractness" of intelligence and also attempt to answer these other questions, but first I shall review some recent work from experimental psychology as well as some historical findings of educational researchers and sociologists that complement the anthropological research just reviewed and provide some of the "scaffolding" of the bioecological framework.

EXPERIMENTAL PSYCHOLOGY

Recently, a number of experimental psychologists have begun to question the traditional notions surrounding intelligence and IQ, arguing for a more ecologically based conceptualization of cognitive complexity (Cole, 1975; Keating, 1984; Neisser, 1979). In the last four years, reports have appeared by experimental psychologists that support the anthropological findings previously cited. These studies demonstrate that even within a given cognitive domain, there are substantial mismatches between performance in one setting versus another. And across cognitive domains, there appears to be far less correlation between one measure of cognitive complexity and another than previously has been assumed to exist. Alone, none of these studies provide incontrovertible evidence for the contextual nature of intelligent behavior, but together they call into question a prior era's unrivaled assumptions regarding the generality of intelligence and the mere adjunctive status of context. They demonstrate the importance of contextual variables as a *constituent* in the perception and solution of complex problems, rather than simply a sociophysical address at which cognition unfolds. We turn next to an examination of some of this experimental research.

Cupcake Baking

Ceci and Bronfenbrenner (1985) have reported the results of a developmental study in which children of various ages were asked to remember to do things in the future, such as remove cupcakes from the oven in 30 minutes or disconnect a battery charger from a motorcycle battery in 30 minutes. While waiting to do these things, the children were invited to play a popular video game. The data of interest concern children's clock-checking behavior while waiting for the 30 minutes to elapse. As can be seen in Figure 3.1, children behaved differently as a function of the setting in which they were studied. When observed in the familiar context of their own homes and in the company of their siblings, children appeared to "calibrate" their psychological clocks through a process of early and frequent clock checking. These early checks permitted the children to synchronize their psychological clocks with

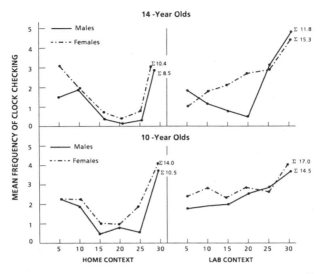

FIGURE 3.1 Children's clock-checking frequencies on the cup-cake-baking task. Late responders excluded Σ = Mean total number of clock checks. (Reproduced Courtesy of *Child Development.*)

the passage of actual clock time. For example, children might begin the waiting period by making several confirmatory checks to ensure that the amount of time that had already transpired was close to their subjective estimate. After several such confirmatory checks, children gained the confidence to allow their psychological clocks to "run" (unchecked) until nearly the end of the 30-minute period, whereupon last-minute incessant clock checking occurred, as evidenced by the rightward scalloping in the figure.

The advantage of using a calibration strategy is that it permits children to engage effectively in other activities (e.g., playing video games), unencumbered by the need to look constantly at the clock. It also allows a maximum degree of precision with a minimum amount of effort. Thus, the use of the calibration strategy does not result in a loss of punctuality: None of the children who gave evidence of employing this strategy burned the cupcakes or overheated the motorcycle battery. Support for the assumption that children were indeed synchronizing their psychological clocks with a nearby wall clock was provided by showing that the U-shaped distribution of clock checking was recoverable even when the wall clock was programmed to run faster or slower than real time (Ceci, Baker, and Bronfenbrenner, 1988). That is, subjects were able to adjust their subjective estimations of the passage of clock time, and once this adjustment was achieved, they were successful at gauging the remainder of the waiting period with only a minimal amount of glancing at the clock.

In the laboratory, however, children displayed no evidence of using a calibration strategy. They, too, rarely burned the cupcakes, but they required nearly a third more effort (i.e., clock checks), with the result being a lessened ability to engage effectively in video game activities during the waiting period. With the exception of older boys who were asked to engage in a traditionally female sex-typed task (baking cupcakes), there was no evidence of calibration in the laboratory setting. These data point to the influence of context on strategy use. Here, context is conceived as not only the physical setting in which the task unfolds (laboratory or home), but the sociocultural features as well (e.g., the sex-role expectations of the task, the age-appropriateness of the task, the presence or absence of familiar persons). Unlike the traditional information processing conceptualization of context as something adjunctive to cognition (i.e., a social/physical address where cognitive tasks are performed), these findings suggest that context should be viewed as a constituent of the cognitive task, influencing the manner in which the task is perceived and the choice of strategies for its completion. Had the investigators assessed children's competence only in the laboratory setting, they would have been led to underestimate the sophistication of their strategies. Conversely, had they observed children's clock checking only in the children's homes, they would have missed the significance of many of the ecological contrasts that the laboratory comparison afforded. A number of other studies by experimental psychologists show similar contextual effects on cognitive strategy use (e.g., Accredolo, 1979).

Capturing Butterflies

In a recent study, Ceci, et al. (1987) extended the cupcake study in a way that allows a direct assessment of the impact of context on cognitive complexity. Their approach was straightforward. They instructed children to predict where, on a computer screen, an object would terminate by placing a cross on the screen at that location (by moving a joystick). The object was one of three shapes (square, circle, triangle), two colors, and two sizes, yielding 12 combinations of features. A simple additive algorithm was written to "drive" the stimuli so that squares would go up, circles would go down, and triangles would stay horizontal. Similarly, dark-colored objects would move right and light-colored objects would move left. Large objects would move on a lower-left to upper-right diagonal, while small objects would move along the opposite diagonal. There were no interactions in the algorithm. Children were given 15 sessions of 50 random trials each to provide probability feedback. Even after 750 probability feedback trials, however, their prediction accuracy was only 22% (lower lines in Figure 3.2).

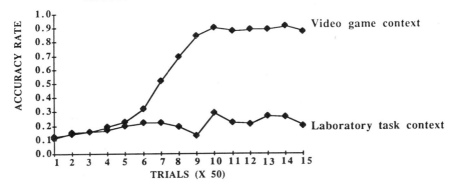

FIGURE 3.2 Children's mean proportion of accurate estimates of a moving object in game versus laboratory contexts (simple main-effects algorithm).

Next, the same algorithm was used to drive a video game. The three geometric shapes were converted to a butterfly, a bumblebee, and a bird. (The same colors and sizes were used.) Instead of placing a cross on the area of the computer screen where they predicted the object would terminate, children were told to place a "butterfly net" to "capture the prey," by moving a joystick. Children were awarded points for each correct capture, and sound effects were added to complete the video game context. As can be seen in Figure 3.2, this shift in context resulted in drastically improved performance. After 750 feedback trials, the children's accuracy was near ceiling.

In an extension of this study, a more complex curvilinear algorithm was used to drive the objects. The following sine function was programmed to determine the distance that either a rectangle would move (in the disembedded context) or the distance a truck would travel before reaching a roadblock (in the video game context):

$$.8 \sin(x) + .6 \sin(y) + .4 \sin(z) + 5\% \text{ error}^2$$

Again, there was a substantial enhancement of performance when the task was embedded in the presumably more motivating video game context, although the overall levels of performance were not as high as those found with the simple additive algorithm (Figure 3.3). These data are important in dispelling the belief that young children are "main effects" thinkers, incapable of grasping more complex, multiplicative models (see Klayman, 1984, for evidence that even college students are unable to assimilate a curvilinear function in a probability learning exercise such as the one under discussion).

CURVILINEAR DISTANCE ESTIMATION TASK

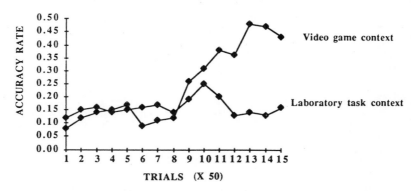

FIGURE 3.3 Children's mean proportion of accurate estimates of a moving object in game versus laboratory contexts (multiplicative algorithm).

Had the children been tested in only the disembedded laboratory prediction context (squares, circles, etc.), a vastly underrated estimation of their competence would have resulted.

These results suggest that children have at least some ability to understand complex (multicausal) models of probability. This may come as no surprise to parents who have spent Saturday afternoons watching their children successfully navigate complex video games at the local mall; the games often require one to consider the influence of one variable on the operation of another. It does give one pause, however, in view of the characterization of American college students as poor at multicausal reasoning, committing a variety of errors, and ignoring or discounting second causes. (See Nisbett and Ross, 1980, for an excellent, though disheartening, review.) It is worth noting that the vast majority of the studies reported by Nisbett and Ross deal with subjects' responses to vignettes that are assumed to be motivationally similar for both the students and the experimenters (e.g., that money is a potent inducement to do something). We shall return to this point later.

City Manager

Dietrich Dörner and his colleagues at Augsburg, Germany, have provided some fascinating data concerning the mismatch between intelligence, as assessed by a standard g-loaded IQ test, and intellectual performance, as assessed by an analysis of one's "everyday" problem-solving behavior (Dörner, et al., 1983; Dörner and Kreuzig, 1983). In one of their studies, subjects were asked to assume the role of a city manager. Seated before a

CRT, the subjects are instructed in the needs of the city, a mythical town called Lohhäusen. Lohhäusen has all of the problems of real cities—inflation, unemployment, the need to raise revenues to build roads, etc. The subjects are confronted with a complex set of competing needs, not all of which are equally important. For example, the need to raise revenues in a manner that has the least impact on the local interest rates may release pressure on local investment, and hence unemployment, but it also may result in less attractive municipal bonds—hence, less revenue generated. Over a thousand variables are identifiable in this task and the subjects' success at managing Lohhäusen is measured in terms of the amount of revenue remaining at the end of the simulation.

Dörner and his colleagues devised a nine-step hierarchy to describe the various cognitive strategies one might use to solve Lohhäusen's problems, from the simplest (trial and error) to the most sophisticated (a systematic hypothesis-testing approach with multiple feedback loops and metacomponents). They consistently have failed to discover any relationship between IQ and level of cognitive complexity in managing Lohhäusen. Similar conclusions have been reached by Dörner and his colleagues on a different task, one involving the number of camels that can be supported on a small oasis—the so-called Sahara problem. Some of Dörner's German colleagues have criticized his approach on the grounds that there is no *a priori* specification of an optimal strategy, but others have disagreed and argued that convergent evidence from other studies strengthens the findings. It is too soon to evaluate the validity of Dörner's claims, but whatever it is that an IQ test measures (speed of acquisition, novelty learning, abstract reasoning, etc.), one would assume that it has some relationship to the ability to deal with novel constellations of variables in cognitively complex tasks like Lohhäusen and Sahara. Tentatively, the work of Dörner casts serious doubt on such an assumption, as does the following study.

A Day at the Races

In a study of expert racetrack handicappers, a sociologist named Jeff Liker and I provided evidence along the lines of Dörner's study. An advantage of our study lies in the *a priori* optimization rule that we used: success at predicting post-time odds at the racetrack. We began by identifying two groups of men who attended the races nearly every day of their adult lives. Although both groups were far more knowledgeable than amateurs about horse racing and were far better at predicting post-time odds than amateurs, they differed greatly among themselves. One group, which we called "experts," was rather amazing at estimating what the odds on each horse would be at post time, while the other group, which we labeled "nonexperts," was less dazzling.[3] We asked the men in these groups to handicap real races as well as hypothetical races that we designed. The latter were included to separate variables that were often too correlated in actual races to determine their importance in an expert's decision. In the hypothetical races, these variables were allowed to covary systematically.

In statistical jargon, we demonstrated that expert handicappers employed a complex, multiplicative model involving multiple interaction effects. By regressing 25 racetrack variables on experts' assessments of odds, we were able to show that simple additive models failed to account for the complexity of their decisions. For readers unacquainted with statistical jargon, in substantive terms, we found that experts not only took into consideration more information (i.e., variables) when handicapping a race, but they did not simply "add up" this information. Rather, they developed implicit algorithms that gave more or less weight to different types of information. And each type of information changed the way they thought about the other types. The following illustration may help clarify this.

To successfully predict the speed with which a horse could run the final quarter mile of a race, experts used what is termed a seven-way interactive model. Use of this complex interactive model discriminated the experts from the nonexperts in the ability to predict odds, with the experts always using it and the nonexperts seldom using it. Here is an example of what is required to estimate a horse's "true" closing (stretch) speed. In the racing program, you observe that horse A's speed statistics for each quarter mile were 29^2, 59^1, $1:30^4$, and $2:02^2$ (superscripts are fifths of a second). Suppose you also notice in the program that A's ordinal position at each quarter mile as well as the number of lengths A was behind the leader were $\frac{1}{4}$ mi., $4^{\circ\circ}$; $\frac{1}{2}$ mi., 5°; $\frac{3}{4}$ mi., 4^6; $\frac{4}{4}$ 6^2. The tiny circles denote how many cart widths from the rail horse A was at the indicated point in the race and signify that A was actively attempting to pass horses along the rail in front of him, moving away from the rail to do this. The superscript numerals for $\frac{3}{4}$ mile and the final quarter ($\frac{4}{4}$) mile refer to how many lengths behind the leader A was at those points in the race, i.e., 6 lengths behind at the $\frac{3}{4}$ mile, with three other horses in front of him (i.e., ordinal position 4), and 2 lengths behind the winner at the end of the race, with five horses who beat him (i.e., ordinal position 6).

Now, here is part of the thinking that goes into computing A's closing speed. The expert handicapper first notes that the winner ran the race in $2:02^2$ and that A finished two lengths behind him in sixth place. Each length is roughly equal to one-fifth of a second, so A ran the race in approximately $2:02^4$, since he was two lengths behind the winner's pace. But this $2:02^4$ time is his official speed and says nothing about A's "true" speed in this particular race. For one thing, A was off the rail a lot, sometimes by several cart widths. Each cart width off the rail means that he was running a longer perimeter at that point in the race. Because the shortest distance around the track is exactly one mile along the rail, A actually ran further than a mile because he was often away from the rail. The expert will also note that A gained four lengths on the leader during the final quarter mile, even while dropping from fourth place to sixth place. Thus, if A's "apparent" final quarter-mile speed is 31^3 seconds (i.e., $2:02^2$ minus $1:30^4$), then his "true" speed was faster by four lengths, i.e., $\frac{4}{5}$ of a second. Thus, A ran the final quarter mile in 30^4 seconds (31^3 minus $\frac{4}{5}$ of a second). The expert handicapper will usually mentally calculate these adjusted closing speeds for each horse in a race, to get some idea of how fast the horse is capable of "closing" under various scenarios having to do with distance off the rail, absolute speed at the time the horse was off the rail, turf conditions, etc. In this example, A was two widths off the rail trying to pass front runners during a brisk 29^2 first quarter mile while in fourth position. Experts realize that this type of move drains a horse of energy more so than if the horse tried to pass front-runners at a slower pace and while closer to the rail. Their opinion of A's actual ability will reflect these and other factors—seven in all, combined in a complex algorithm. While none of these men verbally articulated this model, its validity was evident from their systematic decisions.

The correlation between an expert's IQ score and the b weight for this seven-way interactive term (which is a surrogate for cognitive complexity) was $-.07$. This means that even though the greater use of complex, interactive thinking was causally related to success at the racetrack, there was no relation between such complex thinking and IQ or between IQ and success at estimating odds. Thus, assessment of the experts' intelligence on a standard IQ test was irrelevant in predicting the complexity of their thinking at the racetrack (Ceci and Liker, 1986a; 1988). Within either group (experts or nonexperts), IQ was unrelated to handicapping complexity. Between groups, however, there was an invariant finding: Experts with low IQs always used more complex, interactive models than did nonexperts with high IQs, and their success was due in large part to the use of these complex interactive models. IQ was useless in predicting both how complexly these experts reasoned and the number of variables they considered interactively in their judgments. (Interestingly, the success of experts at making these computations depends, in part, on their skill at doing mental arithmetic and, in particular, at subtractions that cross fifths boundaries. Yet this skill was unrelated to their scores on the mental arithmetic scale of the IQ test, too.)

Taken together, the experimental psychological research just reviewed (baking cupcakes, capturing butterflies, managing a mythical city, and handicapping real races) shows the need to expand earlier notions of intelligence that were developed within the psychometric and information-processing traditions. It is not only that individual differences in IQ test score variance did not coincide with individual differences on these macrolevel forms of cognitive complexity, but the likelihood of using even microlevel cognitive strategies that underpin more complex cognitions was shown to be under the influence of ecological variables such as the sex-role expectations of the task, the physical setting in which the task was performed, the motivational level of the task, and the performance context (game vs. laboratory task). Does this mean that psychometric and information-processing frameworks have nothing to offer the field of intelligence research? No, definitely not. It merely indicates the need to go beyond the present boundaries established by researchers within these traditions if we hope to incorporate the types of results just presented. Expansions of theories of intelligence based on information processing and psychometric approaches should be possible because both traditions provide relevant constructs for thinking about individual differences in intelligent everyday performance. For instance, early proponents of information processing stressed the importance of stimulus "coding." And we have seen that whether an event is coded as a square or as a butterfly is important in activating different semantic pathways and strategies, which in turn determine the level of performance. Moreover, coding is obviously part of what is entailed by the notion of context, but only part (see Chapter 6).

At this point, I have tried to implant the suggestion that something is not quite right with the argument advanced by those who assert that IQ equals intelligence. Assumptions about the singularity and innateness of intelligence were challenged by the cultural anthropological and experimental psychological research just presented. A view of performance has begun to emerge suggesting that cognition is tied to specific knowledge contexts and to motivation. But the argument has a long way to go yet and will not be completed until a fuller exposition of the psychometric and genetic findings takes place in Chapters 7 and 8. At this point, one might argue that the lack of a correlation between IQ and expertise in the various studies described in this chapter (hunting, handicapping, managing a town, etc.) is due to the highly specific nature of the tasks employed. One might suppose that high IQ persons will outperform low IQ persons in the vast majority of domains that lie outside these tiny pockets of expertise. I shall examine the validity of this supposition in Chapter 9 when I take up the issues of abstractness and generality. But the next step in my argument requires me to demonstrate just how labile IQ is in response to contextual and motivational changes—because if one wishes to attach some deep importance to IQ, one needs to be aware of just how labile it is in response to shifts in contexts such as schooling. Before considering educational research in Chapter 5, I want to add a third line of evidence to supplement the anthropological and psychological studies that I have already scanned, namely, that of sociologists. In the next chapter it will become necessary to muddle the picture somewhat and speak interchangeably of IQ, cognitive complexity, and intelligence because the sociologists who have researched this area have focused on the structural antecedents (e.g., social class, workplace complexity, and family size) of cognitive outcomes rather than on the outcome measures themselves.

CHAPTER 4

A Social-Organizational Analysis of Intellectual Development

Sociological analysis is relevant to any discussion of intellectual differences on several levels. At the most general level of analysis, the construct "social organization" has been associated with different categories of thinking. For example, around the turn of the century the classic sociological theory of Emile Durkheim and Marcel Mauss held that "the genesis of the categories of thought is to be found in the group structure. . . .and categories vary with changes in the social organization" (reported in Merton, 1968, p. 519).

Many studies, both historically and more recently, have confirmed that social structure exerts a potent influence on intellectual development. For example, Blau (1981) showed that the proportion of one's neighbors who attended college or whose neighbors had children who attended college was the second best predictor (the first being race) in a multiple regression analysis predicting IQ. This index of social milieu exceeded other indices of SES, such as income and occupational prestige (cf: Mayer and Jencks, 1989, for a more qualified conclusion of the impact of mixed neighborhoods on the attainment of poor youngsters). And numerous investigators have commented on the declining IQ with birth order within a given family (Zajonc and Bargh, 1980). Even critics of this view agree that at least 2% of the intrafamilial variance in IQ is accounted for by birth position—and the concomitant effects of being born into a family in which adult attention is more or less dispersed among few or many offspring (Galbraith, 1982). Therefore, even among a group of children who are otherwise similar, there can be important ecological variations that contribute to their intellectual development. No doubt, some of these variations explain why children within the same family differ, on average, by about 11 points in IQ from each other (and from their parents). In his recent reviews of the literature on the effect of social organization on children's developmental outcomes, Urie Bronfenbrenner (1989; 1990) urged a more intense examination of the features of the organism's ecological niche that contribute to different patterns of intellectual development. By "ecological niche," he meant,

> regions of the environment that are especially favorable or unfavorable to the development of individuals with particular personal characteristics. Operationally, niches are defined by the intersection between one or more social addresses and one or more personal attributes of individuals who live at these addresses. (Bronfenbrenner, 1989, p. 191)

SOCIAL CLASS AND IQ

Despite the long and rich history of theorizing about social organization, I shall focus my brief discussion here on one of its components, namely, *social class*, and discuss its presumed relation to differences in intelligence among individuals and groups. To some extent, this decision is necessary to avoid getting sidetracked (or more sidetracked!), but to some extent the most relevant lesson for cognitive psychologists to learn from sociology is to be found in the narrower literature dealing with class differences and their correlates. This is where the bulk of social research can be found.

Although many sociologists do not use the term *social class* when referring to American groups that differ in wealth, education, or political power, because that term has structural implications that derive from its application to 18th century European society that is not germane to North America, the literature relating to intellectual development is replete with studies that examine the influence of *social class*. In this literature, the concepts of social class, SES, and social status are "carrier variables," that is, concatenations of many other variables. As such, the myriad of components of social class (family income, parental education, occupational prestige, attainment values, housing cost, child-rearing practices, population density, demographic characteristics of the neighborhood, number and type of appliances in the home, etc.) need to be "unpackaged" and examined individually for their influence on intelligence. Some components of social class may be expected to have more of a causal role in intellectual development than others. For instance, from the earliest studies to the present, parental educational level has almost always been a more potent predictor of children's IQ ($r = .50$) than has family income ($r = .30$) (Loevinger, 1940). Perhaps this is why some studies of intellectual development that have used a single, combined index of social class have reported only a very modest relationship with IQ. (The earliest studies of this relationship did report rather substantial correlations between social class and IQ, educational achievement, and later occupational success, but all of these studies suffer from one or more methodological problems that I shall not delve into here.) In a seminal review of the literature linking social status with IQ, Loevinger warned against the use of a global categorization of social status, arguing that if researchers "seek the specific medium of influence [by which social status affects IQ] results more concrete, more conservative, and more fruitful would result from investigation of specific concomitants of social status in relation to intellectual performances" (Loevinger, 1940, p. 205). Recently, Bradley, et al. (1988), taking an apparent cue from Loevinger, showed that global measures of social class were far less predictive of cognitive development in the first three years of life than were specific aspects of the children's home such as the availability of stimulating toys and responsive parents. And Mayer and Jencks (1989) have found that the influence of many social class components shrunk with more sensitive measurement and that the picture regarding the importance of having racially mixed neighborhoods and schools (in terms of increasing graduation rates and college matriculation rates) was not great but was different for various racial and ethnic groups.

And yet, despite the failures to find pervasive effects for global measures of social class, some findings did appear to be robust. By the 1960s, researchers had reached a consensus: When the relationship between IQ and educational attainment was assessed with social class held constant, the reduction in the correlation between IQ and education due to controlling for social class was quite small. For example, both Kemp (1955) and Wiseman (1966) reported correlations between IQ and educational attainment (standardized achievement test scores in arithmetic and English) that were quite high (e.g., rs = .73 to .77). When social class was partialed out, however, these correlations dropped only slightly (rs = .62 to .74).[1] Similarly, Jensen (1972, 1976, 1984), among others, noted that the causal path between IQ and social class was due almost entirely to educational attainment, as opposed to other components of social class (see also Block and Dworkin, 1976). This should come as no surprise in view of the many exceptions to the relationship between some components of social class (e.g., income) and educational attainment. For example, even though teachers, clerics, and social workers frequently earn less than carpenters, plumbers, electricians, and athletes, despite comparable or superior educational attainments (and IQs), they are nonetheless regarded as having higher occupational status:

> We know that test scores play a significant role in determining school grades, in determining how long students stay in school, and in determining what kinds of credentials they eventually earn. We also know that credentials play a significant role in determining what occupations men enter. Occupations, in turn, have a significant effect on earnings. But at each stage in this process there are many exceptions, and the cumulative result is that exceptions are almost commonplace. A significant number of students with relatively low test scores earn college degrees, for example. In addition, a significant number of individuals without college degrees

enter well-paid occupations, especially in business. Finally, people in relatively low status occupations (e.g., plumbers and electricians) often earn more than professionals (teachers and clergymen). . . .The key point is that it doesn't matter much whether IQ differences between Blacks and Whites are hereditary or environmental. IQ accounts for only a quarter of the income gap between them. (Bane and Jencks, 1977, p. 329)

Thus, when social class is decomposed into its constituents, it is clear that some of them do appear to bear a stronger relationship than others to both IQ and educational attainment. For instance, work patterns that are associated with different types of employment have been shown to influence the intellectual development of adult workers (Kohn and Schooler, 1978). Patterns like the degree to which one's job involves tasks of a routine nature and how closely one is supervised bears on "intellectual flexibility," an amalgam of cognitive information having to do with causal reasoning, substantive complexity, and IQ. The Kohn-Schooler model is a theory of inter- and intra-generational change in which early personality and intellectual complexity draws individuals to different types of occupations whereupon, as a result of the work environment in those occupations, the individual's personality and intelligence undergo alterations wrought by the same characteristics that drew the person to that job in the first place!

In a longitudinal study of workers that replicated their earlier findings from a cross-sectional study, Kohn and Schooler demonstrated that even after one statistically controls for the impact of IQ on job selection (in a structural equation model), there is an undeniable reciprocal impact of work patterns on IQ. And the impact appears quite substantial in the sense that for every 10 years on the job, workers systematically shifted several IQ points in response to the demands of their jobs. In turn, these changes in IQ manifested themselves in changes in subsequent work patterns. Recently, Kohn and Schooler have reported seeing this effect in a variety of work contexts, including homework, housework, and professional and blue-collar jobs, as well as in diverse cultures, e.g., Japan and Poland. One complaint raised by some has been that the individuals who are asked to rate the complexity of their jobs are the same persons who are characterized in terms of their complexity, thus resulting in some built-in correlation. It appears, however, that the same results are obtainable even when others are asked to analyze the subjects' jobs.

Finally, it should be noted that many psychologists have reservations about the use of statistical procedures to control or partial social class out of prediction equations. The reason for their concern is that such procedures can obscure the "collinearity" among SES, IQ, and real-world success (e.g., school and work performance), so that controlling for one variable (e.g., SES) can lead to a reduction of the variance associated with another variable (e.g., school or work success). As an example, if SES itself were, in part, influenced by genetics, then removing it from the prediction of work success would obscure any shared genetic variance with the latter and result in an underestimate of the importance of IQ. There are many variations of this argument but it is not necessary to go into them here. Collectively, they are referred to by Jensen and others as the "sociological fallacy," due to their assumption of noncollinearity among the predictors.

MOTIVATIONAL VALUES INCULCATED THROUGH PARENTING

There is an enormous literature that describes the relationship between individual differences in cognition and individual differences in motivational variables such as beliefs, goals, and self-perceptions of one's own ability. Because an adequate treatment of this literature would take us outside the bounds of this chapter, I will say no more about it here other than it does demonstrate the importance of these motivational influences in intellectual development (e.g., Nicholls, Cheung, Lauer, and Patashnick, 1989).

In addition to the world of work, the world of parenting has been shown to exert a powerful influence on intellectual development, mainly through the inculcation of modes of cognizing (e.g., fostering specific types of learning strategies) and through motivational routes that can affect problem solving. In particular, child rearing practices such as the encouragement of autonomy and parental congruence of values bear a significant relationship to children's later intellectual development. For example, "authoritative" parenting (i.e., democratic, yet not permissive) is associated with better intellectual outcomes than alternative methods of discipline, and this effect can be found independently of family income (Baumrind, 1973; Dornbusch, et al., 1987). Apparently, authoritative parents use a more effective form of informal tutoring to achieve their children's often-documented cognitive skill socialization than do permissive, authoritarian, or neglectful/uninvolved parents (Pratt, et al., 1988). Also, there is a substantial literature in the field of child development that documents ethnic, social class, and racial differences in parenting that are linked to differences in intellectual outcomes. For example, lower class parents provide poorer problem solving strategies for their children than do middle-class parents, and they tend to solve problems for their children rather than assist them in solving them (McGillicuddy-DeLisi, 1982). Moore (1986) has shown similar differences among black and white mothers of adopted black children in their tendency to encourage independent problem solving. It is impossible to say whether such parenting differences are responsible for any of the observed ethnic differences in IQ, but the limited reports out of Feurstein's project in Israel (in which immigrants' attitudes regarding the value of their traditions in their children's new lives in urban Israel are found to be related to their children's IQ gains and instrumental success) and some of the social background studies that will be reviewed later suggest that the answer may be in the affirmative.

According to a recent study, parents who hold similar child rearing values appear to foster superior IQs in their children (Vaughn, et al., 1988). Support for this view can be found in any of the preceding works, especially Dornbusch, et al.'s extension of Baumrind's typlogy of parenting styles in a large (nearly 8,000) and ethnically diverse sample, as well as Vaughn's finding that early family socialization experiences are influential in boys' (though not girls') later intellectual development. Interestingly, early studies of the links between parenting behaviors during infancy and children's IQ at age ten also showed that practices labeled "social stimulation" were influential for boys' later IQ but not for girls'. In fact, virtually all parental behaviors were more strongly related to their sons' later IQs than to their daughters' IQs (Yarrow, et al., 1971). The literature in the area of parental behaviors is voluminous, and it is not possible to chronicle it here; I merely point out that it clearly documents the impact of social organization on intellectual development and that the picture is complicated by gender, age, and social status variables.

Earl Schaefer and his colleagues have explored links between parents' attitudes, societal attitudes, and children's own attitudes with the latter's intellectual development. On the basis of this longitudinal project involving the development of low income children, Schaefer (1987) suggested that intellectual competence is mediated by an assortment of "modern" parental attitudes and values. Modern attitudes entail a composite of values about the worth of schooling, technical skills, nonparochial allegiances, opinions about minorities and women, and values and attitudes about child rearing (e.g., the role of punishment). Schaefer suggests that changes in the level of individual intellectual functioning over time can be due to changes in societal and parental modernity, and vice versa. Therefore, Schaefer's environmental model posits that human intellectual capital is not fixed but can be changed by altering parents' attitudes and values:

> Substantial correlations were found of parent values with child academic competence. Parent self-reports of educational behaviors of providing educational experiences, talking with the child, sharing activities, and teaching were correlated with child mental test scores ... Multiple regression analyses showed substantial prediction of child academic competence by parents' beliefs, values, and behaviors and by parent SES indicators. (Schaefer and Edgerton, 1985, p. 301)

Finally, various aspects of the home environment over and above parenting styles and congruent child-rearing values have been shown to be related to IQ and educational attainment, though not to social class *per se.*

Discussion of some of the mechanisms that have been proposed to account for these relationships would also carry us beyond the scope of this monograph; the interested reader is referred to Gottfried (1984), Siegel (1984), Bradley, et al. (1988), and others. Kurtz and her associates have shown that the home environment is a better predictor of educational outcomes than either parental SES, IQ, or cognitive variables, and it remains a strong predictor even after SES was controlled in a regression analysis (Kurtz, 1989).

But there is a caveat to the foregoing conclusions: The link between social class and IQ is usually indexed by verbal intelligence tests, and when the intelligence test is nonverbal, the magnitude of this relation is often greatly reduced (e.g., Butcher, 1970). This suggests that more thought needs to go into exactly what is being measured in these studies. If high social class conveys some esoteric benefit to children, enhancing their power and style of cognizing, this is one thing; if high social class merely means that highly educated parents raise children who mimic their extensive vocabulary, that is quite another. Still, like the studies mentioned on value congruence and parenting style, these studies clearly show that aspects of the home (e.g., its organization) are important independent determinants of later intellectual development. But the demonstration that social class, social organization, or parental styles are contributors to intellectual development should not imply that all development can be accounted for in this manner. Much argument centers around the relative contributions of social versus nonsocial predictors of intellectual development, with some preferring to argue that each makes independent contributions to a prediction equation. Moreover, some sociologists have argued that any influence that parental social class exerts on children's intellectual development is due to genetic mechanisms rather than social ones, through the hereditary transmission of environmental features (Gordon, 1980; 1987; see also Rice et al., 1988).

SAT SCORES AND ACADEMIC ACHIEVEMENT

In a heated exchange of criticisms, Ralph Nader and Nairn Associates, a Washington research consultant group, debated the Educational Testing Service (ETS) in Princeton on the possibility that social class was as predictive of college success (first year grades) as were SAT scores. Historically, the size of the zero-order correlation between social class and children's intelligence, summing across many studies, has been in the range of .3 to .4. Some indices of social class (e.g., parental education) are more predictive of grades than are others (e.g., income); hence, there is a range of correlations in the literature. Children from those families ranked in the highest social class usually achieve IQ scores about one standard deviation above the national mean (i.e., about 15 points above average) while children from the lowest ranking families usually achieve IQs about a half standard deviation below the national average.[2] According to the mathematics of the normal curve (which is how IQ is distributed), if the standard deviation (SD) of a particular IQ test is 15 points, and one group's average IQ is exactly 1 SD below another group's average IQ, then only 16% of its members will have higher IQs than the average individual's IQ from the higher group. Moreover, when it comes to having IQs in the top 2% of the population (130 and higher), the group whose average IQ is 1 SD lower than the other group will have few of its members reach this level—only approximately 1 of its members will reach this level for every 18 who reach it from the higher group. (We can go on *ad infinitum*, extrapolating from normal curve deviates to calculate the ratios of high IQs for groups who differ more or less than 1 SD. For instance, in a group whose average IQ is lower by 1.3 SDs, fewer than 10% of its members will score higher than the average individual's IQ in the higher group and only about 1 of its members will score in the high range (130 and above) for every 60 or so individuals in the higher group.)

Income and SATs:
One Interpretation

The Nader-Nairn report charged that ranking students by their families' incomes produced as good a prediction of success in the first year of college as did ranking them by SAT scores. In several rebuttals of this position, ETS, the makers of the SAT test, have pointed out that the relation between family income, SAT, and college grades, though statistically significant (e.g., the SAT-income correlation is around .30), is far from perfect. Moreover, the relation between SATs and college grades cannot be directly compared to that between social class and grades for several reasons. For one thing, by the time students matriculate in college, there has been "selection" wherein more of those with higher SATs enter college because those with lower SATs are less likely to go to college. Thus, the actual correlation between SATs and grades in college is somewhat reduced by this restriction in the range of SATs among those admitted to college. For another, the size of the relationship may be underestimated in these studies because of an over-reliance on family income as the measure of social class.

Table 4.1 shows the relationships among family incomes of the families of over 600,000 persons who took the SAT in 1973–74. Depending on how one "reconfigures" these data, markedly different stories can be told. For example, William W. Turnbull (1980), then president of ETS, pointed out:

> The relationship of SAT scores to family income is more modest than (Nader/Nairn) implies and is not peculiar to the SAT. . . . About 32% of the students with family incomes below $6,000 rank in the top half of the

TABLE 4.1 1973–74 College Bound Seniors Classified by SAT Average and Family Income*

SAT Average*	REPORTED FAMILY INCOME				Average Income**
	$0–5,999	$6,000–11,999	$12,000–17,000	$18,000+	
750–800	17	117	169	415	$24,124
700–749	239	1,172	1,752	3,252	$21,980
650–699	686	3,994	5,683	9,284	$21,292
600–649	1,626	9,352	12,187	17,992	$20,330
550–599	3,119	17,042	20,822	28,151	$19,481
500–549	4,983	26,132	29,751	37,400	$18,824
450–499	6,663	33,209	35,193	41,412	$18,122
400–449	8,054	34,302	33,574	37,213	$17,387
350–399	8,973	29,762	25,724	26,175	$16,182
300–349	9,622	21,342	14,867	13,896	$14,355
250–299	7,980	10,286	5,240	4,212	$11,428
200–249	1,638	1,436	521	325	$ 8,639
Total**	53,600	188,146	185,483	219,727	
Average SAT Score	403	447	469	485	

*From *College Bound Seniors,* 1973–74.

**The total number of students in this table (646,956) is very slightly smaller than the number (647,031) included in the analyses reported in *College Bound Seniors,* 1973–74. Students on this table must have had both SAT verbal and SAT mathematical scores and have reported family income on the Student Descriptive Questionnaire. Students with only one SAT score were included in *College Bound Seniors.*

total group in terms of SAT scores (above about 450). Other . . . measures of educational achievement would show a similar pattern. There is no evidence that use of test scores *per se* has a dramatic impact on opportunities for low income students. (Turnbull, 1980, p. 9)

Manning and Jackson (1984), the former of whom was affiliated with ETS at the time of writing, made a similar claim:

It is frequently charged that test scores rank individuals according to their family incomes with few exceptions. In fact, the relationship is far more moderate than the critics suggest. . . . Students from each income level obtained the full range of SAT scores. Many students from the top income group ($18,000 and over) earned low scores. For example, 8% scored below 350. Many students from the low-income group (less than $6,000) earned high scores, 5% scored above 600. About one third of.the students in the lowest income category obtained above average scores. (p. 211)

Income and SATs: An
Alternate Interpretation

To place the ETS figures in context, I have compiled the following statistics because they illustrate how slippery such data can be: While approximately 32% of the lowest income students score in the top half of SATs (above 450), there are almost twice as many from the highest income families who do; and although 8% of students from the highest income families receive SATs below 350, more than four times as many (36%) from the lowest income families do.

Some additional figures to illustrate the slippery nature of these data are as follows. According to my calculations, the average SATs of students from the highest family incomes is over 80 points higher than the average SATs of students from the lowest family incomes. And the average income of all students receiving SATs below 350 is about half of that for all students receiving SATs above 700. The point of these statistics is simply that the data in the table support different conclusions, depending where one draws the cutoffs. I believe that most of the points made by ETS in its rebuttal are legitimate, and they should not be singled out for showing that poverty has a deadening effect on scholastic achievement. However, contrary to the gloss that ETS has put on the data, I also believe that they demonstrate substantial asymmetries between the children of various social groups. And as ETS itself has pointed out, the overall relationship between family income and SAT scores may be somewhat understated because of the crudeness of the categories used and the fact that the income figures are self-reports by students. In addition, more effective constellations of social class information that included—but went beyond—family income (e.g., parental education, possession of cultural items in one's home, educational level of one's neighbors) may reveal the asymmetries to be even greater. Finally, my strong impression is that figures such as those in the table are only part of the story: Social class exerts an influence throughout one's entire history of schooling, and probably a vast number of low-income students do not even take SATs because of this. To the extent that this is true, the role of SES is even greater.

In view of what has been said about (1) the reduced relationship between nonverbal measures of intelligence and social class, and (2) the disproportionately smaller role of verbal items on infant IQ tests than on IQ tests for older children and adults, one may surmise that there is a reduced relationship between IQ and social class for infants and young children. This surmise is borne out by the majority of studies on the subject. The relationship between social class and IQ is fairly stable from about 3 years of age through adulthood; however, earlier than 3 years of age, and especially earlier than 18 months of age, there is virtually no relation between social class and IQ. This is undoubtedly because the instruments used to assess the developmental status (IQ) of very young children are less verbal and conceptual and far more psychomotor than those used to evaluate older children, as even a casual observer will note when comparing such infant tests (e.g., the Bayley test) with IQ tests for older children and adults such as the Stanford Binet and the Wechsler Scales.

Before leaving this section on social class and intelligence, let us turn to the seminal and best known study of the predictive validity of IQ. Even here there is evidence of the importance of social class, although this has not been the primary interpretation given by the vast majority of the writers who have commented on this study. As a backdrop for considering the study, keep in mind Figure 1.1, in the first chapter, which depicted social status as the result of IQ, which was itself the result of nervous system efficiency. Historically, the popular thinking has been that IQ, because it reflects a biological reality, is quite instrumental in achieving social status—a form of social Darwinism. A number of older studies indicated that within a single family (i.e., SES held constant), differences in siblings' IQs determined their eventual social status. Those children with higher IQs than their parents tended to move up the SES ladder, while those with lower IQs tended to move down it. Flynn (1988) points out that in the U.S., a boy with an IQ eight points higher (lower) than his father's will tend to attain a higher (lower) social status than his father when he becomes an adult. In fact, over half of all children will move into a different social class than their parents', and the temptation is to think that the bulk of this movement is spawned by native differences in biological intelligence. (I should point out that Flynn does not adhere to this position himself and has provided his own cogent rebuttal of it.)

THE TERMAN STUDY OF GENIUS

In 1921, Lewis Terman began what soon became the best known study of the predictive power of IQ. He sifted through a quarter million student folders of California public school children in the larger urban areas of that state who were between the third and eighth grade and identified 1,528 whose IQs were between 135 and 196, placing them in the top 1% of the population. The children were born between 1903 and 1917. (Actually, 1,470 children were selected, and another 58 of their younger siblings were added six years later, for a total of 857 boys and 671 girls.) Terman used not only the Stanford Binet IQ test, but the Terman Group Test, the Army Alpha Test, and the National Intelligence Test. He wisely, I believe, also solicited teachers' ratings of a large number of temperamental variables such as the children's desire to excel and their persistence at tasks, their school grades, and detailed information about their home life. From the standpoint of today's knowledge, the procedure used by Terman's colleagues to obtain these teachers' ratings were not optimal, as they probably resulted in some degree of built-in correlations, with teachers tending to rate all children who were high on one variable (e.g., school grades) high on others (e.g., perserverance), too. Moreover, teachers probably had a social class bias (as did Terman himself) that tilted the odds against nominating lower class children. Today, nearly 70 years after the inception of this study, Terman and the distinguished team of researchers who survived him when he died in 1956 have produced one of the most remarkable accomplishments in all of psychology: Seven volumes, entitled *Genetic Studies of Genius*, and numerous other analyses by affiliated researchers, have chronicled the lives of these individuals and their children. By the time of Terman's death almost 40 years after the study began, 93% of those who originally participated and were still alive were still participating, an unheard-of accomplishment in survey research! These high-IQ children— called "termites"—were surveyed and tested every seven years, on average. In the 1950s, their children were also tested. The latest analyses I have seen were based on the 1982 wave of interviews with these individuals, many of whom were in their retirement years.

It turned out that "genius" was highly localized among children of the top social classes: Only 6.8% of Terman's high-scoring children came from semiskilled and unskilled parents, while 82% came from professional, semi-professional, and business families. And they resided disproportionately in a few large urban areas. As far as I have been able tell, these figures are nearly the reverse of California's population in 1921, indicating that the 1,528 youngsters who participated in Terman's study were quite unrepresentative. For example, Latin American, Italian, and Portuguese children were underrepresented in the sample, and there was not a single Chinese child, despite a sizable population in the state at that time (segregated schools for Chinese and Japanese children were not canvassed by Terman's team). Also, there were only two black children and one American Indian in the sample, although these groups were far more prevalent in the state as well. And males were disproportionately represented, especially at the higher grade levels. (Terman believed that female intellectual prowess was not maintained after an early age—a biological view rather than a social one.) Of course, you might wonder why I mention these figures at all. After all, Terman selected from *all* California children with high IQs, right? (That is, he didn't reject high-IQ youngsters simply because of their ethnicity, gender, or race.) So if there were disproportionately fewer youngsters from the ethnic and lower income groups, so be it, you might say. But, as you will see, it *is* important to bear these social class figures in mind because Terman himself often based his conclusions about the usefulness of IQ on a comparison of "apples and oranges," in other words, of children who differed in ways other than just their IQs. What is more, countless writers who have commented on his findings have made this same error, as I shall try to show shortly. To do this, some background information about Terman's study is necessary.

I believe that Terman's lack of adequate social controls was a reflection of the hereditarian beliefs he harbored over most of his career. He was a man with a mission, as Minton (1988) notes in his recent history of Terman's project: "He believed that he could contribute to the making of an American society based on the principle of meritocracy. . . .The hierarchical division of labor would reflect (and) individual differences in native ability would thus determine the class structure of inequalities in wealth, power, and status " (p. 139). Toward this end, Terman made decisions that probably biased the selection of his sample against minorities and lower class children. (As I already noted, he did not canvass schools that contained Japanese or Chinese youngsters, nor did his team canvass schools for children who had committed minor infractions.) For Terman, the identification of high-IQ youngsters was more than a research curiosity: It was crucial if society was to plan efficiently. He had already succeeded in persuading the military during the First World War to use intelligence tests to screen nearly 2 million recruits and assign them to military occupational specialties, and during the early 1920s he began to expand the testing horizons:

> There is nothing about an individual as important as his IQ, except possibly his morals. . . . All of the available facts that science has to offer support the Galtonian theory that mental abilities are chiefly a matter of original endowment. . . . It is. . . especially to the top 5%. . . that we must look for production of leaders who advance science, art, government, education, and social welfare generally. . . . The least intelligent 15 or 20% of our population . . . are democracy's ballast, not always useless but always a potential liability." (Terman, 1922, p. 658)

Terman and his colleagues repeatedly assessed this high-IQ group by questionnaires and interviews between the 1920s and the 1980s. In 1945, when these so-called geniuses were 35–45 years old, and again in 1977 when they were in their late 60's and 70's, Terman and his successors conducted careful interviews to determine whether they were excelling in their everyday professional and personal lives. In what follows, I will focus my remarks on the men in Terman's study and say very little about the women. This decision is necessary to give the best test of the hypothesis that having a high IQ conveys advantages in job attainment and earnings during adulthood. This is because women with high IQs would not have been expected to enter professional and business life at rates commensurate with their high IQs because of social expectations and discrimination during this era. And even when they did work outside the home (196 of them had done so by 1940), few of them went into the professions (only 12 by 1940). This stands in contrast to the 262 men who went into the professions by this date.

The Validity of IQ

Most psychology students who are exposed to Terman's study are given the following message: Because the subjects, especially those labeled

Group A, represented a disproportionate number of scientists, persons listed in *Who's Who*, writers, business executives, and other high-status occupations, and because they also tended to be mentally and physically healthier than average and earn above-average incomes, and because their IQs remained superior over the years—in short, because of these and similar things the predictive validity of IQ has been conclusively demonstrated. Even the men in *Group C* (which constituted the 100 or so *least* successful) were earning above average incomes in their 60s. Not only is this the usual interpretation found in undergraduate textbooks, but it is the one most frequently proffered by researchers. Even extremely careful and scholarly writers in the field, like Butcher (1970), Itzkoff (1989), Flynn (1988), Sternberg and Powell (1983), and Brody (1985) have drawn similar, and I believe unwarranted, conclusions from Terman's *Genetic Studies of Genius* as has virtually every undergraduate text that describes this study. Before explaining why I think these individuals have misinterpreted this seminal study, their own statements should be presented to give the reader some idea of the *perceived* strength of Terman's data. More than any other study in psychology, this is the one that has been touted as the single best validator for the IQ = intelligence = real-world success argument. Consider just a few of the statements alleging the usefulness of Terman's study as a validator for the predictive power of IQ:

> This single study remains the most convincing demonstration of how astonishingly well the one much criticised variable, measured intelligence (IQ), can predict level of achievement for decades. . . .An overwhelming demonstration of the value of general intelligence, measured by conventional tests, as the most important psychological variable that can at present be assessed, and (despite large individual changes in IQ) the most stable and predictive over the life span. (Butcher, p. 275)

> The central role of intelligence and its predictive relationship to occupational attainment were prefigured by the results of Terman's classic longitudinal study that demonstrated that school children selected solely on the basis of high intelligence test scores tend to become intellectually eminent in adulthood and tend to have high occupational status. (Brody, 1985, p. 362)

> The strongest piece of evidence (that IQ tests match our primitive notions of the concept of intelligence) comes from Terman's famous study of high IQ subjects. . . . Their life histories are a good match for popular expectations about real world achievements of the highly intelligent. . . . As children they were ahead of their age group at school while none had been held back. As adults an extraordinary number had earned degrees, entered professions, achieved high positions, published books, and articles, and had biographical citations in leading reference works. (Flynn, 1988, p. 3)

> The accomplishments in later life of the (high IQ) selected group were extraordinary by any criterion. By 1959, there were 70 listings in *American Men of Science* and 3 were memberships in the highly prestigious *National Academy of Science*. In addition, 31 were listed in *Who's Who in America* and 10 appeared in the *Directory of American Scholars.* There were numerous highly successful businessmen as well as individuals who were succeeding unusually well in all of the professions. (Sternberg and Powell, 1983a, p. 352)

> Educational and vocational experience, especially in the classic study by Lewis Terman and his 1500 "geniuses," has by now clearly confirmed that intellectual power gives certain individuals a discernable advantage in the social hierarchy as compared with those of their fellows whose IQ scores or educational achievements seem to lag behind. (Itzkoff, 1989, p. 84)

While none of the actual findings regarding the superiority of Terman's sample's income, health status, etc., that Butcher, Brody, Itzkoff, and many others find to be so impressive are wrong, the implications drawn from them are misleading. The fact is that if these 1,528 "geniuses" are compared to children from their *same* childhood social class instead of to economically unselected children, as Terman had originally done, they do not appear to be nearly as exceptional in their later professional or personal lives. I became aware of this when I came across a relatively obscure paper by the sociologist P. Sorokin (1956), writing in *Fads and Foibles in Modern Sociology*. Sorokin conducted an informal comparison of Terman's "geniuses" with children from their same social group on everything from high school and college grades to marital dissolution rates and professional accomplishments. He concluded that the instances of superiority among the "geniuses" in Terman's study are within the expectation of superiority *irrespective of IQ*, in view of their family social background. Notwithstanding the effusive praise that psychologists have heaped on this study as evidence of the predictive power of IQ tests, the IQ scores of these "geniuses" appear to have added little to the prediction of their life outcomes over that gained simply from a consideration of their parental income, education, and occupational status:

> In comparison with the general population, the "gifted group" entered the professional and business pursuits in a notably greater proportion, but in all societies the children of professional and business classes, or the upper and middle strata, enter these occupations in a much greater proportion than the children of the skilled, semiskilled, and unskilled classes that make up the bulk of the general population. When the "gifted group" is compared to unselected (on IQ) college men and women (of the same social class) the difference in occupational distribution becomes slight, practically insignificant. . . . This (same) conclusion is warranted also by the incomes of the "gifted group." . . . In comparison with the median

income of semiskilled, unskilled, and possibly skilled labor, the income of the "gifted group" may have been slightly higher; but they were not tangibly higher than the income of the rank and file from the professional, business, and semiprofessional strata. . . . To sum up: By no stretch of the imagination or of standards of genius is the "gifted group" as a whole "gifted" indeed. It is, rather, a group typical of the professional and business classes of the same age, sex, and other conditions. . . . If instead of a random selection, we selected children from the professional and business classes on the basis of their school marks and teachers' evaluations, but did not use any of the intelligence tests, the proportion of gifted children, and of eventual candidates for *Who's Who*, would probably increase still more. (Sorokin, 1957, p. 80)

Here we see the traditional confrontation between correlational and causal conclusions, and one can well imagine someone with a hereditarian bent arguing that the highest social classes have been "selected" for high intelligence because it is economically adaptive. But the point that I wish to make from this examination is that just as one cannot help but pick up a disproportionate number of high-IQ children by focusing nearly exclusively on children from middle- and upper-class families who reside in urban areas, conversely, one cannot help but pick up a disproportionate number of high school and college graduates by selecting high-IQ persons.

The fact of the matter is that IQ, social class, and school performance are to a large degree colinear and thwart any simple causal explanation of the sort often proffered by Terman when he argued that the school performance of his high-IQ children was superior "in comparison with the general population of whom only about 5% graduate from college and only 30 to 40% graduate from high school." But surely, this is the wrong comparison group! Terman should have compared the graduation rates, college grades, and earnings of his selected students to those of others from their same family background social class. Terman came to this realization rather late and he attempted to construct a more appropriate comparison group *post hoc*. In Volume 4 of the *Genetic Studies of Genius*, Terman and Oden (1947) argued that their high IQ subjects were not only superior in income and occupational status to the national average (on income, by about 70%!), but they were even superior to all college-educated men of that era. Thus, Terman and Oden intimated that IQ, and not family social background, was the cause of the high earnings of their sample since they earned more than a sample of college students of that era, a group they assumed to have provided an adequate control for family background. Yet, a straight comparison between Terman's men and college-educated men of this era who were unselected on IQ shows that the advantage of having a very high IQ is rather slender—$224 per year advantage in earnings for the Terman men when they were between the ages of 30 and 40. And there was no difference at all between Terman's high-IQ men and all college educated men in the proportion who earned in the highest bracket.

And the picture gets even cloudier. The college-educated men that Terman used as his *post hoc* comparison group were undoubtedly different from his high-IQ men in a number of ways, any of which could lessen the force of Terman's conclusions. First, they came, on average, from lower family social backgrounds than did Terman's high-IQ men. This is important if one wants to attribute adult earnings and social status to IQ because it is more likely that family trusts and other forms of inherited wealth were bestowed to the Terman men than on an unselected sample of all college educated men. And it is also more likely that family "connections" opened doors into business for the Terman men than for the unselected sample of college-educated men. Second, the unselected sample of college-educated men had not been raised believing that they had extremely high IQs, nor were their teachers and parents informed they were gifted. There is no way to estimate the effect that this information might have had on the career trajectories of the Terman men. And third, the unselected college-educated men had none of the other advantages of having been in the Terman study, such as having Terman and his staff act as mentors and often write letters of recommendation for the termites and do other things to foster their professional careers. In short, Terman's attempt to prove that having a high IQ conveyed occupational and financial benefits by comparing his high IQ subjects with an unselected sample of college-educated men is unsatisfactory by modern standards of research design. What is needed is a sample of men with unremarkable IQs who came from the same family social background as Terman's high-IQ men (i.e., from the same neighborhoods, with the same degree of parental wealth and education) and who were subjected to similar special treatments (their teachers told them they were geniuses and being mentored by Terman and his staff). If such a sample could be found, would they earn less than those with high IQs? Would they be less likely to become professionals and businessmen? At the end of this chapter I provide tentative answers to these questions using a very different sample, but for now some additional background analysis is needed.

Recently, Schooler summarized a large body of sociological literature that is relevant to what I have been alleging about the Terman results, and before I return to take the steam out of the Terman data even further, I want to consider what this body has to say. Schooler stated that

> Analyses of the social antecedents of adult psychological functioning suggest that having a father with a high level of education, and being raised in an urban area, a liberal religion, and a part of the country far from the South, are all related to having been raised in a complex environment. And each of these characteristics also independently lead to a relatively high level of intellectual functioning. (Schooler, 1989, pp. 23–24)

To drive Schooler's claim home, let us examine the two most extreme groups in Terman's study, group A and group C. These two groups comprised the top and bottom 20% of the sample in terms of life success. At the start of the study, all the individuals in both groups, regardless of which group they belonged to, were about equal in IQ, elementary school grades, and home evaluations. (Depending on the test used, the IQs ranged from 142 to 155 after the start of the study; all groups' IQs had declined by about 10 points after the first wave of testing, due to measurement error and regression effects.) There were approximately 150 persons in each of these groups in the 1940s and about 100 still alive and available for interviews in the 1960s. Although their IQs were equivalent, the developmental outcomes of the men in these groups stand in stark contrast, with group A including a large number of very successful and happy individuals, while group C men are riddled by professional and personal failures that made them indistinguishable from persons with average IQs (see Minton, 1988, for more details about the groups). The remaining group, group B, was comprised of the middle 60% of the sample in terms of life success. Their life outcomes appear to have been distinctly mediocre. (Incidentally, it strikes me as somewhat ironic that while there are a number of talented artists and scholars in Terman's study, there are no individuals who went on to distinguish themselves as notably creative in the sciences or arts; yet there were two men rejected for inclusion in his study for low IQs who went on to win Nobel prizes, a feat that has not been accomplished by any of the individuals included in the study the last time I checked!)

Ninety-nine percent of the men in the group that had the best professional and personal accomplishments, i.e., group A, were individuals who came from professional or business-managerial families that were well educated and wealthy. In contrast, only 17% of the children from group C came from professional and business families, and even these tended to be poorer and less well educated than their group A peers. The men in these two groups present a contrast on all social indicators that were assessed: group A individuals preferred to play tennis, while group C men preferred to watch football and baseball; as children, the the group A men were more likely to collect stamps, shells, and coins than were the group C men. Not only were the fathers of the group A men better educated than those of group C, but so were their grandfathers. In short, even though the men in group C had equivalent IQs to those in group A, they did not have equivalent social status. Thus, when IQ is equated and social class is not, it is the latter that seems to be deterministic of professional success. Therefore, Terman's findings, far from demonstrating that high IQ is associated with real-world success, show that the relationship is more complex and that the social status of these so-called geniuses' families had a "long reach," influencing their personal and professional achievements throughout their adult lives. Thus, the title of

Terman's volumes, *Genetic Studies of Genius,* appears to have begged the question of the causation of genius.

Convergence from Recent Life Course Analyses

Since first realizing the fragile empirical basis for the conclusions drawn from Terman's study, I have been troubled. It is, after all, a classic study that has become part of the ingrained belief system of psychologists. I was concerned that a study so embedded in our thinking as Terman's not be swept out of consciousness so quickly. If not out of respect for Terman's original aims, at least out of deference to the illustrious individuals who took over the project after his death, including Robert and Pauline Sears, Lee Cronbach, and Melita Oden (R. Sears and Cronbach were themselves "termites"), I searched for additional evidence. Fortunately, Glen Elder, a well-known sociologist of the life course, and his colleagues recently reported their own reanalysis of the Terman data and their conclusions are extremely relevant to this discussion. Elder, et al., (in press), noticed something interesting about Terman's sample that had escaped nearly everyone else's attention, namely, that the men who participated could be grouped into two cohorts of equal IQ. The oldest cohort was comprised of men born between 1904 and 1910, and the youngest cohort represented men born between 1911 and 1917. Men born in the earlier cohort were at a disadvantage for future earnings and professional accomplishments because of the double whammy of having reached adulthood during the severe depression of the late 20s and early 30s and then having to enter military service (or related national service) just when their careers were beginning to take off at the start of World War II. The younger men, in contrast, were still in school during the great Depression and they actually appear to have benefited from wartime service, as their careers had not yet gotten under way and the service-related benefits (educational and housing) seem to have helped them.

Elder found that when comparing *below-average* achievers and average achievers in the Terman study, the most potent predictor of life course outcomes (earnings, professional status and awards) was the men's birth cohort and, to some extent, their family background—but *not* their IQ. In comparing those men who had *above-average* career records with those who had only average career records, only one childhood characteristic had any predictive value, namely, whether those familiar with them as children had rated them as "prudent" when they were in grade school. Neither IQ, cohort, nor family background contributed to predicting the differences between the distinguished and average men. (It should be noted that Terman was himself aware that, within his sample, IQ was unrelated to earnings; rather than questioning the validity of IQ as a predictor, however, he suggested that the reason that IQ was unrelated to earnings was due the unreliability due to small numbers of men in some earning brackets—see Terman and Oden, 1947, p. 189–190.)

However, when comparing those men who had average careers with those who had below-average careers, the only characteristic found to be predictive *was* their parental social class (sons of low-status fathers were much more likely to end up in the least successful group), leading Elder and his colleagues to comment that

> Among these men of unusual ability, IQ has little to say about those who ended up with a disappointing career. . . . Men who ended up with careers lacking distinction or even average achievement were likely to be saddled with lower (childhood) family status and birth in the older cohort. . . . For low as well as high achievement, differences in education strongly influenced career achievement. . . . Men who achieved a distinguished career were no brighter than men of average accomplishments and they did not differ on family background. . . . By comparison, both family background and birth cohort are consequential for the less accomplished men. Those who experienced disappointments came from less advantaged homes. . . . The extraordinary promise which the Lewis Terman staff observed in these young men stems from a largely undifferentiated view of intellectual giftedness, as based on an IQ test. (Elder, et al., 1989, p. 20)

I think the bottom line of both the Sorokin and Elder analyses of the Terman data, as well as my own concerns, is that the ecological niche one occupies, including individual and historical development, is a far more potent determinant of one's professional and economic success than is IQ. It was tough being in one's 20s during the 1930s and it was this factor (birth cohort), together with parental social background, that accounts for the adult development of Terman's subjects. IQ totally drops out of Elder's model in all contrasts.

A question can be asked, however, about the generality of the conclusions reached from these reanalyses of the Terman data. Perhaps the predictive power of IQ for life outcomes could in fact be demonstrated in a more intellectually diverse sample. For instance, maybe an individual with an IQ of 135 does not differ from one with an IQ of 196 because both possess more than adequate intelligence to surmount life's important challenges. Putting aside these extremely high-IQ cases, we can ask about the 95% of persons with IQs in the so-called normal range—70 to 130? Does IQ "purchase" anything in predicting their adult development after we control for their childhood family social background and their years of education? Obviously, this is not a question that can be addressed by Terman's data, for it requires a more diverse sample; Terman's mean IQ was around 146 with a standard deviation of about 10.2, rendering it unacceptable for answering this question. The economists Sam Bowles and Valerie Nelson (1974) estimated regression equations from two different data sets in an effort to quantify the extent to which intergenerational economic immobility is due to the inheritance of IQ. They found that earnings during adulthood were strongly related to both years of schooling and family social status but not to IQ:

> We are confident then in concluding that the genetic inheritance of IQ is a relatively minor mechanism for the intergenerational transmission of economic and social status. . . . Evidently, the genetic inheritance of IQ is not the mechanism which reproduces the structure of social status and economic privilege from generation to generation. (Our estimates) suggest that an explanation of intergenerational immobility may well be found in aspects of family life related to socio-economic status and in the effects of socio-economic background operating both directly on economic success, and indirectly via the medium of inequalities in educational attainments. (Bowles and Nelson, 1974, p. 48)

Thirtysomething: To Be Born Rich or Smart

Bowles and Nelson were explicit about the limitations of their estimates because of the lack of individual data. They were forced to estimate the childhood IQs of the subjects in their data sets on the basis of the presumed correlation with educational and economic status. And even here, their measures were less than optimal, as those authors note. So, I decided to undertake a direct test of the thesis that IQ is not as predictive of later occupational success as is commonly thought, once social class and education are taken into account. To do this, I used a data set that is both more diverse than Terman's and less problematic than Bowles' and Nelson's. As will be seen, it turns out that the results with a more diverse sample are similar to those reached by Elder and others.

My colleague Charles Henderson and I carried out an analysis of the impact of IQ on occupational success with the Project Talent data set. This project was begun in 1960 by John Flannagan and his associates at the American Institutes for Research in Palo Alto, California. Unlike Terman, these researchers decided to study a nationally representative sample of over 400,000 high school students, with both intellectual aptitude and parental social class spanning the entire range of the population. In 1960, they interviewed these students and obtained a vast amount of information regarding their intellectual abilities, school performance, and social habits, as well as their parents' social status. By 1974, they had reinterviewed over 4,000 of these persons, most of whom were around 30 years old. During this interview, they obtained, among other things, information about their occupations, including their income. For all practical purposes, this subgroup of

4,000 adults represents a stratified national sample of persons in their early 30s.

Henderson and I ran several regression models, involving years of completed schooling and family social background and a composite intellectual score that is based on reasoning, vocabulary, and mathematics (C-003 in the code book). Excluded from these analyses were all individuals who, at the time of follow-up interviews, were unemployed, either because they were still enrolled in school, were housewives, or were in prison. This left 2,081 cases for our analyses. And in all analyses we included gender as a classification factor, specifiying the regressions separately by sex in view of the large numbers of women who listed housework as their main occupation. The excluded cases were similar to those left in the analyses in terms of their IQ and high school social background.

In one analysis, we looked at IQ as a predictor of variance in adult income and this model showed an apparent impact for IQ. However, when we entered parental social status and years of schooling completed as additional covariates (where parental social status was a standardized score, mean of 100, SD = 10, based on a large number of items having to do with parental income, housing costs, etc.—ranging from a low of 58 to a high of 135), the effects of IQ as a predictor of adult income were totally eliminated.

On the other hand, both social class and education were strongly significant and positive predictors of adult income. Basically, this indicates that the relationship between IQ and adult income is illusory because the more completely specified statistical model demonstrates its lack of predictive power and the real predictive power of social and educational variables. In an additional analysis, we considered three IQ groups (high, average, and low) of approximately equal sizes and examined the homogeneity of regressions of earnings on social class and education within these three IQ groups. Regressions were essentially homogeneous and, contrary to the claims by those working from a meritocratic perspective, the slope for the low IQ group was steepest (see Figure 4.1). There was no limitation imposed by low IQ on the beneficial effects of good social background on earnings and, if anything, there was a trend toward individuals with low IQ actually earning more than those with average IQ (p = .09). So it turns out that although both schooling and parental social class are powerful determinants of future success (which was also true in Terman's data), IQ adds little to their influence in explaining adult earnings.

And the same was true when we statistically controlled years of schooling completed. For example, for every increment of schooling attained by the Project Talent participants, there was a commensurate increase in their earnings. Individuals who were in the top quartile of "years of schooling completed" were about 10 times as likely to be receiving incomes in the top quartile of the sample as were those who were in the bottom quartile of "years of schooling completed." But this relationship does not appear to be

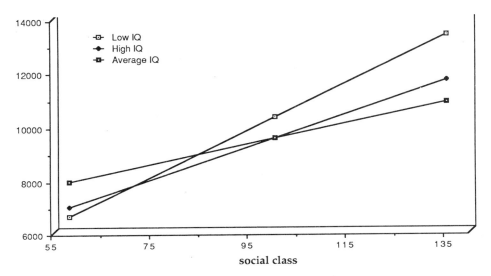

FIGURE 4.1 Regression of Earnings on Social Class for Three IQ Groups, Controlling for Education and Sex.

due to IQs mediating school attainment or income attainment, because the identical result is found even when IQ is statistically controlled. Interestingly, the groups with the lowest and highest IQs both earned slightly more than the average-IQ students when the means were adjusted for social class and education (unadjusted means at the modal value of social class and education = $9,094, $9,242, and $9,997 for low, average, and high IQ groups, whereas the adjusted means at this same modal value = $9,972, $9,292, and $9,278 for the low, average, and high IQs). (Perhaps the low IQ students were tracked into plumbing, cement finishing and other well-paying jobs and the high-IQ students were tracked into the professions, while average-IQ students became lower paid teachers, social workers, ministers, etc.) Thus, it appears that the IQ-income relationship is really the result of schooling and family background, and not IQ. (Incidentally, this range in IQs from 70 to 130 and in SES from 58 to 135 covers over 95% of the entire population.)

So does this mean that the conclusions drawn by many from the Terman data are wrong? I think it does, especially when coupled with the analyses of Elder and his colleagues, Bowles and Nelson (1974), Sorokin (1956), and others. It says nothing, however, about the possibility that genetic sources of variance over and above those that load on IQ may be causative here. But this becomes a tortured argument, requiring more and more assumptions and less and less parsimony. I shall address it later.

Taken together, the preceding analyses support Lewontin's view that if

IQ tests "really do measure intrinsic intelligence, as they are claimed to do, then one can only conclude that it is better to be born rich than smart" (Lewontin, 1982, p. 97).

Recently, Christopher Jencks and his colleagues at Harvard have provided further direct evidence for the role of social class in predicting future success. Jencks (1979) analyzed national surveys and concluded that the key factors to predicting who gets ahead in life are primarily social and personality traits, and, parents' (especially father's) education, income, studiousness, and years of schooling completed. We turn to this last ingredient next.

CHAPTER 5

The Impact of Schooling on Intelligence

How important is schooling in the development of intelligence? It is worth looking rather closely at this question. On the surface, schooling and intelligence appear to be tightly connected. But, as I shall demonstrate, although schooling is certainly linked to the maintainence of IQ scores, there is surprisingly little relationship between school activities and many of the kinds of intelligence that are important in everyday life. In particular, a crucial aspect of intelligence, the complexity of mental functioning, turns out to be unrelated to schooling.

When discussed in this context, we should ask what one means by the term "intelligence." If we mean the rational use of cognitive resources to meet important environmental challenges, the answer is uncertain. In part the uncertainty is due to a lack of understanding of the nature and measurement of these cognitive resources, and in part it is due to a lack of consensus about just what constitutes truly important environmental challenges in people's lives. One individual's perception of an important challenge may not be another's—this much is hardly controversial. But when it comes to defining intelligence as "the use of cognitive resources to achieve an important self-interest," things can become quite problematic. Philosophers and economists have provided many examples of the efficient use of some resource to pursue a self-interested end that does not appear to be in one's "best" interest, let alone in society's interest (see Frank, 1988).

Thus, to yoke the definition of rationality to self-interest presupposes that we can agree on what is in one's best interest and, furthermore, that the behavior of those who do not see the matter in the same way as we do can be relegated to the category of irrational behavior.[1] This relativistic position creates a number of conceptual problems, but since they are not germane to my argument here, they will not be treated further. We can simplify the above question to avoid ambiguities in the definitions of rationality and intelligence with the following rewording: *How important is schooling in the development of a perceived surrogate for intelligence, namely the IQ score?* Having thus circumvented the thorny philosophical issues surrounding self-interest and rationality, an unequivocal answer can be given: Schooling appears to be *very* important in determining IQ performance! Even at the level of individual components of an IQ test, schooling is highly and positively correlated with each scale or component, r's ranging between .35 and .55 for individual subscales on the WAIS (McCardle and Horn, In Press).

But why should we care whether schooling influences IQ if the latter is not identical to intelligence? The answers are many, I believe, but I will mention just six of them. First, whatever it is that an IQ score indexes, it has a certain psychological reality. That is to say, people often know their IQ (or SAT or GRE) score, and their pride and sense of self-worth are sometimes tied to this knowledge, as is their opinion of others whose IQs are known to them. Second, IQ scores (and their surrogates like SATs, GREs, and MATs) play an important gatekeeping function in society, regulating admission to prestigious schools and, occasionally, to jobs and professions. Third, several recent analyses (e.g., Hunter, 1983, 1986; Gottfredson, 1986) have led to the conclusion that IQ scores, and other measures of g, are important and potent predictors of job performance that save the U.S. government and private employers hundreds of millions of dollars each year by virtue of their efficient selection of persons for various jobs. Fourth, a range of social outcomes, such as criminality, mental health, occupational "trainability," and life satisfaction, have all been shown to be related to IQ. It is even argued that any influence that social variables like SES have on delinquency is due mainly to their loading on IQ (Gordon, 1987). Fifth, IQ scores *are* regarded by many persons as identical to intelligence, even if we take issue with that identity here. This has resulted in claims about all sorts of matters that involve schooling.[2] And sixth, if the foregoing five reasons for looking at IQ scores are not sufficiently compelling, it will become apparent that IQ scores are interesting to study in their own right because of their presumed relationship to a number of ancillary issues such as heritability, positive manifold, and cognitive efficiency, all of which will be discussed in the chapters that follow. So we need to explore the relation between schooling and IQ as a means of setting the stage for these subsequent considerations.

CORRELATION BETWEEN IQ AND YEARS IN SCHOOL

Frequently, one comes across very large correlations between IQ scores and the highest grade in school completed. For example, in the Ceci and Liker (1986a, 1986b) study of racetrack handicappers, this correlation is .94. In the largest study of monozygotic twins reared apart, the correlation is .96 (Bouchard, 1984). Even after controlling for SES and other "social variables," the correlation between IQ and schooling has been reported to be on the order of .60 to .80 (Kemp, 1955; Wiseman, 1966). As is true of all correlated variables, there are many alternative explanations for the relationship between IQ and schooling. For example, perhaps genetic selection in some form might be a cause. Genetically intelligent students might be encouraged to begin school earlier or remain there longer. Certainly, the fact that those who remain in school longest also have the highest IQs is not firm evidence of a causal role of schooling on IQ. Jensen (1972) essentially made this argument when he suggested that increases in IQ as a function of schooling might be due to the increased "outbreeding" that has occurred since the turn of the century in America. So a relation between IQ and schooling might be dismissed as an epiphenomenon: Schooling, it might be argued, no more drives IQ performance than the puffs of smoke from a locomotive drive its engine; rather, the amount of schooling that one completes may itself be determined by IQ, just as the puffs of smoke are produced by the engine. Correlational findings are riddled with interpretive difficulties, as social scientists know only too well.

One can also challenge the view that the correlational evidence shows that schooling affects IQ by pointing out that even if the magnitude of the relationship is as large as, say, $r = .7$, that would still leave open the possibility that IQ could correlate as high as $r = .7$ with some other variable (e.g., nervous system efficiency) that is statistically independent of schooling (assuming simple noncollinearity and the square of .7). In other words, a high zero-order correlation between the number of years of school completed and one's IQ score does not rule out the existence of other equally high correlations. These may indicate that other influences on IQ are just as important. (Obviously, to the extent that the other influences are colinear with IQ and years of schooling, the interpretive difficulties are increased.) Bouchard and Segal (1985), who have provided a cogent review of some of the schooling-IQ literature, argue that although the amount of schooling that one receives appears to convey an advantage for IQ, it is quite possible that genetics or some artifactual variable mediates this effect. (For instance, in the well-known reanalyses of the pooled data from many preschool interventions, subsequent IQ and school achievement advantages may have resulted from school officials' differential behavior toward children who went to school versus a control group who did not.)

Despite these interpretive snarls, I will show that when one takes into consideration the entire corpus of correlations that have been reported this century, the very high correlation between IQ and schooling cannot be accounted for on the basis of genetic selection or any alternative explanation (e.g., motivational differences or parental SES). The most plausible explanation for the correlations I shall be reporting is that of a direct causal link: The processes associated with schooling influence performance on IQ tests through a combination of *direct* instruction (e.g., it is in school that most children learn the answers to many IQ questions such as "In what continent is Egypt?" "Who wrote Hamlet?" and "What is the boiling point of water?") and *indirect* modes or styles of thinking and reasoning (e.g., schools encourage taxonomic/paradigmatic sorting and responding, rather than thematic/functional responding, and this happens to be the valued form of responding on IQ tests).[3] This position is not new; many have proffered versions of it in the past (e.g., Cattell, 1963). In what follows, however, I shall review the data that have led me to a rather stronger conclusion about the causal role of schooling in IQ than others have reached. It may also be argued that schools foster disembedded ways of thinking about the world, enabling some to understand concepts are not tied to personally experienced events (an example would be the use of hypothetical situations that defy reality). Perhaps this is why abstractness is thought to increase with schooling. However, reasonable as the suggestion that school experiences promote abstract thinking sounds, there is little firm evidence. I shall return to this matter in Chapter 9, which deals with abstractness.

But we need to remember that although the substantial literature review that follows provides convincing evidence that schooling has an influence on IQ, that conclusion neither justifies nor implies that other variables cannot have an equally large influence on IQ. In a later section, I shall address this possibility.

Of course, there is nothing new about the view that traditional conceptualizations of intelligence are inadequate, especially the notion of a single underlying processor, g, that accounts for a substantial amount of interest variance and is innately determined and immutably fixed. Spearman (1904) termed this underlying factor the *universal unity of intellective functions*. He believed that "all branches of intellectual activity have in common one fundamental function" (p. 284; see also Jensen, 1980, 1981). Spearman considered the causative mechanism to be some sort of mental energy. But others have argued that intelligent individuals do better on IQ tests (and other "g-loaded" tasks) because their superior central-processing mechanism makes it easier for them to glean important information and relationships from their environments than less intelligent individuals. One assumption that is being made here is that this information is available to individuals in all but the most seriously deprived environments (see Clarizio, 1982; Jensen, 1980; Nichols, 1981), but that highly intelligent persons are presumed to be better at detecting, storing, and/or retrieving it. This is precisely what has been assumed: "The published research seems to show that a quite severe degree of environmental deprivation is needed to cause a lowering of IQ, even on a test such as the Stanford-Binet" (Jensen, 1968, p. 12). Psychometric analyses indicate that g-loaded test scores (like IQ) provide the bulk of the predictiveness of large aptitude batteries that are sometimes given to prospective employees, indeed, they provide much better prediction than do scores on specific ability measures (e.g., Jensen, 1986; Thorndike, 1985; Humphrey, 1979).

Although the argument to be presented later in this book is not incompatible with the notion of a central processor g that permeates all cognitive performances, it contradicts the view that (1) an IQ score is an adequate reflection of the operation of such a processor, and (2) the amount of variance accounted for by such a processor is psychologically significant. It accomplishes this by arguing that IQ scores can change quite dramatically as a result of changes in family environment (e.g., Clarke and Clarke, 1976; Svendsen, 1982), work environment (Kohn and Schooler, 1981), historical environment (Flynn, 1987a), styles of parenting (Baumrind, 1967; Dornbusch, et al., 1987; Vaughn, et al., 1988), and, most especially, shifts in level of schooling. Moreover, evidence will be presented later that the shapes of ability profiles change as a function of the child's changing environmental demands (Ferguson, 1956; Wiseman, 1966). That is, patterns of processing may not be isomorphic across ages or generational cohorts.

Contrary to the traditional belief that information contained on IQ tests is potentially available to all children, regardless of environmental conditions, it has been known for many decades that a child's experience of schooling exerts a strong influence on intelligence test performance. Overall, there is an adjusted correlation of .68 between the number of years of school completed and IQ (Jencks, et al., 1972). This relationship is still substantial after potentially confounding variables, such as the tendency for the most intelligent children to begin school earlier and remain there longer, are controlled (Howe, 1972). Thus, to the extent that individuals differ in their level of schooling, they can be expected to differ in their IQs, and this difference may have little, if anything, to do with their underlying ability to detect, store, and/or retrieve information and relations in their environments. In what follows, I shall describe some of the seminal studies that support this interpretation. Although the thesis that schooling has an influence on IQ may sound obvious to readers who are unfamiliar with the psychometric literature, within the latter camp it has always been assumed that IQ influenced schooling, but schooling had little influence on IQ itself: "By intelligence the psychologist understands inborn, all-around intellectual ability ... inherited, not due to teaching or training ... uninfluenced by industry or zeal" (radio broadcast by Burt, 1933, cited in Carroll, 1982, p. 90). Although no single study that will be reviewed is sufficient by itself to make the case that IQ performance is strongly influenced by schooling (even when statistical controls are exerted on social class and motivation), *collectively* these studies provide incontrovertible evidence for such a position.

Influence of Summer Vacation on IQ

A clear illustration of the influence of schooling on IQ performance is seen in the small but reliable decrement in IQ that occurs during the summer vacation, especially among low-income youngsters whose summer activities are least likely to resemble those found in school (Jencks, et al., 1972). This finding has been replicated twice with large samples by independent investigators (Hayes, 1982; Heyns, 1983). The steady increase in both IQ and achievement test scores that accrues as the school year progresses is partly reversed during the summer vacation. Children whose summer activities are closer to those associated with schools (e.g., academic type camps) do not exhibit IQ decrements between June and September.

Continuous Impact of Schooling on IQ

The earliest study of the continuous influence of schooling on IQ was carried out by Hugh Gordon in 1923 (reported in Freeman, 1934). Gordon was commissioned by the London Board of Education to study a group of children who had very low IQs. Some of these children were found in London classrooms, while others attended school very little or not at all—due to either their physical disabilities or their status as sons and daughters of gypsies, canal-boat parents, etc. Gordon argued that the schooled children who had low IQs were mentally retarded in the traditional sense: That is, their low IQs could not be explained on the basis of obvious environmental factors. On the other hand, the children of gypsies and canal boat pilots who had never gone to school had low IQs directly as a result of their lack of schooling. Living on the canal boats meant being cut off from practically all forms of social interaction, except with their own family members, who invariably were illiterate themselves. As Freeman reports,

> Further analysis revealed the impressive and startling fact that the intelligence quotients of children within the same family decreased from the youngest to the oldest, the rank correlation between the intelligence quotients and chronological age being –.75. Not only that, but the youngest group (4 to 6 years of age) had an average IQ of 90, whereas the oldest children (12 to 22) had an average IQ of only 60, a distinctly subnormal level. It is known that children within the same family need not and do not have identical or nearly identical mental abilities; but the marked and steady decrease in intelligence with increasing age suggests that factors other than heredity are at work. . . . The most reasonable explanation seems to be this: the younger children appear to be about "normal" in intelligence, because success in the tests of the earlier years does not depend upon the opportunity for mental stimulus and exercise such as is offered by the school. . . . The results of the investigation suggest that without the opportunity for mental activity of the kind provided by the school—though not restricted to it—later intellectual development will be seriously limited or aborted. (Freeman, 1934, p. 115)

Freeman's conclusion is bolstered by the data from the children of gypsies, who also did not attend school. There was a high negative correlation between IQ and chronological age, as was the case for physically handicapped youngsters.

The next study of the continuous influence of schooling on IQ was carried out in 1932. Sherman and Key (1932) studied children reared 100 miles west of Washington, D.C., in "hollows" that rimmed the Blue Ridge Mountains. Some of the hollows were more remote than others. The ancestors of these "hollow children" were Scottish-Irish and English immigrants who retreated into remote regions of the mountains when their land was deeded to German immigrants in the 19th century. They remained in these hollows for several generations. Sherman and Key assumed that the original genetic pool of the people in the different hollows was very similar. They selected four of the hollows for study, based on their differing levels of isolation from modern communities. (They also studied a fifth hollow, Briarsville, that had been settled by the same Scottish-Irish stock as the others but was situated at the foot of the mountains rather than in an isolated area and had schools in session nine months of the year. Thus, Briarsville represented a sort of baseline against which the effects of isolation associated with the more remote hollows could be evaluated.)

Colvin, the most remotely situated of the hollows, had no movies or newspapers, and virtually no access roads to the outside. There was a single school, but it was in session infrequently, only a total of 16 months between 1918 and 1930. Only three of Colvin's adults were literate, and physical contact with the outside world appears to have been nonexistent. The other three hollows were progressively more modern. They had varying levels of contact with the outside world. Sherman and Key (1932) observed that the IQ scores of the hollows children fluctuated systematically with the level of schooling available in their hollow, the differences being quite substantial. Advantages of 10 to 30 points were found for the children who received the most schooling. Also, there was a dramatic age-related trend in the IQ levels. The older the child, the lower was his or her IQ. Six-year-olds' IQs were not much below the national average, but by age 14 the children's IQs had plummeted into the retarded range. A later study by Tyler, 1965, reached a similar conclusion. She reports that the IQs of children born in a mountainous area of Tennessee were, on average, 11 points higher in 1940 than the IQs of their siblings in 1930. She rejected a genetic explanation for this improvement in favor of one that emphasizes the increased educational and economic opportunities that took place during the decade in question.) Similar "cumulative deficits" in IQ with age have been reported among American blacks and British working-class youths (Jensen, 1980; Vernon, 1969; Wiseman, 1966). Also, Douglas (1964) showed that the average difference between the IQs of differing social classes became larger with age.

Impact of Delayed Schooling on IQ

In another investigation, which was carried out in South Africa, Ramphal (1962, cited in Vernon, 1969) studied the intellectual functioning of children of Indian ancestry whose schooling was delayed for up to four years because of the unavailability of teachers in their village. Compared to children from nearby villages inhabited by Indian settlers of similar genetic stock who were fortunate to have teachers, children whose schooling was delayed experienced a decrement of five IQ points *for every year that their schooling was delayed.* Other studies also have documented the deficit in IQ scores that accompanies delayed entrance into school. For example, in the Netherlands during World War II, many schools were closed as a consequence of the Nazi occupation; and many children entered school several years late. Their IQs dropped approximately seven points, probably as a result of their delayed entry into school (Degroot, 1951).

A half decade later, the situation in South Africa appeared to be essentially unchanged from that reported by Ramphal. Schmidt (1967) reported similar results from his analysis of a different South African community of East Indian settlers. Schmidt measured the impact of schooling on both IQ and achievement, holding constant age, SES, and parental motivation. With age held constant and SES and motivation partialed out, the correlation between the number of years of school attended and IQ was .49 for a measure of nonverbal intelligence and .68 for a measure of verbal intelligence. Additionally, Schmidt reported a correlation between schooling and the Raven's Progressive Matrices of .51, the latter IQ presumably being a measure of g.

Schmidt also found that by the time children had been at school several years, those who began school late had substantially lower IQs than those who began early, another instance of a cumulative deficit. Finally, Schmidt reported that the correlation between the number of years of schooling and achievement test scores (vocabulary and arithmetic) was no higher than that observed between schooling and IQ. This would seem to suggest that IQ scores are just as influenced by schooling as something that is assumed to be explicitly taught in school, namely academic achievement (Howe, 1972). These results, together with others that will be mentioned shortly, strongly suggest that schooling conveys a substantial effect on IQ that is independent of parental motivation or SES (see also Kemp, 1955; Wiseman,1966). Moreover, none of the findings support the proposition that the IQ-schooling relationship can be attributed to intelligent children beginning school earlier or staying there longer or to any form of increased outbreeding.

School Achievement versus Aptitude

Recently, Jencks and Crouse (1982) reanalyzed the Project Talent data reported in the last chapter. After controlling for characteristics of the entering ninth graders, these investigators found that the mean achievement scores (history and social studies) for entire schools varied no more than the mean (intellectual) aptitude scores for those schools (i.e., vocabulary, abstract reasoning, arithmetic, and reading comprehension). They interpreted these results as suggesting that measures of so-called aptitude, such as IQ scores, are as sensitive to variations in schooling as are achievement tests. In a similar vein, Jencks and Crouse report that achievement and aptitude scores have roughly the same correlations with demographic characteristics, suggesting that aptitude tests are most properly viewed as a type of achievement test, too.[4]

Influence of Early Termination of Schooling on IQ

In addition to showing detrimental effects on IQ of summer vacation and of beginning school late, numerous researchers have demonstrated the detrimental effect of dropping out of school before finishing (Degroot, 1951; Husén, 1951; Lorge, 1945). In Harnquist's (1968) study of Swedish males, he selected a 10% random sample of the Swedish school population born in 1948 who, at the age of 13, were given IQ tests. Upon reaching the age of 18 (in 1966), 4,616 of these Swedish males were retested as part of their country's national military registration. Thus, the study is not vulnerable to the usual sampling criticisms. Harnquist was able to compare children who were comparable on IQ, SES, and school grades at age 13 and determine the impact of dropping out of school. He found that for each year of high school (gymnasium) not completed, there was a loss of 1.8 IQ points, up to maximum of nearly 8 IQ points difference between two boys who were similar in IQ, SES, and grades at age 13 but who subsequently differed in the amount of schooling completed by up to 4 years of high school. (Similar findings have been reported by both Degroot (1951) and Husén (1951), using different samples and analytic procedures. In Husén's study, a comparison of 613 Swedish men who had been tested in the third grade in 1938 and at the time of military registration in 1948 indicated that completing junior high school was associated with a three-point advantage, while completing secondary school yielded an eight-point advantage. These studies share several methodological features, thus enabling conclusions to extend over several cohorts.)

Influence of Northern Schooling on Black IQ

A fourth source of evidence for the influence of schooling on IQ is found in the sociological literature, particularly studies aimed at validating Otto Klineberg's original hypothesis about the inferiority of the education of southern blacks in the U.S. (Klineberg, 1935). Lee (1951) analyzed the IQs of blacks who migrated from Georgia to Philadelphia between the two world wars. After controlling for selective migration, Lee found that blacks gained over a half IQ point for each year they were enrolled in Philadelphia schools. Most of this increase could be attributed directly to the quality and additional years of schooling the blacks living in Philadelphia received *vis-a-vis* their Southern counterparts.

Influence of Early School Entry on Cognitive Development

One of the best documented studies of the impact of schooling actually was intended as a methodological demonstration of cohort-sequential analysis rather than a study of the effect of schooling, *per se*. Baltes and Reinert (1969) randomly sampled 630 children from 48 elementary schools in Saarbrucken, Germany. Three crosssections of 8- to 10-year-olds who were separated in age by four-month intervals were administered a German version of the Primary Mental Abilities Test (Thurstone and Thurstone, 1962). Since the German school system at that time required entering children to be six years of age by April 1, it is possible to compare same-aged children who had received up to a year difference in schooling, for example, a child who was born in March and is eight years and two months in May versus a child who was born in April and is eight years and two months in June. The former child would have received an additional year of schooling by the time he or she was eight years old. Baltes and Reinert (1969) found a substantial correlation between the length of schooling and intellectual performance among same-aged, same-SES children. Highly schooled 8-year-olds actually were closer in mental abilities to the least schooled 10-year-olds than they were to the least schooled 8-year-olds! Later in this chapter, I shall describe a recent effort by Frederick Morrison to use the procedure of Baltes and Reinert to assess the impact of schooling on microlevel cognitive abilities.

Influence of Intergenerational Changes in Schooling

With the onset of mandatory schooling through age 16 in the United States during the first half of this century, and the steadily increasing mean number of years of education, one might expect that the entire nation's IQ has been elevated, given the preceding evidence for the influence of schooling on IQ. Although this argument is impossible to confirm in its technical form (the major IQ tests are periodically renormed to insure a mean of 100 and SDs of 15 or 16), there is some evidence that supports the principle that underlies this conclusion. Despite the often-cited declines among high school students' SAT scores since the late 1960s, there are data indicating that this same cohort scores higher on earlier Wechsler and Stanford Binet IQ test norms than did the earlier standardization samples (Flynn, 1984, 1987a, 1987b). Each standardization sample since 1932 has scored higher using norms from the previous standardization. Had the tests not been renormed three times since 1932, children in 1978 would have scored 13.8 points higher than their counterparts did in 1932, a rate of .31 IQ points per year improvement for the cohorts between 1932 and 1978. Parker (1986) reported a similar finding, using different procedures (.25 IQ points per year gain).

Recently, Jack McCardle and John Horn (in press) have argued that the trend toward increases in IQ scores over time, while true at the aggregate "meta" data-analytic level, is not apparent at the individual data analytic "mega" level. In their reanalysis of scores of cross-sectional data sets, they found only a modest improvement in IQ—about one tenth as large as that found by Flynn and Parker. McCardle and Horn mention many procedural differences between their analysis and those of Parker and Flynn—any of which could be the basis for their difference. For instance, McCardle and Horn's analysis is based on a 30 year range of scores from 1950 to 1980 as opposed to Flynn's 55 year range between 1932 and 1978. Since schools changed their practices regarding social promotion and increased their diversification of in-school programs over this period, it could be that the effect of education has been attenuated since 1950.

Along these same lines, Thorndike (1975) reported that both preschoolers and adolescents, but not 7- to 10-year-olds, showed substantial gains on the 1971 renorming of the Stanford-Binet test over the 1936 standardization group. This fairly steady, monotonic increase in IQ test performance is most plausibly linked to changes in society (e.g., more widespread exposure to media and better nutrition), including large increases in the mean levels of school attendance.[5] In cases where only certain age groups have shown gains, as in Thorndike's report, the inference can be made that these were the most likely to benefit from television and the increased presence of preschool programs, as well as educationally oriented science shows. Finally, other countries have reported similar correlations between mean levels of schooling and IQ, including a seven-point gain in IQ in Japan in the 30 years following World War II (Lynn, 1982).

Similarly to Flynn's analysis of cohort changes in IQs, it has been shown that World War II recruits would have scored in the 83rd percentile on IQ if World War I norms had been used to score their IQ tests (Anastasi, 1968). And again, Tuddenham (1948) reported on a sample of 768 World War II draftees who took a revised version of the Army Alpha test (the test used by Terman and others to reach their conclusions that immigrants from southern Europe were predominantly feebleminded and that the average intelligence of their American-born peers was not much better). Compared to the performance of the World War I draftees, those from World War II exhibited an increase of nearly a full standard deviation in IQ. In fact, 80% of the World War II draftees exceeded the median score of World War I draftees. Most researchers attribute such an advantage to the increased average education of the World War II recruits (Anastasi, 1968; Block and Dworkin, 1976; Butcher, 1970; Tyler, 1965). In fact, Tuddenham compared a group of World War I soldiers who had similar educational attainments to those of a group from World War II and found that over half of the IQ differences between the two cohorts disappeared![6]

It seems obvious that the changes described (both gains on IQ tests and drops on SATs) have occurred over too brief a period of time to be the result of genetic changes. Sarason and Doris (1979), in an informative review of the literature on immigrants' IQs, provide an interesting historical footnote to this story. They report that the IQs of first-generation Italian-American children were usually within the low-average to borderline-retarded range (IQs between 76 and 100, with the midpoint around 87). Moreover, this score was not greatly elevated by the use of nonverbal tests or by the careful selection of children who exhibited "mainstream American attitudes." In fact, Italian-American children's IQs were lower than any other ethnic or racial groups except French-Canadians, blacks, Indians, and Portuguese. And commensurate with their low IQs was a high rate of retention in grade and dropping out. For example, in 1911 only 58% of Italian-American seventh graders in the Boston, Chicago, and New York school systems went on to enter eigth grade, compared to 80% of native-born whites. And in 1926, approximately 42% of all New York City students completed high school, whereas even as late as 1931, only 11% of Italian-American registrants graduated from high school.

What Sarason and Doris show so nicely in their review is that as the levels of school completion increased among Italian-American children during the first five decades of this century, so did their IQs. I mention this review because nearly every social commentator and IQ researcher during the early part of this century was convinced that the low IQs of Italian-American children were due to nonenvironmental causes. As a matter of fact, no one (except Sarason and Doris) has bothered to ask what happened to the second- and third-generation offspring of these Italian families and what this development implies about the persistence of low IQs among blacks, Hispanics, and American Indians. Today, Italian-American students' IQs are not considered remarkably low (they are, in fact, slightly above average).

Taken together, the evidence for the influence of schooling on IQ appears quite convincing. (Nine types of evidence for this relationship were reviewed, most involving multiple replications, for a total of 31 separate studies.) While it is not possible to estimate with any confidence the decrement in IQ scores that befalls each month or year of missed or delayed school (Jencks hazards the estimate of one IQ point decrement for each year of missed school), the trend is clearly one of decelerating scores as a function of missed school. All of the studies reviewed accord with this statement and I have been unable to find a single empirical study that argues otherwise. Table 5.1 is a synopsis of these 31 studies, broken down by category of evidence.

TABLE 5.1 SUMMARY OF THE EVIDENCE FOR THE IMPORTANCE OF SCHOOLING ON IQ

	Study
I. *Influence of Total Years of School Completed on IQ*	
r = .96 between highest grade completed and IQ score for MZ twins	Bouchard, 1984
r = .94 between highest grade completed and IQ score for gamblers	Ceci and Liker, 1986
r = .68 highest grade completed and IQ score for children	Jencks, et al., 1972
r = .6–.8 between highest grade completed and IQ score, after controlling for SES	Kemp 1955; Jensen, 1976; Wiseman, 1966
II. *Importance of Delayed Schooling on IQ*	
Large decrements in IQ for every year of delayed entry into formal schooling; the longer one delays entry, the lower their IQ	Schmidt, 1966; Ramphal, 1962
Between 10 and 30 point advantage on IQ tests for children from regions of high educational opportunity compared to low opportunity.	Sherman and Key, 1932; Tyler, 1965
III. *Importance of Summer Vacation on IQ*	
IQ scores decline slightly, but reliably, during summer vacation, especially for low SES youngsters whose summers are least "school-like" (fewer camps, cultural events, or structured activities).	Hayes, 1978; Heyns, 1983; Jencks, et al., 1972
IV. *Continuous Impact of Schooling on IQ*	
In regions of depressed educational opportunities, IQ scores of first graders are nearer to the national mean than are the IQs of their older siblings suggesting this "cumulative deficit" is the result of an accumulation of poor schooling.	Jensen, 1980; Sherman and Key, 1932; Wiseman, 1966; Vernon, 1969; Douglas, 1964
In cases in which children did not attend school (e.g., canal boat children), the correlation between their age and IQ was –.75, indicating that the more years of missed schooling, the lower their IQs.	Gordon, 1923; Freeman, 1934.
V. *School Achievement vs. Aptitude*	
IQ test performance is as sensitive to school variations as is achievement test performance, suggesting that the former behaves in ways similar to skills that schooling is designed to influence (e.g., reading, history, arithmetic).	Jencks and Crouse, 1982

TABLE 5.1 CONTINUED

VI. *Influence of Early Termination of Schooling on IQ*

Between 1 and 2 IQ points are lost for each year of high school not finished, vis-a-vis IQ-matched 8th grade cohort that does finish high school. Similar losses reported for students from Dutch schools closed during World War II Nazi siege.

DeGroot, 1951; Harnquish, 1968; Husén, 1951; Lorge, 1945

VII. *Importance of Northern Schooling for Blacks*

IQ gains of ½ point per school year for Blacks who migrated to Philadelphia from the South, after controlling for such factors as selective migration.

Lee, 1951; Klineburg, 1921

VIII. *Influence of Early School Entry on Cognitive Development*

8-year-olds whose birthdate permitted them to enter school early were closer on IQ scores to 10-year-olds whose birthdates required them to be late entering school than they were to other 8-year-olds who were late entering school.

Baltes and Reinert, 1969

IX. *Influence of Intergenerational Changes in Schooling on IQ*

Commensurate with their below-average level of school attendance, first-generation Italian-American children in the 1920s had below-average IQs, even after controlling for their degree of "assimilation." By the third generation, however, the offspring were slightly above average on school attendance and, consequently, on IQ, too.

Sarason and Doris, 1979

Since the first standardization of the Stanford Binet IQ test in 1932, each new standardization sample has outperformed the prior sample (and would receive a higher IQ score if compared to the prior generation than if compared to their own). Similar improvements have been noted for special populations (e.g., preschoolers, Armed Forces draftees).

Flynn, 1984; Anastasi, 1968; Tuddenham, 1948; Thorndyke, 1972; Parker, 1986; Lynn, 1982

Finally, a number of commentators have noted that educational improvements were the most likely cause of successive gains in IQ from one generation to the next. For instance, Both Burt (1946) and Cattell (1950) studied the IQ scores of large numbers of British schoolchildren and, rather than finding the large expected decline in IQ of two points per generation that was predicted by their genetic theories, they were surprised to discover very small decreases (in Burt's case) or actual increases (in Cattell's case).

Although these investigators continued to hold out the possibility that their findings would be reconciled with genetic theory (e.g., by postulating differential birth rates for the low- and high-IQ parents or by the masking effect of increased schooling), most researchers now agree with Butcher's measured conclusion:

> It is possible, but not very satisfactory, to explain [the increases in IQ] in terms of a real decline in average innate capacity masked and more than offset by gains due to greater test sophistication and to improved standards of health and education. Such might be the view of the convinced heriditarian, but there is no direct evidence for the supposed decline in innate capacity. . . . The studies quoted strongly suggest that environmental changes such as improvements in education have a decided effect on measured intelligence. One would expect this effect to be much more evident in societies and cultural groups where such improvements have been particularly rapid, and Hunt (1961) cites a number of investigations which suggest that this is so. (Butcher, 1970, p. 167)

QUALITY OF SCHOOLING

So far, the evidence reviewed has been in the form of a causal relationship between the quantity of schooling (years of attainment) and IQ. But what about *quality of schooling,* of studying different things in different ways? The answer is mixed: At this time there is no clear trend with regard to qualitative characteristics of effective schools. Some studies have reported no correlation of such qualitative characteristics as amount of expenditure per pupil, class size, teacher credentials, and length of tenure with achievement or IQ (the best known of these studies are the large meta-analyses conducted in the 60's and 70's by Coleman and the Rand Corporation). Bouchard and Segal (1985) have taken the position that the quality of schooling is probably not an important predictor of IQ, given (1) that adult IQ differences are predictable from the IQ scores of children before they even enter school (correlations exist between age 5 IQ and adult IQ around .70), and (2) that major statistical analyses and a few naturally occurring experiments have indicated that variables such as class size, teacher credentials, and access to cultural events were essentially shown to be uncorrelated with later IQ. For example, in the aftermath of the World War II devastation of Warsaw, Poland, there was a near random assignment of children to schools that had been rebuilt under a socialist regime. Yet, the school-related variables accounted for only about 2% of the variation in later IQ scores, prompting Bouchard and Segal to comment that: "While the quality of primary and secondary education is an important concern, there is no firm evidence that it is a major source of individual differences in IQ. It is important to recognize that most of the differences in IQ under discussion here exist before children attend school."

Other studies suggest that more subtle variables may have an impact, such as the alacrity with which a principal sees to repairing broken windows (Edmonds, 1986). Recently, Bruce Fuller (1987) has reviewed this literature and concluded that the indicators of school quality have become more complex and at the same time, less agreed upon. His review indicates that there does appear to be some relationship between quality indicators and educational outcomes, though it is probably still too early to know the magnitude or scope of it. So, while the picture is still unclear about the advantages of some qualitative characteristics of schooling, it is not unclear about the advantage of attending versus not attending school. This has led Vernon (1969) to comment: "But it is clear that sheer amount of schooling, even—in backward countries—of low quality education, helps to promote both school achievement and the kind of reasoning measured by non-verbal tests. . . . if such schooling is unduly delayed, the possibilities for mental growth deteriorate" (p. 350). This conclusion, bolstered by the studies that were reviewed here, represent a rebuttal of Burt's assertion that IQ ". . . is not due to teaching or training."

SCHOOLING AND INFORMATION PROCESSING

Thus far, I have been reviewing evidence for the impact of schooling on IQ. The assumption has been that IQ is identical to intelligence or at least is a surrogate for cognitive complexity. Many cognitive psychologists find this assumption untenable and argue instead that IQ tests are hopelessly heterogeneous in content and blur the distinction between past learning and current intelligence (Gregory, 1981). Some have advocated an alternative approach, one assessing more "fundamental" information processing components presumed to underpin successful performance on IQ tests (e.g., Sternberg, 1985). Many implicitly assume that these microlevel components are less influenced by social variables and play a causal role in macrolevel thinking and reasoning tasks, including IQ. But what about the role of schooling in the development of these more fundamental cognitive processes that presumably underpin successful performance on the various subtests of an IQ test? In what follows, I present a bird's eye view of the relevant research. As will be seen, there is some support for the view that schooling conveys positive effects not only on IQ performance but also on the acquisition and use of even these microlevel information processing components.

Perceptual Abilities

To my knowledge, there have been no experimental studies of the link between schooling and perceptual skill acquisition among American children, although numerous quasi-experimental and correlational studies have been reported in the cross-cultural literature. With a few exceptions, this cross-cultural research demonstrates positive effects of schooling on perceptual skill acquisition and use. Various cultural mechanisms and variations in socialization have been proposed to explain these results, and their elucidation is beyond the scope of this text. (Rogoff, 1981, has provided an analysis of the possible mechanisms that underlie the relationship between schooling and perceptual development.) The important point for our purposes is that *schooling appears to facilitate a number of perceptual skills that underpin successful performance on IQ tests.* For example, the ability to use information such as depth perception cues in two-dimensional pictorial representations, the ability to disambiguate figure-ground relations (Children's Embedded Figures Test scores), and the ability to perceive abstract visual-spatial organizations (Block Design scores) all have been shown to be enhanced through the schooling process (Berry, 1976; Dawson, 1967; Fahrmeier, 1975; Greenfield and Childs, 1972; Hudson, 1960; Kilbride and Leibowitz, 1975; Myambo, 1972; Stevenson, et al., 1978; Wagner, 1977, 1978; Witkin and Berry, 1975). The Soviet developmental psychologist, L. Abramyan (1977), reported that older children's superiority at figure-ground discriminations appears to be due, not to "hard-wired" neural maturational developments, but rather to developments associated with schooling. Abramyan based this assertion on the changes in wording needed to bring younger children to the performance level of older children by changing the context of the figure and the background. For example, if a red circle (figure) was presented against a white background, or a green circle was presented against a yellow background, preschool children found it difficult to avoid the pull of the figure, despite explicit instructions. Even when the stimuli were changed to more meaningful ones (green airplane against yellow sky), children were captured by the figure. It was only when Abramyan told the children to push an "on" or "off" button if the red airplane was going up into a sunny sky (yellow background) or a cloudy sky (gray background) that they were able to respond to both figure and ground changes.

Most of the perceptual skills that have been linked to schooling are indirectly important for the successful performance on commonly used cognitive tasks (e.g., mental rotation and "same-different" judgments), and several of these skills are of direct importance for performance on IQ tests (e.g., both abstract visual-spatial reasoning and the ability to disambiguate figure and ground are important for performance on the Block Design subtest of the WISC-R).

Concept Formation

Numerous studies have documented positive effects of schooling on a variety of conceptual skills, using a variety of tasks (rule learning, free association, multiple classification). Compared to their nonschooled peers, schooled children are more likely to: (1) sort stimuli by form and class rather than by color, (2) group items that belong to the same taxonomic class rather than to the same thematic class, (3) demonstrate greater flexibility in shifting between domains during problem solving; and (4) spontaneously engage in more verbal descriptions of their classifications (Ceci and Liker, 1986a; Cole and Scribner, 1974; Evans and Segal, 1969; Gay and Cole, 1967; Hall, 1972; Greenfield, et al., 1966; Irwin and McLaughlin, 1970; Schmidt and Nzimande, 1970; Sharp, Cole, and Lave, 1979; Stevenson, et al., 1978). In addition to their role in performance on laboratory thinking and reasoning tasks, many of these skills also are relevant to performance on IQ tests (e.g., taxonomic grouping instead of thematic grouping is the basis for successful performance on the Similarities subtest of the Wechsler Intelligence Scales for both Children and Adults).

Memory

Some form of memory assessment has been included on every major IQ test since the inception of Binet's test in 1908. The reason for the popularity of memory among testmakers is obvious: Free recall is linearly related to age, with each successive age group recalling more items than its predecessor until reaching an asymptote during early adolescence. Research has shown that schooling (and the content and procedural knowledge acquired through schooling) conveys benefits not only in the amount of information that can be recalled (with the possible exception of story recall, which is presently being debated), but also in the underlying processes that support memory (Chi and Ceci, 1987; Cole and Scribner, 1977). For example, children who receive traditional Western-style schooling are superior to their peers who receive either non-Western schooling or no schooling at all in the amount of taxonomic clustering in their free recall (Cole, et al., 1971; Sharp, et al., 1979). Other strategies that foster remembering, such as rehearsal and "chunking" have been shown to be enhanced through processes associated with the onset of formal Western-style schooling (Fahrmeier, 1975; Rogoff, 1981; Sharp, et al., 1979; Stevenson, et al., 1978; Wagner, 1978). Thus, direct links between schooling and the processes that support the type of memory made use of on laboratory tasks and IQ tests (e.g., Digit Span and Coding) have been reported.

Recently, Fred Morrison of the University of Alberta reported an "experiment of nature," using a design similar to the one first used by Baltes and Reinert (1969). Morrison selected Canadian children whose birth dates fell immediately prior to or following the Canadian school cutoff date for entrance to first grade, March 1. One group of children was denied entrance into first grade because the children's birth dates fell during the month of March, while the other group got into first grade just under the wire. On average, these two groups of children differed by 41 days of age. At the time of their sixth birthday, they were equivalent on all measures, including IQ and memory span. Morrison followed the two groups for the next two years and discovered that the children who were permitted to enter first grade early significantly outperformed their slightly younger age-mates on strategy use and memory performance. Moreover, they made far greater strides in their use of memory than did their slightly younger peers, suggesting that simple maturational processes could not explain the difference. Finally, in what is perhaps the nicest feature of Morrison's cohort-sequential analysis, he examined the memory performances of children two years later to determine whether there existed an interaction between schooling and maturation. If maturational processes were influential in the mnemonic gains that were associated with schooling, then we might expect the youngest children to outperform their slightly older peers when both groups were observed at the end of first grade. This is because the younger children would have been almost a full year older than their peers by the time they finished first grade, due to their late entry. Morrison found no evidence for maturational influences, at least over the children's age range and on the measures he used. What mattered in his study was only the amount of formal schooling the children received: After one year of schooling all the children were equivalent on the memory measures, regardless of their large age differences—nearly a full year between the groups.

Other Cognitive Skills

The literature on the effect of schooling on other cognitive processes is either unequivocal about the absence of a schooling effect (i.e., schooled and unschooled children do not differ in learning conjunctive and disjunctive rules), inconsistent (e.g., the data on Piagetian attainments is quite contradictory, with slightly more than half of the 30 or so published studies I have read reporting no effect of schooling), or indeterminate (e.g., there are no studies with which I am familiar that have assessed "speed of encoding" or "retrieval efficiency" as a function of schooling, after controlling for chronological age). A scattering of other studies have demonstrated effects of schooling on such matters as children's use of syllogistic reasoning when the premises are couched in terms of events that were not actually experienced by the children (Fobih, 1979; Scribner, 1977; Sharp, et al., 1979). Jeyifous (1985) has reported cross-cultural findings dealing with the "characteristic-to-defining shift" (i.e., moving from an understanding based on a concept's salient or characteristic features to one based on its defining features—e.g., shifting from the concept of a taxi based on the prototypical shape and color of taxis to one based on its essential features of transporting passengers for a fee), as well as with the "identity-preserving shift" (natural objects remain inherently untransformed by surface changes). She demonstrates that both of these shifts occur among the schooled and unschooled Yoruba of Western Nigeria. Her data call into question the importance of schooling for either shift, *provided the materials are culturally derived and personally experienced.* In her study, however, it is not possible to tell whether schooling had no effect on these attainments or whether the effects were too subtle to detect without a longitudinal design. All that is known is that unschooled adults exhibited the cognitive attainments by the time Jeyifous tested them as adults. But it is possible that they made these attainments several years later than schooled adults. Similar cross-cultural results have led several authors to conclude that one of the primary mechanisms by which schooling influences cognitive development is by fostering disembedded modes of thinking that are not tied to personally experienced events. (For excellent synopses of the empirical literature supporting this view, see Cole and Scribner, 1974; Rogoff, 1981, 1982; Wakai, 1985.) An equally tenable interpretation is that schooling fosters thinking about events and concepts as they are learned in school but not necessarily as they are confronted outside of school. Schliemann (1988) demonstrated that Brazilian students' knowledge of school-learned combinatorial rules is quite uneven outside of the school. (Note that this is not equivalent to the former interpretation.)

Taking Stock

There have been many studies of the effect of schooling on cognitive growth, and many writers, policymakers, and educators have expressed opinions about various aspects of schooling. As one looks back over this research, it is illuminating to note how many theories about schooling have come and gone, to be supplanted by newer ones that appear more directed to single events (e.g., declining SATs, failure to master shared cultural knowledge, drop-out rates among minority students) than to the full corpus of data on schooling. I have tried to examine the entrails of each of these new theories for ideas to implant in my own bioecological framework.

I think it is important, though clearly beyond the goal of this treatise, to say something about the job that schools are doing today. It has become almost a national pastime among writers on topics related to education and intelligence to give the schools advice on how to improve the job they are doing. I do not want to disappoint readers familiar with this genre by neglecting to take a shot at the nation's schools myself. It might be (mistakenly) inferred that I think that schools are doing a great job because I have argued that if you want to see how important schooling is for the maintainance of IQ or other cognitive skills, just take a child out of school and see what happens. Lest I be misunderstood, I want to make it clear that just because schooling is instrumental in propping up IQ and other cognitive skills, that is no reason to believe it is working well—any more than thinking that IQ is an index of anything important in and of itself. No, I think that schools have done an increasingly poor job in educating students. But I do believe that one cannot view the performance of schools outside of the context of larger societal issues that impinge upon them. No one ought to diminish the effect of these larger societal issues (e.g., increasing number of children born to single, poor mothers, skeptical work attitudes in the community, and a desire to do the very least amount of work congruent with graduating or getting into college). These are real issues and they present real challenges, the likes of which are fairly new to this country's schools. I think it is fair to say that, until the 1940s, the adage that governed our public schools, *Education for the Common Man*, was both figuratively and metaphorically accurate: Schools were set up to serve the common child, and it was not expected that schools would serve the uncommon child. It was customary for children who lacked motivation to be excluded from formal schooling at an early age. Somewhere along the way, we have expanded our expectation of what schools can and should do. Today, all children must be not just warehoused in schools, but educated! Of course, there are pockets of poor and unmotivated students around the country where the schools seem to be doing a good job inculcating basic skills and cultural knowledge than what has recently been reported for most schools by commentators like Bloom (1987). We ought to study these schools for their ideas (and I believe that we will discover that they have found ways to involve the larger community and families in children's education). Students are adept at figuring out exactly what needs to be done in order to graduate or get into college and the majority of students appear to do no more than this. Twenty-five percent of students do not finish high school, but they are only the tip of the iceberg. Of those who do finish, it has been estimated by Albert Shanker, president of the American Federation of Teachers, that 75% are functioning on what was formerly a junior high school level in basic competencies. This recalls Jencks' assertion at the end of the last chapter that students are not going as far beyond the basics in school today as they once did.

Other than concluding that there is blame enough to go around to all segments of society (schools, parents, communities, policymakers), I have not given serious thought to the solution for this malaise. I suspect that one thing that could be done would be for schools to get across the message that the ante has been raised: To graduate will require more effort, and to get into college will require more effort still. And naturally, the larger community needs to be a player in this enterprise, abetting schools in their effort to raise the ante by inserting themselves into the equation (again). But we all know that there will still be children who cannot cope, no matter what incentives are inserted. They learn early that when the teacher calls on them to answer, they do not do well. And when the teacher offers a reward or competition for the first one who figures out the answer to a problem, it is never they who answer first. Such experiences can prompt youngsters to opt out of the educational enterprise early rather than subject themselves to the continuing humiliation associated with failure. If we expect schools to educate such youngsters, then we need to develop kinder, more humane ways of reaching them and maintaining their interest. Shanker makes a cogent point in this regard: Noting the attractiveness in some sectors for longer school days or longer school weeks or even a longer school year (less summer vacation), he says that if you keep in mind that 25% are not finishing now and of those that do, 75% are not what one would call "stellar students," then increasing the school day is not the answer. Imagine a car manufacturer noticing that 25% of the cars on their assembly line fall off before they reach the end of the line and 75% of those that do reach the end of the line have glitches under their hood. I doubt this mythical car manufacturer would conclude that the remedy is to increase the speed or length of the assembly line. Schools and parents need to come to grips with the fact that they are not producing great products. We at universities who receive these products have known for a long time that this is so. And universities are not the place to begin remedial education, for a variety of reasons.

To close the chapter, I have attempted to provide several types of evidence from several disciplines that pose a challenge to the two dominant psychological approaches to the study of individual differences in intelligence, the psychometric and information processing approaches. In reviewing this evidence I have tried to suggest that the environment (e.g., one's cultural, work, and leisure activities), and in particular schooling, affects not only the IQ score itself, but apparently many of the cognitive processes that underpin successful performance on the IQ test. Both the psychometric (individual differences) and information processing (componential analysis) approaches to this topic, despite their elegance as methodologies, have some obvious shortcomings, most notably their emphasis on isolable subsystems of operations, structures, and processes that are assumed to form the wellsprings of intelligent behavior.

PART TWO

THE BIOECOLOGICAL FRAMEWORK

Up to this point I have focused on the task of defining the question concerning the nature and development of intelligence. Contextual influences on complex performance cast doubt on current conceptualizations and assumptions of intelligence researchers. Beyond describing the dominant arguments and metaphors of today's work, an empirically adequate model of intellectual development would need to take notice of the research carried out by those who are outside the IQ guild. These include anthropologists, sociologists, and educators, as well as developmental psychologists like myself.

In Part II, I shift the level of discourse to a different plane. I shall describe a "first pass" through the rationale for my own framework. Presently, it is a tangle of arguments—entwined as much because of the "work-in-progress" nature of my thinking as because of the inherent complexity of the question being discussed. Thus, the bioecological alternative to current theories of intellectual development is really the presentation of a program of ongoing research, not a completed one by any means. I invite the reader to join me on a short expedition through the web of areas, issues, and assumptions that undergird the bioecological framework.

To presage some readers' reactions to the next half of the book, it is not written with the same authority or smoothness as the first half. This is not because of less time or energy spent on it, but because the argument to be presented is more multidimensional than what was tackled in Part 1. The bioecological framework requires lots of pieces, and it may not be intuitively apparent why each of these pieces is needed. In addition to the obvious ingredients like context, heredity, and multiple abilities, there is a need to say something about generality, abstractness, and the mechanisms that are involved in the transformation of domain-specific abilities across domains of knowledge. The format in which I have chosen to do some of this is by comparing and contrasting the bioecological program of work with dominant theories of intellectual development (Chapter 10). I have tried to make the reasons for including each of the aforementioned pieces as obvious as possible, but readers may occasionally want to skip sections that seem to be going nowhere in their minds. Following is a brief summary of the material in Part II.

In Chapter 6, I discuss the nature and meaning of context and present some conjectures as to how context influences cognitive complexity. Context is central to the bioecological framework, but so are g, knowledge, and multiple abilities. Chapter 6 integrates the last three into the bioecological framework. Chapter 7 presents a preliminary version of the bioecological model of intellectual complexity and comments on the biological, and especially hereditary, mechanisms that might be involved in the transmission of boundary conditions for multiple abilities. The framework is as yet incomplete, though its outline is sufficiently defined to permit a few reactions and evaluations. In Chapter 8, I take on the "singularity of mind" argument, as this poses an important challenge to the bioecological framework. The singularity-of-mind view assumes that there exists a single intellectual energy or nervous system efficiency that underpins all important forms of cognitive complexity. It is the modern successor to Spearman's universal law of unity or g.

In Chapter 9, I challenge the view that intelligence implies abstractness or generality. My view is that so-called abstractness and generality are not inherent aspects of cognitive complexity, but rather manifestations of well-structured declarative knowledge. One can be complex without being abstract or general, and conversely the presence of abstractness or generality implies nothing about cognitive complexity. Finally, armed with analyses of context, heredity, g, and abstractness, in Chapter 10 I review the dominant theories of intellectual development and say what I find right and wrong about them. All of these pieces (context, heredity, multiple abilities, and abstractness) are central to my argument. Any empirically adequate theory or framework concerning intelligence ought to address them.

CHAPTER 6

The Role of Context in Shaping Multiple Intelligences

It was mentioned in Chapter 1 that the bioecological framework is inherently developmental, biological, and contextual. I have already described the basis for the developmental claim, and I shall deal with the role of biology in Chapter 7. In this chapter I want to say why the role of context is crucial and, more importantly, say what I mean by "context."

CONTEXTUALISM AND INTELLIGENCE

The root metaphors that imbue positivist world views of science might be described, in general, as the "world as a machine" and, in the particular topic under discussion, as the "mind as a telephone switchboard or a digital computer." Contextualism, on the other hand, has as its dominant metaphor the ever-changing sociohistorical, cultural, and social milieu in which cognition unfolds. It is manifestly a nonmechanistic world view even though it accepts that regularities of nature can be identified amidst the swirling tides of change. Jaeger and Rosnow (1988) make a crucial distinction between contextualism on the one hand and situationalism and interactionism on the other. In the latter two approaches context is viewed as something adjunctive to the act, whereas contextualist accounts elevate context to a constituent of the act. Situationalism and interactionism, they note

> Regard *context* as a synonym for 'environmental stimuli' that determine a behavioral response. For contextualists, the *context* is an integral part of human actions, to be sure. An act cannot be said to have an identity apart from the context that constitutes it; neither can a context be said to exist independently of the act to which it refers. . . . As Bhaskar said, to think of context as existing in addition to or apart from practices is like imagining smiles alongside or beside faces. (Jaeger and Rosnow, 1988, p. 66)

Thus, a contextualist perspective on individual differences in intellectual development has certain characteristic features. It emphasizes the ecological dependency of cognitive structures, as well as their plurality and spontaneity. Individuals participate in the construction of their own development by virtue of altering their contexts and being altered by them. An implication of such a world view is that those of us who study individual differences need to be sensitive to what our experiments are doing to the subjects and what strategies they are adopting to surmount them (Jaeger and Rosnow, 1988).

If the psychometric and information processing researchers simply viewed their orientations as methodologies, then the criticism made at the end of the last chapter would be unwarranted because both of these methodologies could be used to assess the impact of various forms of knowledge and cultural practices on cognition, and indeed, this is precisely what many of the cross-cultural researchers have done (e.g., Cole and Scribner, 1977; Super, 1980; Irvine and Berry, 1988). But psychometricians and information processors often have assumed that the operations, structures, and processes that have been documented in their analyses are relatively impervious to contextual influences, and, as already noted, many have argued that contextual influences are themselves mediated by genetic action. This imperviousness is part of their lure, because to acknowledge the influence of environmental variables on such fundamental information-handling processes would be tantamount to admitting that the resultant theories are weaker than their proponents often wish them to be. If the environment is a direct contributor to intellectual development, then each theory would need to be qualified by the particular ecology that catalyzed the processes that were instrumental in a particular battery of tasks.

Cognitive Molecules
Out of Context

Giving short shrift to the role of context has been evident from the very beginning of research on individual differences in intelligence, as researchers began presenting their discoveries, not as contextually tied, relativistic depictions of cognitive processes, but as descriptions of powerful, "transdomainal" operations. For example, in the information-processing approach to intelligence, these operations were usually presumed to be the building blocks for higher level cognitions and also were presumed to be insensitive to minor shifts in the environment, even if the higher level cognitions were not. After all, how influential can the environment be in explaining individual differences in basic perceptual recognition speed, attention, memory for digits, and so on, given the simplicity of the tasks and the familiarity of the stimuli used in these studies? (In Chapter 8 this question will be addressed in detail.) Yet it was argued that these elemental processes underpinned all higher level thinking and reasoning tasks, rendering them better or worse as a function of their efficiency. An early researcher in this area, E. L. Thorndike (1924), described an approach to cognition that would seem to be compatible with the information-processing approach:

> Accuracy, quickness, discrimination, memory, attention, concentration, judgment, reasoning, etc., stand for some *real and elemental abilities which are the same no matter what material they work upon;* that in a more or less mysterious way learning to do one thing well will make one do better things that in appearance have no community with it. (p. 272; emphasis added)

Thorndike's molecularization of cognition (which he went on to criticize) resembles current thinking in this respect: There is an implicit endorsement of the view that complex cognitions can be understood by dissecting their components and, furthermore, that the effective deployment of a component in one domain increases the probability that it can be deployed in all appropriate domains. It is precisely because these component processes are assumed to be so fundamental that they are impervious to changes in the environment. In view of the fact that over 90% of the world's psychologists are Western educated, it can be seen how attractive such a view might be because of the assumption that Western schooling fosters context-free, abstract, disembedded thinking (Laboratory of Comparative Human Cognition, 1986).

However attractive this position may have been a half century or even a half decade ago, currently there is evidence that these cognitive component processes are elicited by various aspects of the context. This evidence, some of which was already cited (e.g., the butterfly and cupcake studies in Chapter 3), suggests that the emphasis in modern cognitive psychology on cognitive operations and components practically as forms of disembodied mental activity, is ill wrought because it ignores the crucial role of context in one's perception of a problem as well as in the choice of a strategy for its solution (Lave, et al., 1984). Here the term *contextualism* is used in its broadest sense to include not only the proximal physical and social setting in which cognition occurs, but also the more distal cultural values that help shape cognition and are an integral part of it. As Jaeger and Rosnow (1988) note: "Contextualism requires a shift from viewing 'context' as something external, merely impinging on behavior, to viewing context as integrated within the phenomenon itself" (p. 72).

Culture's Role

One's cultural context is an integral part of cognition because the culture arranges the occurrence or nonoccurrence of events that are known to affect cognitive developments (e.g., literacy and orientation in hostile situations). Each culture and subculture presents a limited number of contexts in which children may interact during development, and many of these could have an impact on cognitive performance. Moreover, culture controls the frequency of occurrence of events, thus dictating the amount of time spent doing some tasks (e.g., weaving) over others (e.g., abstracting and interacting with others).

Similarly, cultures control the level of difficulty of tasks within various contexts. For example, in many cultures it is "uneconomical" to encourage each child's maximal or potential cognitive growth. Rather, these societies endeavor to keep their children in a "zone of proximal development," which is the difference between their level of independent problem solving and their level of problem solving under adult guidance or in collaboration with more capable peers (Vygotsky, 1978, p. 86). The advantage of the latter is that children are exposed to a complete task while only engaging in those aspects found at the limits of their own cognitive competence.

Finally, cultural contexts may control the patterning or co-occurrence of events, thus giving rise to cultural taxonomies. For example, "hungry time" and a trip to the rice market co-occur for the Kpelle people of Western Africa, resulting in their strong conceptual understanding of volume because an error in bartering for a volume of rice could lead to suffering or death (Laboratory of Comparative Human Cognition, 1983).

Contextualists have made more progress on the conceptual plane than on the empirical plane. Many of the mechanisms by which a cultural context is said to shape cognition are inferred *a posteriori* from ethnographies and are, to some, unproven speculations. To date, there are no strong theories of culture that lead to unambiguously correct predictions, hence, theories frequently fit not only a particular cultural setting, but another that is very different. An excellent example of this is the initial discovery that certain African groups had difficulty inferring three-dimensionality from photographs. This led to a rash of cultural explanations that tied this difficulty to aspects of their rural living and traditions, but all these explanations ultimately failed to account for the same perceptual phenomenon in very different cultural contexts, such as among Eskimos (Vernon, 1969). Similarly, despite evidence in some studies that the "uncarpentered" and "open vista" environments of the Kung Bushmen impaired their perception of visual illusions, studies that exercised greater methodological care failed to detect such impairment (Reuning, 1988).

Having made the preceding criticism of contextualism, it is important to note that the cognitive differences found to result from slight alterations in context (e.g., the video game vs. laboratory task reported earlier) provide a measure of support for the contextualist position, because these findings show that even small differences in context can lead to very different cognitive outcomes. This suggests that large-scale cultural differences *are* likely to affect cognition in important ways. One's way of thinking about things is determined in the course of interactions with others of the same culture; that is, the meaning of a cultural context is always negotiated between people of that culture. This, in turn, modifies both culture and thought. Accordingly, we may suspect that Thorndike's list of elemental abilities (or any other list) is in actuality a set of skills that have developed within a specific cultural context and in response to specific material knowledge and that requires, therefore, specific contextual ellicitors for its activation. However, it is one thing to say this, but quite another to prove it. Still, a beginning has been made, and the data that have been gathered are consistent with a contextualist perspective. And the failure of contextualists to provide convincing evidence for their tenets is not a justification for others to ignore context, as so many in the intelligence community have done. This has led Bronfenbrenner to comment:

> It is a noteworthy feature of all preceding (cognitive approaches) that they make no reference whatsoever to the environment in which the person actually lives and grows. The implicit assumption is that the attributes in question are constant across place; the person carries them with her wherever she goes. Stating the issue more theoretically, the assumption is that the nature of the attribute does not change, irrespective of the context in which one finds one's self. (Bronfenbrenner, 1989, p. 207)

To the extent that the foregoing reasoning is valid, then "doing one thing well" may have little relevance for (1) doing other things well that *ostensibly* entail similar cognitive abilities but occur in different contexts, (2) doing other things well that require different knowledge domains for their solution, or (3) doing things well that occur in response to different environmental challenges and possess different attainment values. The results of the distance estimation task involved in "capturing butterflies" demonstrate this: The children's internalization of a curvilinear algorithm was dependent on a specific context (Ceci, et al., 1987). Similarly, Lave and her colleagues (1977; 1984), de la Rocha (in press), Murtaugh (1985), and Rogoff (1978) have shown that competency in using arithmetic operations in completing everyday tasks cannot always be predicted from standardized arithmetic tests that presumably tap the identical operations. For example, Lave (1977) demonstrated that the best predictor in a multiple regression analysis of Liberian tailors' arithmetic solutions to typical school-related problems *in a school context* was the tailors' mean educational attainment (number of years of schooling), while the best predictor of their solutions to these same arithmetic problems *in a tailoring context* was the mean number of years of tailoring experience, despite the formal identity of the two sets of arithmetic problems.

This same point has been made by numerous cross-cultural researchers. For instance, Brazilian "street children" develop intuitive models of probability in order to broker lottery tickets, but these same children appear to be largely ineffective at applying their models to solve the type of probability problems encountered in school (Carraher, et al., 1985). And students entering college who have encountered combinatorial rules to calculate the number of permutations in a set of letters or digits cannot always use these rules in out-of-school contexts (e.g., how many differently colored shirts can be made where each color can be either top, middle, or bottom, Schliemann, 1988). In fact, 45% of the students were at a low level of sophistication in solving this task, whereas bookies with comparable schooling were never at this low level. (I hesitate to press her findings terribly hard at this time because of the small sample sizes, but they are quite interesting demonstrations of the context-dependent nature of at least this type of cognition.)

It is important to note that none of the findings that have been reviewed thus far deny the importance of biological mechanisms in cognition, or individual differences more generally. Most of these studies demonstrating context specificity in fact also demonstrate a fair amount of variation in a given context across subjects who would appear to have had equivalent experiences, schooling and motivation (e.g., Ceci and Liker, 1986a). Biologically based constraints surely exist on all types of cognition, and some persons may be more or less endowed than others in undertaking a particular type of cognition. Although some may be under the impression that the discrediting of Cyril Burt's data on heritability signaled the end of claims of a biological basis to intelligence, nothing could be further from the truth. The field is decidedly as biological today as it was during Burt's day. Cronbach (1975), Bouchard and Mcgue (1981), Plomin (1985), and others all have reviewed the historical and modern behavior genetic findings and remarked how similar they were, despite the generally lower heritability estimates in the modern studies. I shall discuss this evidence in Chapter 7. For now, however, the issue is not the role of biology in accounting for observed differences in IQ test scores, but the relationship between IQ scores, information processing constructs, and cognitive complexity as a function of ecological factors that have been mentioned (schooling, culture, motivation, setting). The evidence presented so far would be congruent with the view that environments differ in their influence on the crystallization of biological potentials (some of which may be indexed by microlevel information processing components). The route this influence takes is reflected in the degree to which one's knowledge structure becomes elaborated or differentiated by different environments. It is an expectation of those holding such views that information-processing efficiency will either be constrained or facilitated by the development and organization of one's knowledge base in response to specific environments, and it is difficult to assess the integrity of the biological underpinnings of such information-processing efficiency in the absence of some understanding of the structure of the knowledge upon which these processes operate. So biology helps influence the efficiency of processing, which in turn helps influence the amount, type, and organization of knowledge that is acquired.

Finally, the knowledge base helps influence the efficiency of the operation of the very processes that were instrumental in its acquisition. This is all another way of saying that context, including domains of knowledge, is important and that processes present in one context may not be operative across others. Following an extensive review of the cross-cultural literature demonstrating the limited generality of cognitive functions, Rogoff (1981) was led to a similar conclusion: "Individual differences may not lie in one's 'ability' to apply certain processes, but rather in the contexts in which those processes are habitually applied" (p. 274).

A BIOECOLOGICAL THEORETICAL FRAMEWORK

The review of the literature thus far has called into question the isomorphism between IQ scores and the type of cognitive complexity needed to manage a city, handicap a race, shop in a cost-efficient manner, etc. Is this because the skills involved in such tasks are inherently unlike those required for successful performance on an IQ test, or because the skills, although similar to those required for successful performance on IQ tests, are operating on inherently different knowledge domains that render them differentially efficient? The answer is mixed, and at present we are unsure as to why this is the case. Successful performance on some of these tasks would appear to require the application of the very same skills that are involved in performance on IQ tests. For example, the type of eductive processes necessary to discern relationships between patterns of variables on a racing form or on a city-management scheme appear similar to those needed to apprehend relationships between items on the Similarities and Picture Arrangement subtests of the Wechsler scales. Yet the use of eductive processes among expert handicappers or city managers was unrelated to their use of these processes on an IQ test (Ceci and Liker, 1986a,b; Dörner, et al., 1983). Therefore, it is more than simply postulating that different types of cognition are being measured by IQ tests and these forms of real-world problem solving, with those required for successful IQ performance being perhaps more abstract than those required for successful performance in some real-world contexts. Were it this simple, extant theoretical accounts would suffice, provided we added a statement about the relatively greater abstractness and predictiveness of those skills indexed by IQ tests. In Chapter 9, I shall argue that the concept of abstractness is problematic on two different levels, precluding its invocation in the preceding argument.

To drive home the point that some of the evidence that has been reviewed appears to implicate the same cognitive processes in both IQ tests and everyday forms of cognition, even among persons who fail to display them in both contexts, let us revisit a couple of the studies described in Chapter 3.

Recall that the men in the racetrack study appeared to be able to use mental arithmetic operations in making handicapping calculations (e.g., $2{:}03^2$ minus $1{:}30^4$) yet they had difficulty with the Arithmetic subtest of the Wechsler Adult Intelligence Scale. Similarly, unschooled Mano adults appeared unable to use the taxonomic classification scheme that is rewarded with bonus points on some IQ tests until they were presented an isomorphic problem that involved classifying containers of rice, an activity that holds some importance in their life. At the same time, it must be noted that there are many instances in the cross-cultural literature wherein an indigenous person's "cognitive deficit" refuses to disappear when the problem is couched in familiar terms or stimuli (Super, 1980). The reason for these failures can be due to any or all of the three components to be discussed next.

Introducing the Bioecological View

The fundamental observation that cognition is frequently context sensitive, along with the evidence for the importance of various ecological influences, leads to the following proposal: Individual differences in intellectual development are best understood within a bioecological framework consisting of three broad components, the first two of which are influenced by the timing of experiences during sensitive periods of development. First, there exists not one cognitive potential or central processor g underlying most or all cognitive performances, but multiple cognitive potentials. Second, the role of context, broadly defined to include motivational forces, social and physical aspects of a setting or task, values inculcated through various types of parenting, and the elaborateness of the knowledge domain in which the task is embedded, are important not just in the initial period of development of the cognitive potentials, but also in their later evocation during testing. And third, there exists a fundamental inseparability of knowledge and "aptitude" such that cognitive potentials continually access one's knowledge base in the cascading process of producing cognitions, which in turn alter the contents and structure of the knowledge base. The roles of these three components will be examined in the following discussion about the multiplicity of intelligences and will be carried over to the next chapter as well.

Multiple Cognitive Potentials

Existing evidence suggests that there is not one intellectual force g underlying most or all of the variance on intellectual tasks, but multiple, and in part genetically determined, cognitive potentials. Support for this statement, excluding the part about "in part genetically determined," comes from three types of research: *cognitive, neuropsychological,* and *psychometric.* Concerning the latter, it is apparent that a relatively large number of cognitive abilities are statistically independent, if not in the sense of being orthogonal, at least in the sense that the internal consistency (r_{xx}^2) of each is statistically larger than the squared multiple R of its prediction by a linear combination of the rest of the abilities, thus suggesting that the failure to account for unique factor variance is not a result of low reliability of the measures (Horn, 1978).

Naturally, all factorists recognize the existence of specific abilities, differing mainly in the proportion of common variance (size of g) and the placement of the axes. (See Sternberg, 1977, for a discussion of differing criteria for rotation of factors.) And while factorists are not in agreement over the actual number of such specific cognitive abilities or about the magnitude of g, less than a dozen so-called primary mental abilities have been repeatedly revealed by independent researchers (see Carroll, 1941; Horn, 1978; Horn and Donaldson, 1980; Willis and Schaie, 1986; Thurstone, 1947). Willis and Schaie showed that these specific abilities, in combination, account for over half of the variance on solving everyday tasks. Some have noted, however, that specific mental abilities are themselves capable of being factored into a modest g (Cattell, 1971); but virtually all researchers agree that they represent unique contributions to explaining variance in individual differences (Royce, 1988).

Using somewhat different test batteries, conceptions of intelligence, and subject populations, theorists like Gardner (1983) have arrived at a number of different specific mental abilities. While the specific number and types of these independent cognitive abilities are not settled, and are indeed beyond the scope of this book, it is pretty clear that more than one type of intellectual ability emerges in these factor analytic studies, and they are differentially predictive of many forms of everyday problem solving. For example, Bray and Howard (1983) found that the level of managerial rank attained by workers during their 20-year longitudinal study at AT&T was somewhat predicted by these workers' specific cognitive abilities. Also, Camp, et al., (1988), demonstrated that a specific cognitive ability (inductive reasoning) was somewhat predictive of everyday problem solving tasks in the home. The model to be described later is predicated upon this assumption. The lists of cognitive abilities shown in Figures 7.1 and 7.2 are a composite of data from a variety of such investigations (e.g., Horn, 1978; Sternberg, 1985) and reviews of literature (e.g., Gardner, 1983). The ephemeral nature of these lists cannot be overemphasized; as more is known about cognitive abilities and their interrelationships, it is expected that these lists will change. For the time being, any list of abilities ought to be viewed as illustrative and highly tentative.

The second type of evidence in favor of multiple cognitive abilities, as opposed to one large first factor or g, comes from cognitive studies involving (1) observations of subjects who perform complexly in situations outside of a formal testing situation in which they may have appeared deficient (e.g., Cole, 1975; Scribner, 1976; the various anthropological studies cited earlier), (2) the demonstration of more cognitive complexity by an individual in one domain than he or she exhibited in another domain, and (3) a nonlinear relationship between cognitive complexity (indexed by a greater use of interactive thinking) and IQ, at least in some domains (Ceci and Liker, 1986b; Dörner, et al., 1983). Concerning this last, a case in point is Kohlberg's demonstration that the relationship between level of moral reasoning and IQ is nonlinear, with essentially no relationship between the complexity of one's moral reasoning and IQ beyond about 95. A similar nonlinearity can be found in Streufert's study of executives: There is no relationship between his measures of cognitive complexity and IQ beyond an average IQ. Such "inflection points" in the data suggest that a minimum level of IQ is needed to reason complexly, but beyond this minimum, the size of IQ has no predictive power.

The second type of evidence suggests that sometimes the same form of intelligence is needed for performance on an IQ test as on some nonacademic task, but it is differentially deployed as a function of the context. At other times it is clear there are forms of intelligence that, although conforming to our common usage of the term "intelligence," are statistically unrelated to those forms required for successful performance on an IQ test (e.g., Cornelius, et al., 1989; Sternberg, et al., 1981; Wagner and Sternberg, 1985).

The third type of evidence, which in many ways I regard as the weakest, comes from the neuropsychological literature. Clinical studies have repeatedly shown that discrete cognitive functions are lost as a result of lesions to local areas of the cortex. For example, aphasics with inferior frontal lobe lesions or lesions of the basal ganglia are severely impaired on long-term memory tasks but not on short-term memory tasks. By contrast, patients with posterior temporoparietal lesions do not appear to have problems on either type of task. Such findings have led researchers to conclude that specific neural connections in the inferior frontal lobe and basal ganglia underpin long-term, but not short-term, memory processes (Risse, et al., 1984). Other studies suggest specific neuroanatomical bases for many discrete cognitive processes (see Gardner, 1983; Geschwind, 1979). As one example, Schlotterer, et al. (1983), have demonstrated the functional separation of processes that support the encoding of information using an energy mask (e.g., a burst of light that covers the viewing field, following the offset of the stimulus) versus a patterned mask (e.g., a burst of densely packed letters or digits in the viewing field). The type of higher order interruption that results from a patterned mask, which can be evidenced by its interference with categorical processing, as when the subject tries to decide whether a stimulus belongs to a pre-specified group (Rawlings, 1986), is thought to implicate the pyramidal cells of the hippocampus and layers III and V of the association cortex. And, finally, there is the *laterality* literature. It has been known for some time that certain cognitive functions are subserved primarily by one hemisphere or the other. When a task is administered that requires processing from one hemisphere and another task is introduced which also requires resources from that same already engaged hemisphere, there is greater interruption of the ongoing cognition than if the newly introduced task requires resources from the non-engaged hemisphere (e.g., Hellige and O'Boyle, 1989). This and other laterality evidence that I will not delve into here (based on both interruption and facilitation paradigms) suggests that there are separate functional resource pools for each major hemisphere rather than a single resource pool that pervades all processing.

Yet despite the fairly persuasive evidence for the neuropsychological modularity of intellectual functions, there has been the suggestion in some quarters that the brain contains a unitary central processor that can be indexed by evoked cortical potentials. In fact, throughout this century there has been a debate between "locationists," who argue that specific forms of cognizing are tied to known neural structures, and "equipotentialists," who argue that processing is diffusely distributed over much of the cortex and proceeds as a "mass action." Hendrickson and Hendrickson (1982), Eysenck (1982), and Schafer (1987) have each argued that certain amplitude-based parameters of the electrical potential generated in response to simple auditory stimuli (clicks) are highly correlated with IQ scores. The task used in the studies does not require any deliberate response on the part of the subject other than simply listening to the clicks. Amplitude-based parameters derived from evoked potentials nevertheless are claimed to correlate *extremely* highly with the most *g*-loaded subtests of the Wechsler Adult Intelligence Scales—in excess of .80. This is an enormous correlation, about as high as some tests of intelligence correlate with others! Presently, the evoked potential–IQ literature is empirically "lean," and therefore, many of the alternative interpretations cannot be ruled out until replications and extensions of the experiments are conducted.

To appreciate this position, consider Schafer's recent finding that high IQ subjects who listen to "regularly timed" presentations of clicks emit less evoked potential (EP) response than when they listen to clicks that are randomly presented, whereas lower IQ persons emit disproportionately larger EPs to regularly timed clicks. From this difference, Schafer infers that the brains of high-IQ individuals are adapted to ignore trivial events in the environment, such as highly predictable clicks, while for unpredictable events, there is survival value to staying tuned in:

> A measure of neural adaptability derived from EP amplitude ratios correlates .66 with WAIS IQ obtained on 74 normal adults (corrected to .82 for range restriction). . . . Results suggested that the brain that efficiently inhibits its response to insignificant inputs and that orients vigorously to unexpected, potentially dangerous stimuli, is also the brain that manifests high behavioral intelligence. Neural adaptability as indexed by the temporal expectancy effect on evoked cortical potentials appeared to provide a biological determinant of g factor psychometric intelligence. . . . given

these observations and those of other workers studying EP correlates of IQ (Hendrickson and Hendrickson, 1982), we can agree with Jensen that Spearman's g is not an empty mathematical artifact of factor analysis but rather a construct possessing biological significance. (Schafer, 1987, p. 240).

What are we to make of such findings, and how are they to be reconciled with the previous findings of specificity of cognitive loss following neural insult? Do they not suggest that *g* is an abstraction of a singular global (i.e., nonspecific) cognitive resource that is biologically determined? I shall have more to say about this "singularity" claim in Chapter 8. But in this chapter I want to raise alternative interpretations of the neuropsychological findings. During the course of researching and thinking about this area, I have come to the conclusion, after intermittent bouts of infatuation and disappointment with neuropsychological findings, that they are not very relevant to a discussion of the nature of the mind. But to be fair, it must be said that the neuropsychological findings are just as irrelevant to the discussion of the singularity issue when they are used to refute it as when they are used to support it. Here is what prompts me to this assertion.

On logical grounds, it would seem that the use of neuropsychological findings to *refute* the singularity of (and support the modularity of) intelligence is problematic. Although the results of clinical studies are more congruent with a modular view of the mind, these studies can be explained in another manner, too. An analogy from athletics may be informative. Suppose that performances on various track-and-field events tend to be correlated, with running speed, javelin throwing, long jumping, and pole vaulting all tending to go together. In other words, athletes good at one event tend to be good at the others, too. Some underlying general factor could be invoked to explain this positive manifold of correlations, perhaps physical strength. That is, it could be suggested that the reason all of these track-and-field performances tend to be correlated is because they all require some degree of physical strength.

But this inference does not mean that other factors are not also involved in the track-and-field performances. Localized damage to the nondominant hand might diminish pole vaulting, as both hands are needed in that sport, but leave the remaining track-and-field skills unimpaired. Or localized damage to the dominant hand might impair javelin throwing (as well as pole vaulting), while leaving running and jumping unimpeded. And so on. The point of the analogy is simply that one could imagine a model that would postulate different neural bases for specific as against general factors and damage to the former would tend to mask the existence of the latter, especially if the latter represented a necessary but not sufficient condition for some specific performance. Thus, neuropsychological results from patients with brain lesions may indicate only that the specific functions that are impaired are subserved *in part* by specific neuroanatomical systems—not that these functions are independent of some general neural factor. In short, that specific functions are lost as a result of localized brain insult is pretty clear; but that this implies the absence of a global biologically based cognitive factor (g) is not so clear, at least not until evidence is found that rules out the existence of some underlying cortical unit that performs a uniform operation but varies from region to region and area to area, according to extrinsic connections. So even when the neuropsychological findings are used to argue in favor of the hypothesis of multiple cognitive potentials (and against the singularity of intelligence), they are problematic.

When the neuropsychological findings are used to *support* the existence of a singular processing component, their usefulness is no less compromised by alternative interpretations. Consider another analogy, this time with heart rate measurements. Suppose that subjects were asked to listen to regularly timed and randomly presented clicks while their heart rates were recorded. And suppose that we were to discover that persons with high IQs gave larger heart rate responses to randomly presented clicks than to regularly presented clicks, whereas lower IQ persons gave relatively higher heart rate responses to regularly presented clicks. What might we conclude from this? I think that many behavioral scientists might hypothesize stylistic differences in listening. Persons with high IQs may be more inclined to try to detect some sort of pattern or puzzle in the randomly presented series of clicks (they are adept at problem solving in such contexts, even when the experimenter informs them that there is no pattern or regularity). And persons with lower IQs might exhibit more anxiety during these tasks, given their history of testing and poorer school performance and this could modulate their EPs. In that case, then higher or lower heart-rate responses to clicks might be the result of cognitive modulation.

Obviously, I am not claiming that this cognitive modulation mechanism is the one that produced Schafer's and the Hendrickson's evoked potential results. Rather, I am claiming only that the conclusions that are drawn from their work are not without an alternative explanation that moves away from the straightforward view that some singular processor in the brain is at work while the organism is at rest and this is the same mechanism that accounts for IQ performance. When people listen to auditory clicks, their brain activity could be modulated by their psychological expectancies and plans, even if the task may not appear on the surface to call for such mental activity. The only parts missing from this argument, and crucial ones if the hypothesis is to carry weight, are that (1) the propensity to engage in surreptitious mental activity during this task must be more common among higher IQ subjects, and (2) if we were to alter the instructions given to subjects in a way that preserved the timed versus untimed dimension, but tried to equate the groups on their suspicion that there may be a hidden task embedded within the instructions, then the EPs of low IQ subjects would mimic those of high-IQ subjects.

Recently, neuroscientists have tried to address the issue of g more directly by searching for evidence of some type of repetitive cortical unit, of known anatomical organization, that might serve as the "cortical glue" for holding together specific types of cognition (Gilbert and Wiesel, 1979; Martin, et al., 1983; Phillips, et al., 1984). Since, to date, this research has not found evidence for the existence of such a neural equivalent of g, the foregoing evoked potential evidence ought to be viewed as undefinitive, in light of its indirectness as a measure of neural functioning.

Clinical studies have shown that specific forms of brain damage are associated with specific patterns of performance on IQ tests. In fact, clinical neuropsychologists frequently view IQ subtest profiles as diagnostic of various types of brain damage (Beaumont, 1983). Implicit in this research is the assumption that there are multiple neural systems or columns, each subserving independent types of cognition, and little, if any, g accounting for substantial variance across types of cognition, as damage to an underlying central processor would be expected to result in deleterious effects across all cognitive tasks. Thus, the evidence for the biological bases of discrete cognitive functions seemed fairly persuasive to Gardner when he wrote:

> This much is known to all followers of biological science. What is less widely appreciated is that specificity of cognitive function can be tied much more precisely to finer regions of the human cerebral cortex. . . .We find, from recent work in neurology, increasingly persuasive evidence for functional units in the nervous systems. There are units subserving microscopic abilities in the individual columns of the sensory or frontal areas; and there are much larger units, visible to inspection, which serve more complex and molar human functions, like linguistic or spatial processing. These suggest a biological basis for specialized intelligences. (Gardner, 1983, p. 57)

PITTING g AGAINST A MULTIPLE POTENTIALS PERSPECTIVE

A great deal of misunderstanding in psychology surrounds the issue of whether a statistically defined g and the psychological and physiological mechanisms used to explain g, such as "central-processing-capacity" are identical, and whether the existence of these mechanisms is tied to the discovery of their role in producing individual differences. As already mentioned, g is a statistical concept, referring to the magnitude of the linear variance accounted for by the first unrotated principal component, or to the magnitude of common variance obtained when the first-order factors are themselves factored. None of the widely used IQ tests provide a factorially pure, unidimensional measure of *g*. Either they were developed too early to benefit from factor analysis (as in the case of the Stanford-Binet test) or they combine the variances associated with discrete subtests, thus ending up with a total score (Full Scale IQ) that can be due to qualitatively different patterns of performance on the component subscales, as for example, when verbal and performance variances are combined into a single IQ score (McNemar, 1964). Thus, some subtests load far higher on *g* than others.

Although *g* is a well understood statistical concept, the psychological and physiological mechanisms underlying it are not. It could be that *g* is the result of any of the following factors:

1. Environmental variables (e.g., enriched environments tend to be supportive of a wide range of cognitive achievements, just as surely as disadvantaged ones tend to dampen a wide range of achievements, and therefore, environmental influences may underpin statistical *g*).

2. Biological variables (e.g., the biological capacity of some "central processor" such as *attention* or *memory capacity* that is involved in virtually all cognitive performances may be responsible for statistical *g*).

3. Metacognitive or executive processes (e.g., individual differences in insight into the inner workings of one's cognitive system may underlie the observed consistency of individual differences across tasks simply because it is hard to imagine a task that is unaffected by the operation of executive routines like planning, revising, anticipating, etc.).

4. Motivational and temperamental variables (e.g., individual differences in the level of intrinsic motivation or impulsiveness could lead to across-the-board performance differences; so could a parent's endorsement of a "modern world view," as seen in the work of Schafer and his colleagues).

5. A highly interlocked system of independent cognitive abilities that are not easily separated through exploratory factor analysis.

Detterman (1982) makes the latter (#5) case cogently when he states that "Since every part of the system is in some way related to every other part, the operation of any part will reflect every other part. . . .Therefore, it must be expected that all parts of the system will be interrelated in operation" (p. 107). Closely related to Detterman's hypothesis is one that claims that *g* results not from a single microlevel cognitive resource that runs through all of the tests, but because each of the macrolevel tests in the battery samples some (but not all) of the microlevel cognitive resources or "bonds" (Thomson, 1948) that are shared in the solution of at least one other macrolevel problem.[1] For reasons that are beyond the scope of this treatise, there is a fundamental difference between Spearman's g and Thomson's g that will not be described here other than to point out that although both of these measures of g reflect the difference between all zero correlations in a matrix versus the magnitude of all non-zero ones, Thomson's g makes no assumption of singularity of process.

Central Processing Capacity

Obviously, the preceding candidates for understanding the nature of g are not mutually exclusive. For example, some have argued that motivational differences may themselves be inherited; they produce differences in "experience-producing drives" which, in turn, lead to differences in learning (see Hayes, 1962). Others have argued that central processing capacity sets limits on the influence of both environmental and metacognitive differences. Moreover, the concept of g itself may prove inherently unfalsifiable (Sternberg, 1985). Despite these question marks, it may be enlightening to examine the various combinations of *statistical g* and *nonstatistical g*. For the purposes of such examination, we shall focus on the biologically based central-processing capacity (CPC) explanation, as it is at the root of what many researchers believe to underpin statistical g. (See Chapter 8 for numerous statements by researchers who endorse this view.)

Central-processing capacity (CPC) is conceptualized as a biologically imposed constraint on cognitive processing, due to some undefined limitation in an underlying resource pool, such as limited storage space and/or attentional capacity that restricts the number of operations that can be performed simultaneously on a information tht is retrieved from long-term memory or newly received from our senses (Eysenck, 1988). Logically, it is possible to identify four combinations of statistical g (i.e., statistically intercorrelated performances) and central-processing capacity limitations that are analytically orthogonal: high g–high CPC; high g–low CPC; low g–high CPC; and low g–low CPC. Let us let our imaginations run loose with these combinations in order to help illustrate the interpretive problems surrounding the nature of statistical g.

One can readily imagine a person who possesses high g–high CPC; many psychometric theorists see the first principal component in just this way. Individuals tend to occupy the same relative ranks on a range of cognitive tasks (thus exhibiting a large first principal factor) due to the presumed dependence of all tasks on CPC. Thus g and CPC can, in this instance, be seen as identical. But we can also imagine a person who is high g–low CPC. For such a person, individual differences on each of the tasks would result from non-CPC sources, such as motivation, culture, exogenous fatigue, and style of processing (e.g., reflective or impulsive), and these latter sources of variance would swamp that which resulted from central-processing capacity limitations. Note that this condition need not posit an absence of central-processing capacity in shaping task performance, but only that its role contributes relatively little variance compared to that contributed by non-central capacity sources. For instance, many biological functions are presumably involved in shaping all sorts of behaviors (e.g., breathing and walking) but contribute little in the way of individual difference variation to these behaviors. If these biological functions were impaired, however, performance would be dramatically affected. The important point here is that although g might be based on the contribution of a single source to variations in individual difference, not all important sources of performance may contribute to such variations. For example, while pulmonary muscles may underpin the ability to breathe, and removing them would surely render this ability impossible, individual differences in the amount of expiration may be unrelated to these muscles because for the most part they exist in adequate amount in the entire population.

It is easy to imagine the remaining two combinations, low g–high CPC and low g–low CPC. The former could result if the central-processing capacity was relevant across a wide range of tasks but the knowledge contexts associated with these individual tasks differed greatly, resulting in a lack of consistency in CPC efficiency. Here the importance of CPC is not denied, only the uniformity with which it is accessed by the various testing contexts. Much of the cross-cultural research is suggestive of this combination, as is some of the "policy-relevant" research, which assumes that all cultural groups have the necessary CPC to achieve, but they differ in the match between the level of their background knowledge and that which is required for successful performance on the task in question. Finally, it is possible to postulate an absence of intercorrelations among task performances that could result if each task required a unique type of mental energy instead of a unitary CPC. Some factorists such as Guilford (1967) have reported smaller consistencies across a range of tasks and have been led to emphasize the importance of unique cognitive resources rather than a single central processing mechanism. (More recently, Guilford (1985) has reported some degree of second-order correlations among his first-order factors, and others have found that second-order factors tend to intercorrelate themselves and yield a g that accounts for twice the variance of a linear combination of the specific factors.).

The "multiple intelligences" advocates like Gardner (1983) would seem to be firmly entrenched at low g–low CPC. According to Gardner, at least seven discrete types of processing capacity can be identified and this is why there is a small g (provided, of course, one selects the appropriate battery of tasks). The bioecological framework being put forward here also is perched at low g–low CPC, but for reasons that differ in an important way from those of Gardner and others. In this framework, there is no desire to prove the nonexistence of a CPC that permeates all cognitive performance, as long as one acknowledges its relative insignificance *vis a vis* specific cognitive resources in evoking variations in individual differences. It may very well be that a CPC not only exists, but also contributes to variations in individual differences—even more than the traditional 10% barrier. However, to whatever extent such a CPC reflects the *g* that has been documented by factorists thus far is anyone's guess. A test of the hypothesis that *g* is a reflection of a biologically based CPC requires much more diversity in the test battery and testing environment than has heretofore characterized psychometric approaches. Ultimately, it would require training studies, the use of multiple markers for each cognitive trait, and a diversity of contexts. If this sounds laborious, it is. Detterman (1982) commented that

> Unfortunately, I can think of few, if any, real attempts to answer these calls (for systematic experimental attempts to explicate factor analytic findings). It is certainly much easier to modify theory than to modify tests or tasks. These are the only two alternatives available when confronted with a pervasive source of common variance (*g*). Only the modification and refinement of tests or tasks have the potential of producing independent parameters explanatory of the common source of variance. Certainly, we will not know if such a task can be accomplished until someone makes a serious effort to try it. (p. 105)

Since the arguments concerning *g* versus a central-processing capacity are couched in the correlational tradition and not in the information-processing tradition, it is prudent to refer to *g* in strictly statistical terms rather than make the inference that a biologically or psychologically based CPC, as opposed to individual differences in motivation, environment, metacognition, etc., is the source of observed intercorrelations. It would seem that the componential analyses carried out by information processing psychologists have one important conceptual advantage over correlational analyses that form the basis of factor analytic solutions of "off-the-rack" ability tests. In the latter, one is constrained by the nature of the tests and the context in which they are administered, to say nothing of the researcher's theory of intelligence that leads to choices between number, placement, and rotation of axes. A specific test's *g*-loading can, and has, fluctuated dramatically—from .2 to .8, depending on the other tests included in the battery. For example, in referring to another researcher's analysis of differences between blacks and whites on 13 tests, Jensen commented that

> Thus, the Spearman hypothesis (i.e., that Black-White differences are most pronounced on *g*-saturated tests) seem to be borne out by these data. . . . yet there is a question, as the *g* loadings on some of these tests are not in marked agreement with their loadings where they have been factor analyzed in other contexts. (Jensen, 1980, p. 536)

Most standard ability tests are quite heterogeneous in their content, requiring a merger of several basic cognitive skills for performing successfully on them. Moreover, as will be addressed subsequently, such performances are the result not only of cognitive aptitudes, but also of the nature and complexity of one's domain-specific knowledge that is instantiated by those aptitudes.

The Problem with Factors

In the information-processing approach, however, it is possible to begin by identifying so-called well-springs or kernels of cognition, proceed to analyze the performance of individuals on tasks designed to tap these components, and finally examine the relationships between these findings and more molar forms of cognition (Detterman, 1980). By following such an approach, it is possible, for example, to directly assess the relationship between some basic cognitive-ability task and IQ scores, SATs, or any other presumed surrogate for g and this is the approach that some have followed (Fredericksen, 1986). It is also possible, at least in principle, to assess the link between g and the cognitive processes that were assumed to have been instrumental in task performance. By contrast, the factor-analytic approach seems to maximize the likelihood of superimposing analogy upon such observations. It is scientifically unsatisfying to wade through the numerous analyses that show that part of the shared variance among some group of ability factors (e.g., the primary mental abilities) can be accounted for on the basis of some singular underlying component; rather, the ability factors themselves are in need of scrutiny to understand their nature.

Historically, evidence for the existence of a sizeable g derived from two facts: (1) Only a limited range of abilities were assessed, and even those were assessed through batteries of academic-type tasks that often made it impossible to determine the basis for the intercorrelations. (2) The contexts in which these limited abilities were assessed were themselves limited. As one example of the former, consider the frequent finding that verbal and quantitative abilities are correlated, sometimes very highly. (On the SATs, they are correlated at .70 and on the WAIS the arithmetic and vocabulary scales are correlated at .62—McCardle and Horn, in press.) Upon examination, it is obvious that the correlation between these seemingly disparate abilities could have as much to do with the similar wording in questions that invoke them as with some underlying common factor g. That is, the quantitative scores may be correlated with the verbal scores because they both require for their solution the understanding of what is frequently complex and ambiguous sentence structure in questions on tests. Commenting on this same point, a journalist who wrote a recent critique of ETS concluded bluntly that the quantitative SAT questions ". . . .are verbal problems, not mathematical ones" (Owen,

1985, p. 131). Many others have made this same point (Guilford, 1967; Jensen, 1972; Cronbach, 1970).

For example, Jensen demonstrated that a particular test's *g*-loading is a function of the other tests in the battery; it can range anywhere from being a pure measure of *g* to being virtually no measure. In this same vein, Cronbach has remarked that the correlation between verbal and quantitative abilities may be an epiphenomenon of an individual's being jointly trained in both: "The high correlation between verbal and numerical abilities is due in part to the fact that persons who remain in school are trained on both types of content" (Cronbach, 1970, p. 479). One can easily imagine that a skill that is suddenly deemed important enough by the dominant culture to be included in its schooling will correlate highly with other skills taught concurrently, such as verbal and numerical skills. Thus, if cooking, computing, and cartography were suddenly inserted into the school curriculum, they, too, would tend to intercorrelate. But such intercorrelation would be unrevealing about the nature of the nervous system, unless researchers knew a great deal more about the latter than they do at present.

This very obvious point reveals the folly of some factorists' claims that factors reflect the nature and segmentation of the nervous system. The remark of a well-known hereditarian that the structure of the intellect revealed through factor analysis is context-dependent deserves mention: "The ascending layers, essential to the idea of a hierarchy, emerge only from statistical procedures that, to some degree, assume their existence in the first place" (Herrnstein, 1973, p. 97). The magnitude, nature, and number of factors are not invariant, but change with changes in the test battery, research hypotheses, sample characteristics, and statistical choices. Recently, McCardle and Horn (in press) and Loehlin (1989) presented impressive evidence that different structural models of intellectual development can be fit to the same matrix of means, standard deviations, and correlations and these various models can be quite dissimilar, despite their near equivalent fit.

To give one final example, Lorenz (1987) performed an interesting analysis of *spatial ability* as part of her doctoral dissertation research. She collected data on a battery of 19 tests of so-called spatial ability, in addition to recording the subjects' SAT scores. Some of the tests were "ecological" measures of spatial ability, such as distance estimation, giving directions, orienting oneself in a strange building, and locating landmarks. Other tests were conventional psychometric measures of spatial ability such as the "space relations form" of the Differential Aptitude Test, the Flaggs test (a test of spatial thinking), and the Cut the Cube Task (CCT). (CCT instructions are as follows: "Imagine a cube 3"×3"×3, " painted red all over. Now imagine that it has been cut into smaller cubes, each 1"×1"×1 " by making two horizontal cuts, two vertical cuts from front to back, and two vertical cuts from left to right." Subjects are asked how many of the smaller cubes have 0, 1, 2, 3, and

4 red sides.) In addition to these measures, Lorenz asked her subjects to give directions to someone unfamiliar with various campus buildings, e.g.,

> Pretend that you want to tell someone how to get from the rear entrance of the post office in the new student union building to the psychology text book section in the book store. Your directions. . . . should allow someone who has never been in the building before to locate the psychology books without getting lost. *You may use words or pictures or both.* (Lorenz, 1987, p. 34)

Based on a principal component analysis with a varimax rotation of subjects' scores, Lorenz extracted four interpretable factors: a "spatial manipulation" factor which reflected most of the conventional psychometric measures (as well as the SAT quantitative score), a geographical direction factor that reflected locating landmarks and giving directions, a landmark memory factor, and a route memory factor.

Had Lorenz stopped there, she would have provided an interesting demonstration of several points. Her findings made clear that "spatial ability" is multidimensional and not adequately captured by a single measure or even a single type of measure. But Lorenz did not stop there. She noticed that 62% of the subjects, when asked to give directions, preferred to do so verbally; the rest preferred to use pictures, or both pictures and words. She wondered whether these two groups differed in organization of spatial abilities as reflected by spatial structure. So she reanalyzed the data, separating the 80 or so subjects who preferred to give written directions from those who used pictures. Interestingly, she found that subjects who preferred to give written directions were characterized by a different factor structure than that originally extracted, one containing a single ecological factor (four out of the five ecological variables had loadings greater than .60 on this factor), a psychometric factor that included all of the conventional measures, and an aptitude factor that included SAT verbal *and* quantitative scores. Note again the difference between the two findings: Unlike the original analysis, in this one a single ecological factor emerged, suggesting that subjects who preferred to give written instructions approached the tasks differently from those who used pictorial directions. Further statistical examination of the data confirmed this difference. Subjects who preferred to give directions by drawing pictures used the same approach on SAT quantitative and psychometric tasks that they used on the ecological tasks. Subjects who gave written directions used a different approach for the ecological and psychometric tasks. By lumping the two groups together, one arrives at a potentially misleading understanding of the nature of spatial ability and its relationship to other abilities. To the extent that subjects are heterogeneous with respect to their approaches to problems, the resultant factor structure may not be an adequate characterization of anyone's abilities. As Lorenz (1987) concluded:

The major finding of these studies is that this structure is not the same for all individuals. . . . for those subjects (who gave written directions), a clear distinction between psychometric and naturalistic measures of spatial ability was apparent. This distinction was not present in the subgroup of subjects who preferred to provide directions in the form of maps, however. Thus, it is not possible to discuss the structure of the spatial domain, since different structures appear to exist for different subgroups of individuals. (p. 61)

A question I asked myself upon reading Lorenz's analyses was, Where does it end, this reconfiguration of results on the basis of sample characteristics? At what point can factorists cease subdividing their sample, cease searching for differential approaches to solving what are *prima facie* identical problems, but what may be represented quite differently by part of their sample? The answer is that there is no absolute index for deciding when to probe and when to stop, other than simple statistical heuristics. No one should rest content with a factor structure as long as alternative interpretations can be generated!

On the basis of the empirical literature already reviewed and the logical arguments of Guilford, Jensen, Herrnstein, and Cronbach (among others), it may be argued that to the extent factor-analytic studies have (1) relied upon a small set of fairly homogeneous psychometric-type skills (e.g., spatial ability and verbal fluency), (2) have narrowly conceived models of the information-processing system, (3) have assessed the presence or absence of the aforementioned skills with disembedded stimuli (e.g., nonsense shapes and linguistic analogies), and (4) have not taken into consideration the myriad routes that can be taken to arrive at the same answer, they have tilted the odds in favor of discovering a large first principal component, i.e., g. This is, of course, not an original point, many others having remarked similarly (e.g., Detterman, 1982; Gardner, 1983; Neisser, 1976; 1979). As test batteries are expanded to include a wider range of both contexts and problems, and as more complex models of information processing are subjected to experimental confirmation, the evidence for the existence of additional independent factors is likely to increase and the size and significance of g decrease.[2]

Moreover, the selection of homogeneous tests means that it could be possible to perform well on them using only a subset of the pool of discrete biological potentials. A corollary of this argument is that to the degree a particular battery of tests includes tasks that entail a wide range of cognitive potentials (as found in Lorenz's study) *and* assessment contexts, the first principal component will be diminished and the presence of specific factors will be enhanced.

The term "assessment contexts" takes on a rather specific meaning and does not simply refer to the ecology of the testing situation, important though that may be. "Assessment contexts" also refer to the domain-specific knowledge that underpins successful task performance. If one is interested in assessing inductive ability, for example, then problems from a number of content domains should be sampled, under the assumption that content knowledge needed to engage in induction is not equally elaborated and differentiated in all domains for a given individual. For instance, if one has a relatively inelaborate verbal content domain, then the exclusive use of word analogies or sentence completions may lead to an ungenerous estimate of one's inductive "aptitude," especially if different results were to be obtained in the solutions to problems from more embellished pictorial or action do-mains (e.g., gambling, planning retirement income, or work-related knowl-edge).

CONTEXTS OF CRYSTALLIZATION

In addition to postulating multiple biologically determined cognitive poten-tials, I have emphasized the role of developmental contexts (including moti-vation) in each of their crystallizations. By the term "crystallization," I refer to a firming or erection of the cognitive muscle. Muscles can be deflated through disuse, and therefore, the connotation of crystallization as "turning into stone" is not intended here. To use a metaphor, cognitive potentials, like muscles, must receive exercise to develop maximally and to stay developed. Moreover, different developmental contexts are probably differentially im-portant for the development of various cognitive potentials. Thus, a home filled with age-appropriate reading material and parents who engage their children in word games, etc., may be more valuable in the development of one type of cognitive potential (verbal reasoning) than in another (visual-spa-tial reasoning). Obviously, environments often contain contexts supportive of growth across many cognitive potentials. For example, some families may provide their children with complex grammars, a sensitivity to social details, and various types of puzzles and mechanical toys that can foster the crystal-lization of quite different biologically constrained cognitive potentials. It is equally clear that an extremely deprived environment may provide little or no growth-fostering contexts for *any* cognitive potential (Edgerton, 1981). Such a state of affairs could easily lead to the mistaken notion of a large, single underlying intellectual force (*g*), as performances across ostensibly disparate cognitive domains will tend to be correlated because of correlated environments.

Motives as Crystallizing Agents

Thus, individual differences in cognitive complexity can be expected to result from differences in any of one's cognitive potentials (which are in part biologically determined), from differences in domain-specific knowledge needed to demonstrate the potential and/or from differences in appropriate rearing as well as testing contexts, including motivation. The literature that was reviewed demonstrates that it is not sufficient for one to be biologically endowed with a cognitive potential and even to be exposed to appropriate opportunities for its crystallization: One also must be motivated to benefit from this exposure. Performance is influenced by learning, refinement, shaping, etc., and the role of motivation cannot be ignored in such matters. Extrinsic motivators (such as the value that one attaches to attaining success on a task), as well as intrinsic motivators (inculcated through various parental styles, such as fostering autonomy, valuing schooling, and adopting a modern world view, that were described in Chapter 2), are equally important in shaping cognitive outcomes (Schaefer, 1987).

Environmental Challenges

Throughout history it is easy to find examples of individuals whose most motivating environmental challenges, although related to aspects of the environments in which they were reared, were unrelated to educationally relevant contexts, as well as vice versa. (See Gardner, 1988, for some interesting biographical descriptions of some persons who have excelled in various domains.) As a rather obvious example of the first part of this statement, an individual born around the turn of the century and forced to leave school at an early age to contribute to the family's welfare may not have been motivated to learn vocabulary by reading. Such people may have perceived more important environmental challenges than performing well in school (e.g., earning money through gambling or becoming a local resource for other gamblers). Despite the availability of books in their homes, many of them may not have been motivated to benefit from their presence. Thus, there is no easy way of discerning how high their verbal IQ scores could have been because whatever underlying potential or "muscle" for verbal fluency they may have been born with, may be largely unexercised. (Cf. the study of expert handicappers by Ceci and Liker, 1986a.)

Although it is not clear how much importance one should attach to it at this time, there is some evidence that the IQ performances of minority students can be enhanced by some simple and brief exercises designed to make them feel more rapport with the tester (Bernal, 1984). This demonstration is very promising, but needs to be expanded and replicated in view of the mixed results in the area. Also, lest one think that motivation operates only on this macrolevel, there are many studies that indicate that it can lead to improvement in microlevel skills like rehearsal, too. When kindergarteners are given incentives to rehearse cumulatively, they continue to employ this strategy a week later, unlike a control group taught the strategy but without incentives (Ward and Toglia, 1987). This is all pretty well known, and most do not find it surprising that positive reinforcement affects learning.

KNOWLEDGE VERSUS INTELLIGENCE

The final prong of the bioecological framework deals with the inseparability of *knowledge* and *intelligence*. As used in this discussion, knowledge is broadly construed to refer not only to the accumulation of factual information, but also to the accretion of heuristic rules and strategies (including such things as shortcuts, rules of inference, and even learning to sit still to concentrate) that may be learned and that are highly useful. Knowledge can be created by an individual—through the concatenation of previously learned information and thinking about this information in new ways—or it can be created externally by other agents of society and passed down from generation to generation through media, education, and language. Traditional theorists distinguish between *knowledge* and *intelligence;* the latter has historically been conceived of as the underlying capacity to acquire knowledge, detect relationships, and monitor ongoing cognitions in order to adapt to ever-changing demands. Thus, intelligence, as traditionally indexed by IQ tests, refers to one's aptitude for using cognitive processes in situations where knowledge either is of minimal importance or is so basic that it is thought to to be shared by all persons (Gregory, 1981).

But it is illusory to imagine that performance on IQ tests or, for that matter, even microlevel cognitive tasks like the encoding of àlpha-numeric stimuli, is devoid of knowledge. Nor can it be said that the detection of relationships can, even in principle, be estimated without reference to past learning and to the elaborateness of one's knowledge base (i.e., cognitive complexity). As Scribner (1986) has recently concluded on the basis of her analysis of everyday problem solving, the ability to solve problems is intimately tied to the amount and quality of relevant knowledge one possesses. She argues that "From earlier assumptions that problem-solving can be understood in terms of 'pure process', a consensus has arisen that problem-solving procedures are bound up with amount and organization of subject matter knowledge" (p. 29). If one wishes to locate the most complex thinkers at a racetrack, the soundest advice is to begin by testing for factual knowledge about racing. High levels of cognitive complexity at the races are almost never found among unknowledgable patrons, regardless of how high their IQs may be (Ceci and Liker, 1986a, 1986b; 1988; Liker and Ceci, 1987).

Some Illustrative Anecdotes

A few examples of the interplay between knowledge and process may help illustrate this point. Let us consider some of the feats accomplished by three well-known mathematicians. First, there is the oft-reported story about the renowned mathematician, J. von Neumann, as this strikes at the very essence of the argument being made here. When von Neumann was given the following problem to solve by a friend, it is reported that he pondered for a time before coming up with the answer:

> There are two cyclists, a mile apart, cycling toward each other and each going at 10 miles per hour. A fly flies from the nose of one cyclist to the nose of the other, backwards and forwards between them, until the cyclists meet. The fly flies at 15 miles per hour. How far has the fly flown when it gets squashed by the cyclists' noses meeting?

By not seeing the easy way to solve the problem, von Neumann performed quite a remarkable feat: He mentally computed the distance that the fly flew as the limit of a mathematical series (it is quite beyond the ability of even most skilled mathematicians to do this mentally), instead of realizing that the time it would take the two cyclists to meet, with each travelling 10 mph, was three minutes; therefore, the fly would have flown $\frac{3}{4}$ mi in three minutes (at 15 mph). This raises the question as to whether von Neumann was being incredibly complex in solving the problem with his complicated method, or whether he was being cognitively unsophisticated in not perceiving the simplicity of the problem. The point is that even on logical grounds alone, cognitive complexity is inseparable from knowledge. For most of us, the requisite knowledge to calculate the limit of a mathematical series is lacking and any measure of complexity that depended on such a calculation would thus also be wanting. This is a nice example of the distinction between knowledge and problem solving. In science, the simplest solution is usually preferred.

As another example of the role that knowledge plays in our perception of a problem, including its "blinding" influence, we can consider the case of Alan Turing, the mathematical logician. Turing owned a bicycle, on which he rode to work each day. The mathematician Ian Stewart wrote in *Nature* that:

> At intervals the chain (on Turing's bicycle) would fall off. A methodical man, Turing kept a bottle of turpentine and a rag in the office to clean his hands on arrival. After some time it dawned on him that the chain fell off at very regular intervals. On counting the revolutions of the front wheel he found exact regularity: after a certain number of turns, the chain fell off. He took to counting the revolutions so as to be able to execute a manoeuvre that kept the chain on. Tiring of this, he fixed a counter to the wheel. Later he analysed the mathematical relation between the number of spokes in the front wheel, the number of links in the chain, and the number of cogs in the pedal: he discovered that the mishap occurred for a unique configuration of wheel, chain, and pedals. On examining the machine he found that it happened when a particular damaged link came into contact with a particular bent spoke. The spoke was duly straightened and the turpentine and rag removed from the office. This tale illus-

trates both the power and perils of logical reasoning. A cycle mechanic would have solved the problem in five minutes. (Stewart, p. 1987, p. 26)

In reading this and similar anecdotes, it is pretty clear that the amount and type of knowledge we possess cannot only be recruited to aid our understanding of a problem, but at times can also blind us to more efficient ways of solving it. From what I know of Turing, I suspect that he approached many real-world problems with the same formal algorithmic processes demonstrated in the quote. But as we all know, there are times when the problem can be solved more efficiently by stepping outside of our highly entrenched ways of thinking about the world.

Finally, consider the case of the famous Indian mathematician, Srinivasa Ramanujan, who collaborated with G. H. Hardy on some of the latter's most imaginative papers and who, at a recent AAAS symposium, was hailed as one of the three greatest mathematicians of all time—together with Archimedes and Gauss. Although he died at the age of 32, Ramanujan left over 600 discoveries that remained unproven until recently (Borwein, 1987; Forrester, 1987), because despite his many insights, he was lacking in the formal mathematics education needed to provide proofs for them. In C. P. Snow's forward to Hardy's A Mathematician's Apology (1963, p. 37), he describes a visit Hardy paid to his Indian friend as the latter lay dying in a British hospital in Putney:

> It was on one of those visits that there happened the incident of the taxi-cab number. Hardy had gone out to Putney by taxi, as usual his chosen method of conveyance. He went into the room where Ramanujan was lying. Hardy, always inept about introducing a conversation, said, probably without a greeting, and certainly as his first remark: 'I thought the number of my taxi-cab was 1729. It seemed to me a rather dull number.' To which Ramanujan replied: 'No, Hardy! No, Hardy! It is a very interesting number. It is the smallest number expressible as the sum of two cubes in two different ways.' That is the exchange as Hardy recorded it. It must be substantially accurate. He was the most honest of men; and further, no one could possibly have invented it. (p. 37)

Knowledge and Process in Symbiosis

The anecdotes about von Neumann, Turing, and Hardy reveal an important feature of high levels of complexity: They almost always co-occur with high levels of declarative and procedural knowledge. Complex cognitive processing and elegantly structured knowledge are in symbiosis. It is fascinating to speculate as to how these men's knowledge of numbers was organized. For Ramanujan to be able to specify that 1,729 was the smallest sum of two cubes in two ways reveals an extremely differentiated, yet integrated knowledge structure. (He was purportedly able to identify similar features for other numbers, a feat that must have seemed arcane even to other mathematicians.) But the point of this is simply that, given a restricted level of complexity in their domain-specific knowledge, something complex for one individual may be simple for another individual possessing a more extensive knowledge structure. This has led Nisbett and his colleagues to remark that "even quite young children readily reject invalid arguments when they have world knowledge that is helpful, whereas even adults accept invalid arguments when their world knowledge encourages it" (Nisbett, et al., 1988, p. 5). Thus, there is no objective measure of complexity that allows for its estimation exclusive of knowledge (given that the definition of the latter includes such things as strategies, shortcuts, factual information, heuristics, and the like). The philosophers Block and Dworkin (1976) have made a similar argument:

> We have been arguing that people of equal intelligence probably differ in the knowledge and skills they demonstrate on IQ tests simply because they probably differ in knowledge and skills. A more insidious fault of IQ tests is that people with the same knowledge and skills (and equal intelligence) probably differ markedly in their ability to *demonstrate* this knowledge and these skills on an IQ test for reasons whose connection with intelligence is tenuous. (p. 450)

For the reasons given, it is logically unsatisfactory to speak of cognitive complexity or abstraction as the *sine qua non* of intelligence in one breath and equate IQ with intelligence in the next, in view of the crucial role played by prior knowledge (broadly defined) in the performance of both. To say, as some have, that complexity or abstraction is one of the hallmarks of intelligence is to acknowledge the importance of prior learning and knowledge in intelligent behavior. An expanded version of this logic has led Gregory (1981) to remark:

> What is it to think without knowledge? If we ignore, for now, specific factual knowledge: are not rules for problem solving knowledge—knowledge developed by education and experience? Again, strategies for even the simplest tasks may surely be gained from the environment and by education, such as the skill of concentrating on the test, being confident and yet self-critical—and guessing what kind of answers are needed. . . . If education did not have such effects, it would be useless. . . . So, the claim that IQ tests can be freed of education and other biases by suitable choice of tasks seems ill-founded. It is indeed deeply misconceived, for intelligence requires and surely is in large part effective deployment of knowledge. (Gregory, 1981, p. 304)

Earlier, a number of studies were reviewed that were difficult to explain in terms of the dominant psychological theories of intelligence. For instance, the findings from the studies of town managers and racetrack handicappers are hard to reconcile within either the psychometric or information-processing traditions, as individuals with low IQs sometimes developed more cognitively complex models of these tasks than did high-IQ individuals, even when both groups were equated on experience and motivation. Of course, one can assert that such instances of complex, abstract thinking are congruent with existing theory—perhaps they evidence an occurrence of some overdeveloped specific ability found among individuals who are otherwise quite concrete in their thinking. However, not all of the corpus of findings can be explained in this way, as will be seen in the next two chapters.

Before leaving this chapter, it is important to recap what was and was not argued in the preceding pages. An argument was made that it is logically impossible to separate differences in (1) the efficiency with which cognitive processes operate and (2) the level of intellectual performance from (3) differences in knowledge. Because motivation, personality, and cognitive processes are all important determinants of knowledge acquisition, and because the contents and structure of one's knowledge base are inseparable from intellectual performance, it follows that motivation, personality, and cognitive processes are axiomatically involved in intellectual performance, too. The problem with viewing IQ as an indicant of intelligence, even a highly imperfect one, is that it is a *product* of unknown combinations of the other processes. It is logically akin to accepting someone's *TOEFL* (Tests of English as a Foreign Language, given to foreign applicants to American universities to determine their command of the English language) as an indicant of underlying aptitude for expressing and understanding English. Most of us would not make this leap from a score to an aptitude, as it is crucial to know something about the applicant's prior level of training in English, interest in learning English, and even level of understanding of his or her native language, if our interests are of a theoretical nature, as opposed to simply prediction. (Actually, it is also important to know these things for predictive purposes: Imagine trying to predict the future productivity of a nation on the basis of its most recent measure of productivity, in the absence of knowing anything about relevant historical forces that might be at work!) Simply accepting an end *product*, whether it is a TOEFL score or an IQ score, tells us very little about the underlying processes that produced it. Of course, this does not mean that these scores are without value. As expressions of someone's current level of functioning, they are often quite valuable in directing remedial efforts, suggesting diagnoses, and so forth. Cognitive psychologists entered the aptitude testing field in the last decade precisely with this goal in mind, to attempt to come to grips with the cognitive components that underpin IQ performance and specifically in response to what they saw as the shortcomings of past correlational analyses (Pellegrino and Glaser, 1979).

So much for what *was* argued. What *was not* argued was that intellectual performance is without a biological component. Indeed, my own view is that it is hard to find any psychologically interesting characteristic that does not have *some* biological component. Along with many readers, I assume, without empirically adequate data, that very few persons who are exposed to rearing environments like the ones that Ramanujan or Mozart experienced will end up with the same competencies as these men.[3] Yes, biology has its place—just as surely as nonbiological forces are deeply implicated in producing expertise. Otherwise, heritabilities would be much higher than have been reported and we would not have so many instances of one member of a monozygotic twin pair being heralded for a great contribution and the other being described as unremarkable. We also would not expect to find the unevenness in cognitive abilities that is found among gifted individuals: As a rule, they are outstanding in a single area and average in the rest.

Thus, while not dismissing the importance of biology, my argument was against assuming that cognitive performance could be attributed in any direct manner to biological differences and, moreover, that IQ scores are an index of biological differences. Michael J. A. Howe put it nicely, when he stated:

> It is worth emphasizing that the importance of genetic factors is not being challenged here. For many individuals, genetic influences do undoubtedly place constraints on what can be achieved. Moreover, the notion that anyone, even a mentally retarded person, could attain the highest levels of performance if only the right environmental circumstances were provided, is firmly refuted by the factual evidence. What is being challenged is the assumption that genetic mechanisms operate in ways that make it impossible for someone to aspire to the highest levels of performance unless he or she happens to be one of the rare individuals who possess a particular combination of genetic potentialities. The latter view, that progress is completely mapped out in advance by each individual's genetic make-up, is as wide of the mark as the one that attributes all differences to variations in physical environments. (Howe, in press. p. 13)

IQ tests are heterogeneous with respect to cognitive processes and personality traits, reflecting some more than others. Biology probably plays different roles in the development of each of these processes, and they are differentially instrumental in producing differences in IQ, so it is never possible to say what the basis is for an individual's performance on an IQ test. And it never will be, unless test developers are guided by far more explicit theories of task performance than have heretofore been seen, theories of both the nature of the items and the individuals who attempt to answer them— often by taking diverse routes (Whitely, 1980; 1983). Moreover, even if the test items were homogeneous with regard to some theoretical process, the persons who answered them might not be. And if the theory that has guided the test construction postulates that cognitive variables are not general across tasks, then the quality of traditional IQ tests can be questioned.

CHAPTER 7

A Model of Cognitive Complexity

The models depicted in Figures 7.1 and 7.2 can be developed into a variety of theoretical frameworks that are able to account for much of the research reviewed thus far. Both models are comprised of (1) structures that receive and make sense of environmental inputs, that is, the traditional sensory registers and what Horn (1978) calls "simple associative memory processes" that perform contiguity analyses; (2) the knowledge base, a reservoir of an individual's knowledge, divided into specific domains of varying elaborateness and differentiation (i.e., complexity); and (3) multiple cognitive operations that instantiate information located in the knowledge base in an effort to solve a problem. The cognitive processes in the model depicted in Figure 7.1 differ from those represented in Figure 7.2 in the following important respect. In Figure 7.1, processes are represented as "transdomainal," that is, operating with equal efficiency across all content knowledge domains. It is assumed that these processes are transdomainal from the time of their acquisition or very soon thereafter. On the other hand, processes in Figure 7.2 are tied initially to a particular domain of knowledge and operate on information only within that domain. It is hypothesized that with development, these processes gradually become powerful and general algorithms capable of being applied to information across domains. (This is represented by the extent to which the processes have emerged from their specific knowledge domains into the center circle of the figure, indicating transdomainal applicability. The mechanisms responsible for the course of development of the processes in Figure 7.2 will be discussed in Chapter 10, when the work of structural/knowledge theorists is covered. In both models, the roles of experience, processing efficiency (in part biologically constrained and in part constrained by the organization of the knowledge base), and motivation can result in changes to the knowledge base, which will have repercussions on processing efficiency as well as on the other components.

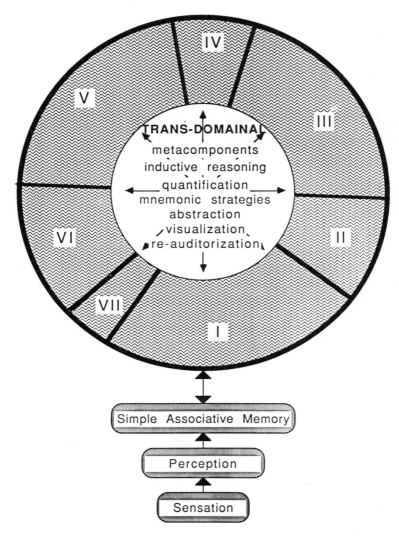

FIGURE 7.1 Schematization of disembedded model of cognitive complexity. Domains of knowledge are represented by Roman numerals, and cognitive processes are not embedded in specific domains but assumed to be operative across all domains of knowledge.

As can be seen, both models make the traditional assumption of *reciprocity* between the sensory/perceptual processes, the knowledge base, and the cognitive operations. For example, in the process of perception, events must be identified (by some look-up process in the knowledge base), and this

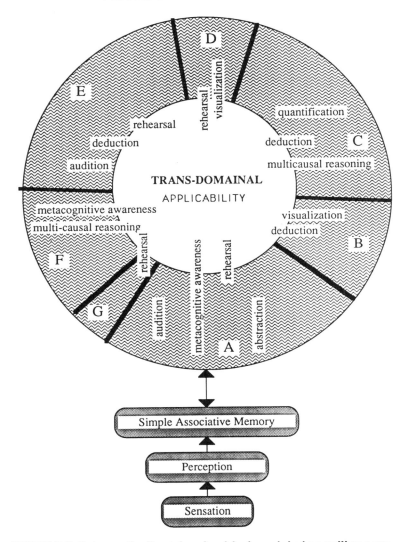

FIGURE 7.2 Schematization of embedded model of cognitive complexity. Domains of knowledge are represented by Roman numerals, and cognitive processes are tied to specific domains, to varying degrees (e.g., rehearsal has become nearly transdomainal).

involves a "give-and-take" wherein the elaborateness of the relevant domain-specific knowledge determines the speed and nature of the perception. Both figures illustrate seven content knowledge domains within the total reservoir of knowledge. This is a completely arbitrary number, and no significance should be attached to it, as the actual number of content knowledge domains depicted would undoubtedly be far greater than seven if we included as domains all of the regions of declarative knowledge (e.g., sports knowledge, culinary knowledge, kinship knowledge). On the other hand, domains can be defined in terms of their abstractness, by their modality (visual, spatial, verbal, kinesthetic), or by neuroanatomical substrates involved in their execution. Others have addressed these different definitions of *domain* (e.g., Marini and Case, 1989; Keil, 1984) and I shall say no more here other than that the definition used throughout this monograph refers to declarative knowledge regions (after Chi and Ceci, 1987). Such a definition is in line with the central goal of showing that individual differences in intellectual development can arise out of differences in either processing efficiency or knowledge elaboration, themselves existing in a type of symbiosis.

Several features of the models are not represented in the figures. For instance, nothing is implied about the role of specificity of modalities. Numerous researchers have written on this topic and the results of their studies seem to indicate an important role for modality to play in any adequate model of intelligence. To date, it is not known with any confidence whether the cognitive processes themselves are different in some fundamental sense for linguistic and nonlinguistic modalities, or whether the processes are the same but the knowledge structures on which they operate are different—in other words, whether knowledge is represented differently in each modality. Recently, Lansman, et al. (1983), Jarman (1984), and Ceci, et al. (1980), have reported differences in the manner in which information is represented and/or accessed in linguistic and nonlinguistic modalities. Their research indicates that even microlevel information-processing components (in the Lansman study, focused versus divided attentional searching; in the Ceci study, the use of retrieval cues) are functionally independent in the different modalities. Thus, the models depicted in Figures 7.1 and 7.2 need further differentiation according to modality-specific pathways to the knowledge representations or even modality-specific representations themselves.

Another omission of the figures is the degree of elaboration of the content knowledge domains. At any given time, some domains are likely to be more elaborately developed than others, even though different degrees of elaborateness are not shown in the figure. This is an important feature of both models because various cognitive operations must access or "operate on" some type of knowledge, and if the knowledge is insufficient in some sense (e.g., either lacking in quantity or not organized efficiently for a given task), optimal cognitive processing (in the sense of its biologically constrained potential) can be impeded. The sole difference between the two models—the assumptions they imply about the origins and transdomainal nature of processes—may be difficult to test because of the confounded role of elaborateness. Some assessment of elaborateness within a given domain always is necessary in order to make inferences about relative superiority of process-ing.

Finally, the models are underdetermined with regard to *g*, multifacto-rial arguments, and the hierarchization of the processing system. Obviously, there is nothing in the models themselves that forecloses the possibility of the existence of a single source of common variance: *g* could exist within either of them, without much redrawing. Similarly, it is possible (even desirable, had my goal been different) to add more description about the models' account of microlevel and macrolevel linkages. Are microlevel processes fully embed-ded within macrolevel ones? Do common links exist between one or more processes at each level? Because the models make no differential demands on these issues, they will not be dealt with here.

As mentioned earlier, it is possible for an individual to appear to lack a certain cognitive potential (e.g., the ability to deduce relationships) when the real problem may be a mismatch between the task at hand and the subject's relevant content knowledge required to actualize that potential. By exploring a variety of contexts and materials, it is possible to demonstrate the existence of a cognitive potential even in the face of deficits in its operation on a particular task due to the lack of elaborateness of that particular content domain. More will be said about this later, but first it is instructive to review some of the research cited earlier that was said to be inconsistent with existing theories of intelligence. The three-pronged models shown in Figures 7.1 and 7.2 are, for the most part, capable of handling these difficulties.

In Ceci and Liker's (1986a) study of race track handicappers, the differ-ences between experts and nonexperts in cognitive complexity could be due to differences in their underlying cognitive potential for educing patterns of relationships from race track data—that is, thinking in a multiplicative man-ner. (All of the subjects had been attending races regularly for much of their adult lives, so motivational and experiential variables *per se* are less likely explanations.) Alternatively, the observed differences could be due either to differential degrees of elaborateness of the knowledge bases of experts and nonexperts (with the former's racing knowledge being somehow more differ-entiated, thereby permitting them to separate correlated variables and recom-bine them into higher order interactive terms) or to context-sensitive elicitors (i.e., perhaps for some reason the nonexperts possessed an equivalent under-lying potential for deducing principles from patterns of relationships, but racetrack handicapping did not serve to elicit it). The first option implies that the sort of complexity exhibited in race track handicapping is unrelated to the complexity measured by, for example, the *Comprehension* subtest of the Wechsler IQ test; the second option suggests that this type of thinking also could be valuable for performance on certain subtests of IQ tests, but that the nature of the material and task demands associated with IQ testing are poorly suited for eliciting the relevant potential in individuals whose life courses have strayed from courses that pertain more to formal schooling. This argument assumes that it is possible for low-IQ race track experts to reason multiplicatively whenever the material to be used is elaborately represented in their knowledge domain. While it is clearly impossible to localize the source of the differences between expert and nonexpert handicappers, it can be seen that the model in Figure 7.2 is not lacking for possibilities. It is sufficiently "operation-able," and there are several experiments that could be carried out to test the alternate explanations. Currently, my colleagues and I are involved in an intensive case study of two experts. We are asking them to do various tasks that either bear a structural similarity to race track handicapping, but are couched in a different domain (the stock market), or involve the identical domain, but require a different multiplicative model to be developed for their successful performance. It will be some time before we have a handle on this issue, unfortunately, because the tasks need to be presented over prolonged periods of time.

The anthropological studies of hunting and shopping are also open to alternative explanations, as are the German studies of managing a mythical city and a desert oasis. The important point here, however, is to acknowledge that high levels of cognitive complexity appear to exist independently of one's IQ score (see also Charlesworth, 1976). They appear to be domain specific (or tied to knowledge development in such a manner to permit it to surface in several, but usually not all, domains simultaneously) and often to be useful in performing laboratory tasks and even in answering IQ questions. Neither information-processing nor psychometric theories of intelligence can adequately account for these observations, the former being a descriptive methodology that takes no a priori stand (inherently, that is) on such matters and therefore has been associated with multiple, contrary explanations, while the latter appears to construct an edifice without an established foundation in explaining such results. (Lest one think this criticism too one-sided in favoring information-processing, it can be pointed out that information-processing approaches may give the illusion of establishing a foundation (i.e., the underlying cognitive processing components), while in reality the underlying processes are products themselves (of the knowledge base), but on a somewhat more microscopic level than factors.) In contrast to the model depicted in Figure 7.1, the inductive-theoretical framework described in Figure 7.2 appears better equipped to understand these instances of intelligent behavior, at least in principle. Individuals may be expected to differ in intellectual performance whenever they differ (1) in the efficiency of underlying cognitive potentials relevant to such performance, (2) in the structure of the knowledge relevant to such performance, and/or (3) in the aforementioned contextual/motivational factors that initially served to crystallize the underlying potential and later (during performance) served to elicit the relevant knowledge and cognitive processes.

EVALUATING THE BIOECOLOGICAL FRAMEWORK *VIS-A-VIS* CLASSIC FORMS OF EVIDENCE: THE CASE OF HERITABILITY

Theories are rarely discarded because of anomalous findings; rather, an alternative theory must exist that can be shown to handle the anomalous findings as well as the seminal findings upon which the existing theory is based. Thus, before the bioecological framework can be considered adequate, it should go beyond the demonstration of its ability to incorporate diverse findings that are not easily accommodated by existing theories of intelligence (information-processing, individual differences, and genetic) and also demonstrate its ability to incorporate the very evidence mustered to support those theories. To do this, we must shift the level of discourse and consider what is regarded as the bedrock evidence in support of current theories of intelligence—that is, research pertaining to heritability, the positive manifold, and generality. In this chapter, we shall focus on the first of these matters, addressing the others in the following chapters.

If IQ really is inherited to a substantial degree and really is related to a manifold of very basic decision processes, then several contentions being made here would appear to be undermined. For example, a contention made earlier was that IQ reflects only one of many types of intelligence. If so, and if this type of intelligence is highly heritable, then are the other types of intelligence also highly heritable, and do they also correlate with other basic decision processes? Also, in Chapter 3 it was argued that even when the same processes that are indexed by IQ tests are also required for successful performance on some everyday tasks, performance on one sometimes (though not always) cannot be predicted from performance on the other, leading to the contention that contextual elicitors and motivation interact with domain-specific knowledge to produce this disjunction. Obviously, if IQ and the other types of intelligence are highly heritable, then the bioecological framework requires additional explanatory constructs. Because of its importance, then, we shall examine the heritability argument in some detail in this chapter and examine the positive manifold argument in the next.

Genetics

Even if one accepts the view that an IQ test is not an all-encompassing index of one's cognitive abilities (as most psychologists do) as well as the related suggestion that multiple forms of intelligence exist independently of those tapped by IQ (as some psychologists do), it is still necessary to confront the high heritability of IQ scores. The reason for this assertion is that if IQ scores are highly heritable, wouldn't this seem to suggest that at least those

forms of intelligence that are reflected by IQ scores are outside of the bioecological framework? After all, throughout this monograph, it has been alleged
that social and cognitive experiences like those associated with schooling
provide an environmental context appropriate for the crystallization of cognitive skills and the construction of domainal knowledge that is related to IQ.
Intelligence test scores were shown to change systematically and substantially with the amount of schooling attained and the type of parenting received. If the heritability of IQ is as high as some have alleged it to be
(somewhere between .45 and .8), how might this be reconciled with the
bioecological framework?

Actually, high heritability itself is not problematic for the bioecological
framework, unless it approximates unity (h^2 = 1.0), a figure that no serious
theorist has so far proposed. After all, this is a theory that acknowledges a
role for biology, as the name indicates: The bioecological framework posits
multiple intelligences arising out of the interplay between relevant learning
regimes and specific genetically influenced predispositions for cognizing,
each with its own polygenic potentiator in the form of a synaptogenetic
schedule. Moreover, there are so many reports in the behavior-genetics literature of substantial heritability coefficients, calculated from an enormous
variety of sources, that it is unlikely they could all be wrong. Researchers on
the Colorado Adoption Project alone have produced over 50 published reports during the 1980s, each documenting substantial heritability for a variety of cognitive skills, including IQ (see Plomin, et al., 1988; Rice, et al., 1986;
Rice, et al., 1988). I see no point in trying to discredit these studies, as some
have done. They were exceptionally well designed, and their path analyses
are among the most sophisticated I have seen. I do believe, however, that the
exact size of the heritability coefficients and, more importantly, the meaning
that one attaches to them, are grounds for serious debate. Here, there is deep
disagreement within the scientific community. Wahlsten (1989) noted the
dramatic differences that exist between human and animal researchers in
their beliefs about gene-environment interaction. Heritability analysis requires an absence of interaction, and most human behavior geneticists have
concluded from their analyses that gene-environment interaction is indeed
negligible, with additive environmental and genetic components being
uniquely assessed. On the other hand, animal geneticists have expressed
doubt over the alleged lack of interaction between genes and environment in
these analyses. On the face of it, the animal researchers would seem to have a
powerful advantage, namely their ability to randomly map the full array of
genotypes over the full range of environments:

> At the 1987 Behavior Genetics Association meeting in Minneapolis, the
> concurrent sessions on human and animal studies were almost like two
> separate worlds in terms of attitudes towards interaction. Many human
> behavior geneticists dismissed interaction and cited heritability estimates

with great confidence, while most of those studying mice, rats, and fruit flies documented one case after another and expressed skepticism about heritability coefficients. (Wahlsten, 1989, p. 11)

On logical grounds it would seem to be impossible to conclusively settle the issue of interaction without some currently unanticipated (statistical or biological) technological innovation that allowed scientists to study human heritability in new ways. In what follows, I shall explain why I believe that although biology plays a role in shaping each of the multiple intelligences, it does so in a way that does not translate into the heritability differences we are so frequently confronted with in studies of intellectual development. And I shall try to cast these heritability studies in a different framework, one that minimizes their explanatory power. But first, some basics are required.

Heritability is defined as the proportion of total variation in a trait that is caused by genetic variance (e.g., variance of IQ = variance of heredity/total variance). If heritability (h^2) is the shared variance between one's genotype (or genetic potential) for intelligent behavior and one's actual level of intelligent behavior, then the square root of heritability would represent the size of the correlation between the genotype for intelligent behavior and intelligent behavior itself. This much is copy-book maxim and noncontroversial.

Numerous methods of assessing the heritability of a trait exist, but all have one thing in common: They are based on the extent to which *differences* in a trait are the result of *differences* in genetic makeup. I have underscored "differences" for a reason. Heritability is not equivalent to demonstrating that the expression of a trait is the direct result of genetic action; rather, it is a demonstration of a genetically caused *variance* in the trait. What one means by the statement that the heritability of IQ = .8 is that 80% of the *differences* between people's IQs are due to genes, not that 80% of their total IQ scores are due to genes. And, importantly, the size of the heritability estimate will vary directly with the amount of environmental variation (as well as phenotypic variation) such that if there were a sharp increase in poverty in a community and as a result many of its children were denied access to educationally relevant activities, then the size of the heritability estimate for that community would be reduced. This is because genes are relatively more important in producing differences between children who are from identical environments than they are in producing differences between children from vastly different ones.

Take, for example, the number of ears we are born with. This trait of "earness" is due to genetic action. But because there is essentially no variation in "earness" in this or any other human community, the heritability for having two ears is approximately zero despite the obvious role that genetics plays in ear formation. (Hunt, 1984, makes the same point in an example he gives of the number of legs influencing athletic ability.) Heritability is always relative to a specific population at a specific point in individual time (i.e., the developmental stage of the organism) and historical time (i.e., "period" effects associated with a particular cohort). There is nothing magical about the size of heritability estimates: They can and do change, and they change mightily when researchers move from sample to sample or make different statistical assumptions. Recently Robert Plomin (1989), one of the foremost researchers in this field, noted that the size of h^2 appears to have shrunk in the past two decades from what was typcially reported in the first half of this century. In the first half of this century, the values were, on average, around .7 while for studies conducted after 1970, the average is around .5. (I find this decrease in the size of heritability mildly perplexing because a genotype is most likely to be fully expressed in a good environment and certainly the environment in the first part of this century was not as conducive to development as more recent environments (i.e., diet, educational opportunities, and cultural events are all more widespread today than they once were, thus permitting someone with potential to express it). Heritability is a fragile and often evasive concept. As Sandra Scarr noted,

> The most common misunderstanding of the concept of "heritability" relates to the myth of fixed intelligence. If h^2 is high, this reasoning goes, then intelligence is genetically fixed and unchangeable at the phenotypic level. This misconception ignores the fact that h^2 is a population statistic, bound to a given set of environmental conditions at a given point in time. Neither intelligence nor h^2 estimates are fixed. (Scarr, 1981, p. 73)

Biological influences at one point in time may produce differences in psychological traits that are not present at another time and vice versa. Earlier it was pointed out that a developmental perspective was critical to a theory of individual differences in intellectual development because of the timing of events that impinge on the various cognitive muscles at their various points of unfolding. For example, Rice, et al. (1988) reported that adopted children's home environment during the first year of life exerts a more potent influence on their later intelligence than it does during the second, third, or fourth years of life. This conclusion should be qualified by examining the differences found for each of the specific cognitive abilities that were used to derive the measure of general intelligence, but as a generalization, it is certainly accurate to claim that aspects of the home environment were more important earlier than later in these children's lives. On the other hand, Yeates, et al. (1983) demonstrated that aspects of the home environment (controlled for maternal genetics through IQ partialing) was *not* a significant predictor of children's IQs at age two, but was an important predictor of their IQs at ages three and four. Unlike the Colorado Adoption Project children, who were reared in white, middle-class families, the children in the Yeates, et al. study were chosen because of their risk for sociocultural mental retardation, perhaps accounting for some of the disparity with Rice's conclusion.

Even physical traits that currently have very high heritabilities, such as skin color or height, could have low heritabilities at some future time or place. The heritability of skin color among all Caucasians in New York City is probably quite high, with color running along family lines. On the other hand, the heritability of skin color among Jews or Italians who live in New York City and those who live in Miami is probably much lower, even though the total amount of variation is probably similar in these two groups. This is because much of the variance in skin color found between Jews or Italians living in Florida and those living in New York is due to their relative exposure to the sun whereas in the group of all Caucasians in New York the differences would, to a greater degree, be the result of genes. As Angoff (1988) recently commented, "In an environmentally homogeneous group, heritability coefficients are likely to be inflated; in an environmentally heterogeneous group, the opposite is likely to be true" (p. 715). The upshot of all of this is that if everyone were genetically identical, then variation in skin color, IQ, or anything else could not be due to genes, and hence, its heritability would be zero.

Methods of Estimating h^2

Heritability of IQ can be estimated from the correlations among relatives' IQs, with IQs being more similar as a function of the degree of genetic overlap between the relatives. Although estimates of the contribution of heredity relative to the contribution of the environment should always be couched in terms of a particular population at a particular point in both historical and personal time, and *in a particular environment* (Scarr, 1981), this has not always been the case. When Arthur Jensen wrote his 1969 article in the *Harvard Educational Review,* "How Much Can We Boost IQ and Scholastic Achievement?," his answer to this question was "not much," because he believed that most of the variance in IQ scores was due to heredity. Although his estimate of heritability was high, his thesis was not a new one, and it was hardly controversial within the field of genetics. (See Provine, 1986, for an excellent historical review on this issue in which he points out that Jensen's position in the 1969 paper was actually less hereditarian than that of the majority of geneticists up to 1953.)

What *was* controversial in Jensen's reasoning was his extrapolation from documented sources of variation in IQ scores *within* particular racial or social groups to the cause of differences *between* groups. Even if the heritabilities are equivalently high for a trait within two different groups (e.g., $h^2 = .8$ within a group of whites and $h^2 = .8$ within a group of blacks), one cannot infer, as Jensen did, that any IQ *differences* between these groups must also be the result of heritability. (Jensen, it should be pointed out, was well aware of this fact and defended his genetic extrapolation as "consistent" with various lines of evidence, even if not rigorously proved.) Because the fallacy of Jensen's extrapolation is well known, and the examples used to illustrate its faulty logic can be found throughout the scientific literature, I shall say no more about it than this: It marked the height of the controversy within educational psychology and raised the collective consciousness of educators and psychologists about the issue of the relative importance of heredity and environment and the assumptions that underpinned Jensen's calculation.

The proportional contribution of heredity to IQ scores can be assessed in a variety of ways, from a comparison of the similarity between identical twins versus fraternal twins, or from a consanguinity analysis of the degree of IQ similarity as a function of genetic similarity, or simply as the degree of IQ variation when heredity is held constant divided by the degree of IQ variation when both heredity and environment are allowed to vary. Heritability is an important concept in the development of modern theories of intelligence because a highly heritable IQ suggests that whatever it is that an IQ test measures, it is something real because of its genetic transmission. So if we take the position that IQ is not identical to intelligence, the question then becomes, *If it is not intelligence, then what is this genetically transmitted trait that is moderately predictive of both school and work success?*

More than anything else, the heritability findings, in which IQ is almost always claimed to run along family lines, is the single most important barrier to escaping traditional theories of intelligence. The reasoning is simple: *If* the potential for having a high-IQ is genetically transmitted, and *if* IQ is highly correlated with very simple sensory functions that would appear to be impervious to differences in the environment (see Chapter 8), and *if IQ* is predictive of real-world success, *then* any theory of intelligence must take the heritability of IQ seriously. There is, however, a plausible alternative to this line of reasoning that allows one to profitably step outside the IQ = intelligence = real-world success = heredity camp. Part of it was addressed earlier in the argument that the efficiency of basic sensory mechanisms was susceptible to environmental influences. Therefore, if these mechanisms are not direct reflections of nervous system efficiency, as many have touted them to be, it would seem to call into question the presumption that IQ reflects nervous system efficiency because of its relationship with such mechanisms. The other part of the argument, part of which will not be presented until Chapter 8 and part of which will be presented momentarily, is that the heritability concept itself is fraught with both conceptual and statistical problems.

As I stated in Chapter 1, there has been a quiet revolution during the 1980s in the field of intelligence research, and many of my colleagues in social, cognitive, and biopsychology are unaware of it. The revolution has occurred in the field of intelligence research, but its implications run deeper than that, indeed throughout all of the literature on individual differences. It posits that individual differences in intelligence are *substantially* determined by individual differences in biological (CNS) efficiency. In order to appreciate the import of this view, let us briefly reconsider several of the strands of the argument that was outlined in the first chapter.

First, IQ is highly heritable (around .5 for adult IQ), as seen in the many reports from the Colorado Adoption Project. Second, microlevel information-processing skills, such as the speed to detect simple stimuli, are themselves intercorrelated and somewhat heritable. Third, individual differences in these microlevel information-processing skills are correlated with individual differences in IQ (e.g., Larson, et al., 1989). Fourth, simple neurological measures (e.g., the amplitude of EEG waves while subjects listen to "clicks") are highly correlated with both microlevel informationprocessing skills and IQ (e.g., Hendrickson and Hendrickson, 1982; Schafer, 1987.) And fifth, real-world success at school and work (e.g., quality of work samples, grades, supervisor ratings) is reliably correlated with IQ (e.g., Itzkoff, 1989; Hunter, et al., 1984; Gottfredson, 1986), as are group differences in delinquency rates and a host of other social conditions (see Gordon, 1976, 1987, for extensive analyses of IQ as a predictor of delinquency over and above SES).

Figure 1.1 depicted a simplified path diagram for these lines of argumentation. Central nervous system (CNS) efficiency, indexed by some rate measure (e.g., Hick's law or a rate of transmitting information), is biologically determined, at least to a sizeable extent. This singular biological resource (or constraint, depending on how well endowed with it we are) is assumed to flow into all of the microlevel cognitive processes that form the basis for higher level achievements required on IQ tests. In turn, school performance and IQ are thought to be influenced directly by the efficiency of these microlevel processes (e.g., how quickly stimuli can be coded, transformed, retrieved, etc.). IQ and school performance are also thought to be bidirectionally linked: The more one is capable (biologically) of learning in school, the better one is capable of scoring on IQ tests. And the better one is capable of scoring on IQ tests, the more one is capable of learning in school. All of the paths between CNS and IQ are depicted as being moderated by microlevel processes.

The inference that falls out of the preceding tenets is that individual differences in both IQ and microlevel information-processing skills are probably due to (inherited) individual differences in central nervous system efficiency because the correlation between these measures and the EEG measure is very high (depending on the study, somewhere between .7 and .9!). To some, this helps explain why a ranking of individuals by the speed with which they can decide which of two vertical lines is longer is correlated with a ranking of these same individuals by their IQs. Thus, the reason IQ is said to be predictive of school and work success is because it is a surrogate for the physiological underpinnings of the information-handling mechanisms that support success on the microlevel tasks and that are involved in success in academic and occupational contexts. As I already noted, these physiological underpinnings are assumed to be inherited.

I said that the revolution in intelligence research was a quiet revolution because I do not believe that many psychologists are fully aware of it. Since the studies that support the argument for a genetic basis of individual differences in intelligence have not been tied to racial or ethnic group differences, there has been little clamor inside or outside the intelligence community about what could be seen as an overly deterministic perspective. And although there is nothing inherent in the concept of heritability to warrant it, this line of thinking could lead some to minimize the importance of social-contextual research done by those who deal with matters other than intelligence. I say this because if the EEG studies can be believed, then there is not much nongenetic variance left to be explained in the IQ-work relationship or the IQ-school relationship because of the $r = .9$ relationship between EEG and IQ. However, since the IQ-work relationship is itself only of moderate magnitude (accounting for about 12% of variance), that still leaves plenty of room for contextual factors to moderate work success. Even more important, the concept of heritability needs to be understood to appreciate that it is a highly labile and relativistic entity, capable of dramatic changes in magnitude as a function of both statistical changes and changes in sample. Since most social researchers do not know about this line of reasoning, I shall devote some attention to it in the remainder of this chapter and the next. I shall begin by explaining some of the concepts needed to evaluate the merit of the argument depicted in the first chapter (Figure 1.1) and then I shall attempt to use these concepts to unravel it.

The Stability, Nature, and Meaning of h^2

In a recent analysis of the impact of the early home environment on later IQ scores of the children in the Colorado Adoption Project, it was demonstrated that most of the evidence for a direct environmental impact on children's IQ could be explained as mediated by parental genetics (e.g., parents create environments that are congruent with their own genotypes and it is really their genotypes, and not the environment, that have a direct causative influence on the children's IQ). Yet Coon, et al. (in press), noted that the detection of a direct environmental impact depends on the heterogeneity of the sample. The standard deviation of environments in the Colorado Adoption Project (and all other adoption projects, due to the practice of placing children in stable, middle-class homes) may be too small to provide a sensitive test of the environmental influence on later IQ. And, recently, Loehlin (1989) has shown that the very same data from an adoption study. All of this just goes to show that the concept of h^2 is quite relative, changing dramatically depending on (1) demographic features of the cohort, (2) individual and historical forces that impinge upon the cohort, (3) assumptions one makes about the magnitude of gene-environment interaction and reciprocol causation between parents and probands, and (4) sampling characteristics of both entry variables and process variables.

Because of the "relativistic" nature of heritability, it should be possible to find fluctuations across different samples and even across the same sample at different points in time. This is precisely what has been found in both human and animal studies. For example, Rice and her associates (1986) have reported that general intelligence has a much higher h^2 in adulthood (about .50) than in childhood (about .25), and the h^2 for one-year-olds is .08 and increases to .21 at age four and .35 at age seven (Rice, et al., 1988.) Coon (in press) has demonstrated that specific spatial and verbal abilities show increasing heritability with age, too, but this pattern is not found for abilities such as memory and perceptual speed. Elsewhere, however, the h^2 for children's IQ has been estimated as being higher than h^2 for adult's, IQ, although the procedures used to compute the two quantities were quite different (Morton, 1974; Rao, et al., 1974.) And in another study, the estimate of h^2 from different familial relationships led to significantly different values for the same population: When male sons' IQs were compared with their fathers' IQs as well as with the IQs of their fathers' identical twin brothers, with whom the fathers share all of their segregating genes and with whom the sons themselves are as genetically similar as they are to their own fathers, h^2 was different in the two comparisons (Rose, et al., 1979.) Still more, the heritability of IQ fluctuates depending on social class and race: Among whites living in Britain, it is 20% greater than among whites living in America (Bouchard, 1986.) And as another study shows, the heritability of IQ scores increases with increasing social class (Fischbein, 1980).

Finally, Jensen himself reports that the similarity between a child's and parent's IQ changes with the age of the child, as does the correlation between two IQ scores for the same child. This is quite interesting because *if* the magnitude of h^2 is assumed constant over various stages of development, *and* if the size of the correlation between two IQ scores gets smaller as a function of the elapsed time separating them, *then* there is only one conceptual remedy if one wishes to retain the concept of heritability in the equation: IQ = intelligence = heritability. It is necessary to posit a changing genetic makeup for different stages of intellectual development, at least for some children! Humphreys (1984) arrived at a similar conclusion, commenting that the "assumption of a constant genotype during the period of maturation may represent a logical inference from genetic theory, but a construct that cannot be estimated is useless in any scientific theory, and nothing is gained in the prediction of a practical behavioral criterion by bringing in an estimate of genotype" (p. 227).

For all these reasons, it should be clear that the heritability concept bears close study. Many similar differences in h^2 as a function of age, sample characteristics, etc., could have been cited to illustrate the relativistic nature of the concept, but the examples given should suffice to disabuse one from inferring that some hard-wired parameters of information-processing of an established magnitude can be estimated from the heritability literature. Certainly, heritability is a relativistic concept, and estimates of the heritability of IQ appear to be especially labile. This much all researchers in the area would agree with. But even if one accepts the role of genetics in producing individual differences in IQ scores at a given point in one's individual and historical development, in a given sample, and under given conditions, how one interprets this finding will still depend on a variety of other factors, including whether the inherited trait in question (i.e., IQ) in this sample (at this time and under these conditions) bears any relationship to what a layperson identifies as "intelligence."

What if a genetically transmitted trait that was itself unrelated to intelligence influenced educational success through some indirect route (e.g., inherited temperament influenced children's activity levels, and this in turn affected their educational adjustment)? One possible consequence might be that genetically overactive children end up with lower IQs because teachers find it difficult to maintain their attention. But the path from what is inherited to one's IQ is far from direct; and the basis for these children's lower IQs would be *nonintellective* in nature, namely, a temperamental disposition. Lest one imagine that this is a completely far-fetched argument, it repeatedly has been demonstrated that temperament is related to concurrent IQ scores, as well as sequential short-term memory, achievement test scores, and grades (e.g., Keogh, 1982; Klein and Tzuriel, 1986).

Obvious extensions of this kind of reasoning can be imagined. For instance, what if what is inherited affects a child's physical appearance (e.g., attractiveness, skin color, myopia), which in turn influences others' reactions to him or her, including those of the child's teachers, peers, and parents? The genetic effect on the child's appearance could then foster or hinder his or her school progress, which, as was pointed out in Chapter 5, is a major context for crystallizing some cognitive potentials. This in turn could affect IQ. Thus, a highly heritable trait that is itself unrelated to one's underlying cognitive potentials could ultimately affect IQ by influencing the contexts needed to crystallize those potentials that are tapped by IQ. To the extent that genetics plays such an indirect role in producing IQ, it could give the false impression that the IQ score itself (or the processes that support it) is directly inherited. In other words, genes may affect IQ scores in two ways. First, one may inherit constraints on a certain cognitive capacity (e.g., one may inherit a limited memory-storage space) that influences the amount that can be learned. Ultimately, this capacity limitation would work against IQ. On the other hand, one may inherit the types of environments to which one is exposed, as for example the color of skin one wears. Imagine if inherited facial features influenced the way parents, peers, and teachers responded to a child. There is a venerable literature in social psychology that amply documents the fact that children whom adults view as attractive receive more positive reactions than do children who are viewed as unattractive; that a child's physical resemblance to authority figures influences his or her identification with those figures; that tall children evoke different responses from their environment than short children; and that body build affects the way children are treated by peers and teachers (e.g., Hartup, 1983.) Indeed, it has been known almost since the beginnings of IQ testing in America that attractive students score slightly higher than unattractive students (Mohr and Lund, 1933.) It is one thing to score low on an IQ test (or on any other achievement test) because of a genetic limitation on our cognitive abilities. It is an entirely different matter to score low on tests (or in life, in general) because of the genetic transmission of nonintellective features. The basis for the heritability of IQ in a given sample at a particular point in time could be due to either mechanism. The findings that were discussed in the earlier chapters would argue in favor of the latter route, i. e., the inheritance of nonintellective factors, including the type of environments to which one is exposed. Other findings also are congruent with the importance of this route. For instance, numerous training projects have shown IQ to be fairly malleable, even showing large gains over brief training periods (e.g., Feurstein, 1979; Gardner, 1987; Herrnstein, et al., 1986.) Of special interest has been the finding that the largest IQ gains in intervention in a high risk sample are found among those families at greatest risk (Landesman and Ramey, 1989.)

The Assumption of Additivity

Finally, on top of these concerns about the labile nature of h^2 and its meaning and malleability are those concerns having to do with its actual computation. To understand this last concern, consider what would happen if the influence of a genetic disposition is different in different environments. That is, what if there is no *constant* genetic action of the same magnitude across all environments, but instead a multitude of actions? Then any assessment of the proportional contributions of heredity and environment would have to specify the amount of variation in both of these factors for the population in question as a function of their *interaction*. Yet heritability analyses require the absence of interaction among genetic and environmental contributions to IQ. In a careful examination of all of the known genetic studies of IQ available to her nearly 50 years ago, Loevinger (1943) criticized them for their failure to confront this additivity assumption, i.e., that an IQ score is the sum of heredity and environment, each acting uniformly over the diversity of variation in the other, and that intelligence could be described simply as an additive result of heredity and environment in the absence of any interaction between them:

It is rather remarkable that although all of the techniques for analyzing proportional influences of heredity and environment on intelligence depend on an additive assumption, only one of the papers utilizing the techniques with actual data has contained an examination of the appropriateness of the assumption. . . . If the amount of variation in a trait when heredity is constant depends on the particular constant value of heredity, we shall have more than one answer, and thus no answer, for the proportional contribution. (p. 742)

Loevinger's criticism proved to be a harbinger for modern researchers studying intelligence, and the issue of additivity and the constancy of genetic action across environments has become an important point of contention today despite major advances in quantitative methods. A non-IQ analogy proves interesting: A $3,000 increase in family income, from $12,000 per year to $15,000 per year, may have a greater impact on a trait that is, in part, genetically determined among that family's offspring than would a $3,000 increase for a family earning $90,000 per year. *Mutatis Mutandis*, genotypic changes of a specified magnitude, could result in different levels of expression at different levels of environmental enrichment.

In a recently completed analysis, Douglas Wahlsten made a telling criticism of the behavior genetic studies based on their failure to test for the additivity assumption in a statistically powerful manner. His argument is fairly elaborate and statistical and need not be presented here, except to note that he demonstrates that the same weak tests of interaction in an analysis-of-variance model that Loevinger had criticized, and which have propelled some to argue that h^2 is substantial, would lead to distorting many of the phenomena that we know to be true because of similar additivity assumptions and weak tests of interaction. A case in point is Newton's law of universal gravitation, $F = Gm_1m_2 + d^2$, where F is force, and m_1 and m_2 are the masses of two bodies, and d^2 is the square of the distance between the centers of the bodies. Wahlsten notes wryly that

Conclusions from the ANOVA (on gravitational data) would be that both mass and distance are important for force, although the internal factor (mass) is rather more important and accounts for more variance than the external factor (distance), and that mass and distance are additive because the interaction term is not even close to significance. One might even proclaim that Sir Isaac, lacking a good understanding of the principle of parsimony, made up a formula that was too complicated for practical situations!. . . . The inescapable conclusion is that. . . . ANOVA is relatively insensitive to the presence of real non-additivity of the kind considered plausible by many investigators. (Wahlsten, 1989, p. 14)

Addressing the higher correlation between family income and SAT scores for black, Hispanic, and Asian-American students than for white students, Manning and Jackson (1984) note that "these data are consistent with the suggestion that, for a given increment in family income, minority families are able to convert the increase into proportionately larger gains in scholastic aptitude (and school achievement) than are white families" (p. 214). Although modern techniques for assessing the "heterogeneity of regression" were not in wide use at the time Loevinger reviewed the literature, their use today provokes a continual source of debate among researchers, especially when the topic concerns the basis for ability differences as a function of social class, race, or ethnicity. The Genetics Association of America, the best known and most respected group of researchers in this area, concluded that the available evidence up to 1976 on group differences in IQ was insufficient to specify a genetic component as the source of these differences. Over 90% of its members endorsed the following statement:

> The interpretation of IQ scores is especially troublesome when comparisons are made between different cultural groups. These limitations must be borne in mind in any genetic analysis. . . . Although there is substantial agreement that genetic factors are to some extent responsible for differences in IQ within populations, those who have carefully studied the question disagree on the relative magnitudes of genetic and environmental influences, and on how they interact. Moreover, in general, even if a trait is largely genetic, this does not mean that the degree of expression of that trait cannot be significantly altered by environmental manipulation. Nor does a large environmental component in variation necessarily imply that we can easily change it. . . . It is particularly important to note that a genetic component for IQ score differences *within* a racial group does not necessarily imply the existence of a significant genetic component in IQ differences *between* racial groups. . . . Similar although less severe complexities arise in consideration of differences in IQ between social classes. It is quite clear that in our society environments of the rich and poor and of the whites and blacks, even where socioeconomic status appears to be similar, are considerably different. In our views, there is no convincing evidence as to whether there is or is not an appreciable genetic difference in intelligence between races. (Russell, 1976, p. 101)

DIFFERENCES BETWEEN CORRELATIONS AND MEANS

But let us return to the question of whether evidence for a biological basis for cognitive differences presents problems for the bioecological theory. So far, I have argued that a biological basis for IQ has been inferred from its relation to basic sensory mechanisms like encoding, but that it ought not to be until the objections that were raised in earlier chapters are satisfied. I also have argued that the biological basis that has been inferred from heritability estimates is labile, varying over sample, conditions, and time. And finally, I have argued that it is unclear what the underlying mechanism of the hereditary transmission of IQ is: It could be the transmission of a basic cognitive deficit or it could be the transmission of something that is fundamentally *non-intellective*, such as temperament, skin color, or attractiveness, which subjects a child to an environment that is not optimal for gathering the talents needed for successful IQ performance. The latter explanation would help explain why some persons who do poorly on IQ tests can do well on other tasks that rely on the same processes but not the same knowledge structures. None of the foregoing, however, means that I believe that there is no biological basis for intelligence. I believe that there is not one intelligence, but many, and each of them has a biological basis as well as a set of environmental contexts that foster or impede their crystallization.

Since the bioecological approach posits a number of biologically based cognitive potentials, it is to be expected that some genetic basis for their transmission is operative. Where this theory departs from the traditional hereditarian view held by some intelligence researchers is in its guarded objection to the mapping of IQ onto these cognitive potentials in any direct way. It is not that such potentials are irrelevant to IQ performance—only that they cannot be directly inferred from such performance because of the moderating roles of both developmental contexts (which in turn influence knowledge representations) and motivation. While heritability (h^2) is a measure of covariation among relative ranks on cognitive tests and, therefore, is indexed by correlation, the schooling and social class studies reviewed in Chapters 4 and 5 focused on overall mean levels of performance on such tests, and their impact is indexed by the absolute magnitude of the test scores. Even when h^2 is extremely high, as it is for some physical traits like height (h^2 for height is nearly 1.0), environment still can exert a very powerful influence, as attested to by the surge in the heights of second-generation Japanese raised in the U.S. Even though h^2 among this group remained over .90, meaning that American-reared sons of tall Japanese fathers tended to be taller than both American-reared sons of short Japanese fathers and, more importantly, taller than the Japanese-reared sons of tall fathers. The American-reared offspring were over five inches taller than they would have been if they had been reared in Japan (Greulich, 1957). And Tanner (1962) showed that both American and British teenagers were a half-foot taller, on average, than their predecessors a century earlier. Finally, Angoff (1988) reported that the heights of young adult males in Japan were raised by about three and a half inches since the end of World War II, an enormous gain in such a brief period of time! If something as highly heritable as height can fluctuate so dramatically in such a relatively short period of time, then surely traits like intelligence can be altered, too.

We can extend this line of reasoning about gains in height to the area of cognitive potentials. High heritability is silent as far as the absolute magnitude of cognitive potentials are concerned. All intelligence researchers, no matter how hereditarian their bent, would agree with this. The dispute on this particular issue concerns the status relegated to IQ.

As already mentioned, an enormous number of studies have reported substantial heritability estimates for IQ and other so-called aptitude test scores, using designs that permit the comparison of similarities between adoptive parents and children, half-siblings, identical twins reared apart, and grandparents and grandchildren. Findings from such studies are reviewed in Bouchard and McGue, 1981; Bouchard, 1984; DeFries, et al., 1981; Dixon and Johnson, 1980; Erlenmeyer-Kimling and Jarvik, 1964; Hoffman, 1985; Horn, 1983; Loehlin and Nichols, 1976; Loevinger, 1943; Plomin, 1985; Rice, et al., 1988; and Scarr, 1982. Taken together, these studies have led many researchers who span the spectrum in terms of their genetic orientations, their views on the efficacy of early enrichment, and their theoretical perspectives to conclude that the issue of heritability is no longer in dispute; only the size of the heritability index remains in question. Consider Jensen's (1981) conclusion first, and then similar conclusions by a number of leading authorities in the field of behavior genetics and intelligence. These quotes reveal just how dramatically the thinking in this area has shifted since 15 years ago, when statements like Jensen's were viewed as exaggerated and were roundly attacked by environmentalists:

> There is a great deal of agreement among scientists regarding the heritability of intelligence. The experts are not concerned with arguing about any particular value of h^2 within the whole range of most empirical studies; that is, between about .50 and .80. They all recognize the reasons for the variation. . . . They are, however, generally in agreement concerning the substantial heritability of intelligence and IQ. (Jensen, 1981, p. 104)

Along the same lines, Bouchard (1984), after carefully reviewing recent criticisms of behavior genetic findings by Kamin and others, concluded that most of these criticisms were invalid. Similarly to Jensen and others (e.g., Plomin, 1985; Gordon, 1980), he concluded that "there is compelling evidence that the heritability of IQ is well above zero and probably between .50 and .80, depending on the population being studied" (p. 170).

Although long a proponent of the view that the environment is extremely important in shaping cognitive outcomes, Plomin (1985) also noted that the significant role of genetics could hardly be disputed:

> The convergence of evidence in the area of IQ is impressive. . . . the data converge on the conclusion, remarkable as it seems, that fully half of the variance in IQ scores is due to genetic differences among individuals. . . . There is no finding in the behavioral sciences that begins to approach the magnitude of this result. (p. 304)

Even with the exclusion of discredited or otherwise problematic data (e.g., Burt's data, which many believe to be fraudulent), heritability estimates remain quite high, a point made by Gordon (1980): "Contrary to impressions promoted by journalists and idealogues, heritability estimates still remain substantial, mostly in the range .5 to .7" (p. 176).

Itzkoff (1989), in the final volume of his four-volume series on intelligence, entitled *The Making of a Civilized Mind*, comments that "the hereditary, biological roots of (IQ) differences seem to be clearly established. . . . as well as the by now well established 80/20 heredity-environment variance ratio, today noted even behind that erstwhile Maginot Line of environmentalism, the Soviet Union and the East Bloc" (p. 84). And Royce (1979), a noted factor analyst, concluded that ". . . it has been estimated that 60% to 80% of the observed variation in general intelligence is due to genetic variation" (p. 614).

And finally, Zigler (1988), a noted proponent of early intervention and a founding member of the *Head Start* movement, noted that "until we have further evidence, our best estimate (of h^2) is probably at least .50" (p. 7).

So the picture that has emerged in the intelligence research community is decidedly biological—and not just in some restrained sense, but with seemingly unrestrained biological exuberance!

And the picture is somewhat the same in studies that have examined the heritability of specific cognitive abilities rather than global IQ scores, although there are exceptions. For example, Rose, et al. (1979), have shown that biological limits exist for many specific cognitive abilities, such as short-term memory and perceptual speed, as have DeFries, et al. (1981) and Rice, et al. (1986). (In the latter analysis, short-term visual memory appears to have a substantially lower heritability estimate than do vocabulary, spatial skills, and perceptual speed. But the reasons for this are not clear and the exact size of the estimates are quite labile and likely to change dramatically across sample and test battery.) An example of the biological constraint on a specific cognitive ability is seen in the finding that unless one uses recoding "tricks," the limit on digit span is about a dozen chunks of information. Staszewski, 1989, has trained subjects to surmount this biological constraint by recoding digits into sports records, and consequently, these subjects are able to store and correctly recall over one hundred digits.

The argument derived from the literature reviewed earlier is that individual differences in biological constraints on specific cognitive abilities are not necessarily (or even probably) directly responsible for producing the individual differences that have been reported in the psychometric literature. I shall return to this claim later and provide some evidence for an indirect role of genetics, suggesting at the same time that some of the classic evidence for the direct influence of genetics is problematic.

It is clear that even if one were to agree with Jensen (1980, 1986) and others that the heritability of IQ is high, and further, that this is evidence that some genetically transmitted quality of the central nervous system (neural efficiency; speed, rate, and extent of myelination; signal-to-noise ratio of transmissions; etc.) is responsible for the continuity in mental development (see also Eysenck, 1988, 1982; Jensen, 1986; Razz, et al., 1983), both within an individual over time and across members of a family, it still would be possible to incorporate evidence for the heritability of IQ into the bioecological framework. Put simply, one could argue that contexts such as schooling serve to elicit certain underlying genetically transmitted cognitive potentials that are relevant to success on the homogeneous battery of tasks typically used by factorists. In the absence of a relevant context, the particular cognitive "musculature" that relies on it will remain undeveloped. In support of this view, most of the studies that have documented the high heritability of IQ scores also have documented the importance of the environment; that is, *absolute* IQ scores are nearer to adoptive parents' and siblings' scores than to those of biological parents and siblings, despite the higher *correlation* with the latter (e.g., Skodak and Skeels, 1941).

Nonetheless, I do not endorse the preceding position that the heritability of IQ scores is evidence that some genetically transmitted aspect of the central nervous system has set limits on the manner in which or the rate at which one learns new information. Perhaps for some persons with a known organic insult or chromosomal aberration, such as individuals with Down's syndrome, this is a fair summary of the cause for their lack of success. But for the vast majority of others, it does not seem plausible to me, given their cognitive abilities in other domains. An equally plausible view exists, and there are actually more data to support it than there are to support Jensen's view. According to this view, the highest heritability estimates reported are seen as more illusory than real. Naturally, some degree of correlation between genes and IQ would be expected, because one basis of IQ performance *may be* genetically influenced cognitive potentials that are relevant to IQ test performance. But even though such potentials may be genetic in origin, this does not mean that they contribute to variance: Everyone may possess them to the same degree, and the variance may be due to environments and/or motivations that led to their differential crystallization. We need to go beyond the fact of correlations to assess these possibilities, and when we do, the evidence for very high heritability is quite perplexing, leading some researchers to estimate a much lower h^2 than others (e.g., Jencks and his colleagues believe the best prediction is closer to .40 than to .80).

For example, Johnson, et al. (1977), analyzed the collateral resemblances of aunts and uncles who were biologically related to their nieces and nephews versus the collateral resemblances of those who were related only through marriage. As with the Colorado Adoption Project analysis, there was considerable variability in their findings: Relatives resembled each other more on some cognitive measures (e.g., vocabulary and matrices) than on others (e.g., memory and mazes). As expected, biological aunts and uncles were more similar to their nieces and nephews than were nonbiologically related aunts and uncles (r = .18 with the first-principal component of the test battery for biologically related aunts/uncles vs. r = .04 for nonbiologically related ones). On the other hand, cousins, who share only 12.5% of their genes, resembled each other more than they did their aunts and uncles, with whom they share 25% of their genes (r = .29 between cousins). As many have remarked before, findings such as these demonstrate the importance of genes and environments on cognitive task performances, and more so for some tasks than others (see, for example, Bouchard and McGue, 1981; McClearn and DeFries, 1976). They do not, however, demonstrate extremely large genetic effects that are independent of environmental variation. In fact, short of identical twins reared apart, these correlations between collaterals are probably the best source of behavior genetic evidence available because, unlike families who adopt or give up children for adoption, aunts, uncles, and cousins have not done anything unusual (they constitute an ideal, "unselected" sample), and in principle it should be easy to assess the degree of environmental similarity between them. Moreover, these pairs of related individuals do not suffer from the uniqueness of being physically identical. Another advantage of collateral analyses is the frequency with which one encounters all combinations of collaterals. For example, an uncle who is related by blood to a nephew he has seldom seen should resemble him on cognitive tests more than he should resemble another seldom-seen nephew on his wife's side of the family. Unfortunately, very few studies of this kind have been reported, and those that have do not permit any certainty in establishing the size of the heritability quotient, beyond the claim that while its magnitude may be nontrivial, it is not nearly as high as some might have supposed.

Confounding Genetic and Ecological Sources of Shared Variance

Nongenetic sources of IQ variance may be contextual, motivational, and/or due to differences in the knowledge base that disappear when the format of the test or of the material is changed. It is difficult to gauge the importance of these nongenetic variables because they may be correlated with genes in a manner that render their independent influence difficult to disentangle. And the size of h^2 can be expected to change with the developmental status of the child, generally increasing with age. With the exception of the handful of data pertaining to identical twins reared apart from birth and the collateral analyses already mentioned, the claims of high heritability of IQ are based on data that confounds schooling, social class, and ecology. It is logically unnecessary to accept as evidence that a genetically transmitted quality of the nervous system directly or indirectly accounts for the continuity or similarity of the mental development of individuals who happen to share ecologies as well as genes. (On the basis of a different philosophical analysis, Block and Dworkin (1976) have come to a similar conclusion.)

A most suggestive clue in this regard was reported by Daniel Goleman, science writer for the The New York Times. The Burakumi of Japan are a group of ethnic Japanese whom we in the West hear little about. Until Goleman introduced me to them, I had never heard of them, which might not be surprising except that I collect comparative national and international data of the sort at issue here. I have come to realize that the low profile of the Burakumi is a result of the secrecy and care exercised by Japanese researchers and policymakers: No one seems particularly interested in talking about the Burakumi to Westerners, if indeed they are even discussed publicly in Japan. Apparently, the Burakumi have a long history as an underclass. Historically, they did the dirty work (public executions, garbage collection, leather tanning, etc.) and congregated among themselves in several large cities in western Japan. Although they are racially Japanese, they are apparently identifiable by their past, clothing, behavioral patterns, and neighborhoods. Their children have every deficit known to African-American and Hispanic children, including a 10–15 point gap on IQ. Social efforts have so far resulted in little improvement in their IQ scores, from what I have been able to tell—although my information comes from a handful of Japanese colleagues, and none of them is an expert in this area. So it is highly interesting to hear Goleman (1989) report that "one study shows that Burakumi children in America, where they are treated as any other Japanese, do as well on IQ tests and in school as other Japanese" (p. 22).

This sort of reversal of what has come to be regarded as an almost immutable, genetically based problem reminds me of the analyses carried out by Flynn (1988) and Eyferth (1961). Flynn reported an interesting analysis of the intellectual development that occurred over a 17 year period in Scotland. First, he noted that in America, white 11-year-olds often attained similar IQ performances to black 13-year-olds, and this suggested to some that real differences in intellectual ability were involved. Then, Flynn reported that on two subtests of the Wechsler Intelligence Scale for Children, Scottish 11-year-olds in 1982 mimicked the performance of Scottish 13-year-olds in 1965, suggesting that similar black-white differences in America may be environmentally based. The 1965 and 1982 Scottish cohorts are separated almost entirely by environmental differences, and yet their relative performances are along the lines observed in transracial comparisons, prompting Flynn to comment: "So, the generations appear to be separated by real ability differences and yet, unless evidence of extraordinary achievement is forthcoming, that ability difference cannot be identified with an intelligence difference" (Flynn, 1988, p. 26). Hence, the large IQ differences observed between 1965 and 1982 among Scottish adolescents seems to be due entirely to improvements in the 1982 cohort's environment, and not to genetic improvements. This is why Flynn said that it represented an "ability difference not identified with an intelligence difference," since the genetic potential remained unchanged over this 17 year period.

Eyferth (1961) studied the children left behind in Germany after the Second World War whose fathers had been white and black occupation troops. While the children of racially mixed marriages in the U.S. are known to possess below-average IQs, this was not the case with the German children. The average IQ of white German children was 97.2 and the average of the racially mixed children was 96.5, a difference that is neither substantively nor statistically significant. Of course, in Eyferth's study, those with a hereditary bent can always raise alternative explanations (e.g., black recruits had higher-than-average IQs, while white recruits may have had lower-than-average IQs). However, one can also imagine that the children of the black G.I.s had an enormous disadvantage growing up in post-war Germany, their skin color being evidence of their mothers' "affiliation with the enemy." In any event, because no single study is likely to be definitive, it is good to have studies such as the ones reported by Goleman, Eyferth, and Flynn to add to the empirical base.

Before analyzing the most rigorous evidence for the genetic hypothesis, namely, the data on identical twins, it is instructive to examine what is known about national IQ differences among Caucasians outside the U.S., Canada, Australia, New Zealand, and northwest Europe (i.e., the British Isles, Scandinavia, France, Germany, Switzerland, and Northern Italy.) Richard Lynn (1978) reviewed the pertinent literature, and Table 7.1 summarizes his data.

I think these data are quite interesting for the following reason: If the data are aggregated, the mean, median, and midrange of these 12 studies of Caucasians is 85. Furthermore, as pointed out by my colleague Richard Darlington, if you exclude the six studies that either have small samples ($n \leq$ 25) or unreported sample sizes, the remaining six studies are all within the range 83–87. This is pretty compelling evidence that the mean IQ for this kind of sample is 85—exactly the same as that reported for American blacks! And the offspring of immigrants from these same countries are known to have IQs well above that figure, indicating that although genes may influence the rank ordering of individuals, in a correlational sense, it is the ecology in which they live that dictates their average IQ levels. This distinction between correlations and means was originally made (I think) by Woodsworth (1941), but it recently has been reiterated by Scarr and others because it is so important for our understanding. To see just how important it is, try the following exercise. Based on the famous (or infamous, in light of Longstreth's 1981 criticism of its design and assumptions) study of Skodak and Skeels (1949), ask yourself which is a more important determinant of IQ, genes or ecology. In their study, a group of mothers who gave their offspring up for adoption had a mean IQ of 85.7. The children themselves were tested during early

TABLE 7.1 Mean IQ of Caucasians Outside North Anerica, Northwest Europe, and Anglo Pacifica

Location	Test	N	Year	Type of Subjects	Mean IQ
India	Stanford-Binet		1926	Students at University of Calcutta	95
Zagreb, Yugoslavia	Raven's				89
Thessalonika, Greece	Wechsler Performance Scale		1972	Schoolchildren	89
Spain	Raven's	113,749	1954	Army conscripts	87
India—17 locations in several states	Raven's	5,000+	1968	Children aged 9–15 86	
Southern and Eastern Europe	Mostly non-language group tests	1,291	WWI to USA	Children of immigrants	85
Italy	Nonlanguage group tests	500	1919–1938	Immigrants to USA 85	
Italy	Binet	500	1919–1938	Same sample as above	84
Portugal	Binet or group	671	1919–1938	Immigrants to USA	83
Shiraz, Iran		1972		low	
					80's
Baghdad, Iraq	Draw a Man		1965		80
India	Raven's	25	1968	Graduate students 75 at university	

85 = Mean, median, midrange of the 12 values.

adolescence and found to have an average IQ of 107. Is this 21.5-point IQ superiority over their biological mothers incompatible with high heritability?

The answer is *no*. The 21.5-point gain over their biological mothers' IQ is consistent with a heritability as high as .6 (narrow) and .7 (broad.) You can see this if you regress IQ on their offspring as follows.. First, make the assumption that assortive mating occurs at .39. (Likes mate with likes, so the mean IQ of the biological fathers of these children was probably similar to the mothers' 85.7.) But since there must be regression toward the mean, the fathers must be closer to the mean of 100 than were the biological mothers, given the mothers' low scores. (Statistically, such upward regression is a necessity; if the mothers' IQs were over 100, then the fathers' IQs would have had to regress downward toward the mean and, as a result, be lower than the mothers' IQs.)

Anyway, if there is assortive mating as high as .39, then we would expect the biological fathers of the children to have had IQs around 94.5 (the population mean minus .39 × the difference between the mothers' mean IQ and the population mean, or 100 minus .39 × 14.3 = 94.5). So if the biological mothers' average IQ was 85.7 and the biological fathers' average IQ was 94.5, then their midpoint average IQ was 90.1 (85.7 + 94.5 ÷ 2 =90.1). Now, to regress IQ on the offspring, simply take the population mean for this trait (100) and add it to the product of h^2 and the biological midpoint IQ minus the mean population IQ, or:

est. $IQ_{offspring} = IQ_{population} + h^2 (IQ_{midpoint} - IQ_{population})$ or 100 + .6 (90.1 − 100) = 94.2

So we expect the children's IQ to be 94.2, not 85.7 like their biological mothers. But 94.2 is not 107, and doesn't this imply that high heritability (.6 to .7) is incompatible with these data? No, because 94.2 is the IQ that would be expected if the trait were entirely due to genes, that is, if it were 100% genetically determined. The difference between the IQ expected on the basis of 100% heritability (94.2) and the observed IQ of 107 is well within a heritability as high as .6 or .7 if we examine the homes in which the children were raised. From the account given by Skodak and Skeels, the adoptive homes were far above average, with the heads of households usually professionals. How much above average these family environments were is hard to gauge, but it seems reasonable to suppose that they were between one and two standard deviations above the national norm. So let us take the superior status of the adoptive homes to be 1.5 standard deviations above the norm during that era. How much of an increment will such an environment produce to the observed value of 94.2? This is a straightforward calculation: Since the standard deviation of the IQ test was 15 or 16 (let us be conservative and say it was 15), the total variation to be explained is 15 squared, or 225. Thus, the proportion of the variance that may be due to nongenetic sources (environment and measurement error) is the reciprocal of h^2 multiplied by total variance, or .3 × 225 = 67.5. To get the amount of enhancement due to environment, take the square root of this value, or $\sqrt{67.5}$ = 8.25, and multiply it by 1.5. Since 1.5 × 8.3 = 12.33, we expect that even with an h^2 as high as .6 to .7, children reared in superior homes ought to have an IQ of 106.5, almost what was actually observed. And if the higher standard deviation figure was used (16), the amount of enhancement due to environment would be higher still! So, it is hard to imagine an IQ differential between a child and her biological parents that would be incompatible with even h^2 values as high as .6 to .7. (This exercise should also make it clear that if we are interested in promoting human potential, even quite high values of h^2 still leave plenty of room for environmental enhancement.)

Against this backdrop, we return now to the heritability literature. The most rigorous test of the genetic hypothesis concerns the handful of data pertaining to separated identical twins, no more than 47 pairs known to scientists, many that do not meet rigorous scientific criteria. (At this time, new twin data is beginning to be collected by several teams of researchers but their results are not yet known, e.g., Vernon, 1986). Bronfenbrenner's (1974) reanalysis of these classic data showed that the correlation between the IQs of monozygotic twins reared apart was .87, *if one considered the entire pool of twins together*. If, however, one differentiates the pool of twins according to the ecologies of their childhood, an interesting finding emerges: In the case of twins who were reared in dissimilar environments (urban vs. mining or agricultural settings), and who attained different levels and quality of formal schooling, the correlation dropped to .27. Thus, when one comes closest to randomly mapping genotypes over environments, the evidence for heritability becomes substantially attenuated.[1]

Thus, there is strong evidence that nongenetic factors exert substantial influence on IQ scores, even in those studies that have reported very high heritability estimates. But does the importance of education and social experiences mean that genes are unimportant? Clearly, it does not. What is at issue is not really the role of genetics in IQ (surely, genes play a nontrivial role in virtually all important aspects of mental functioning and development), but, rather the degree to which the high heritability associated with IQ is interpretable as evidence for the high heritability of *intelligence*. As mentioned earlier, many of the biological bases for behaviors are distributed so pervasively among persons that, despite their critical role in producing the behaviors, they are associated with little variance in individual difference.

There exist too many reasonable and yet contradictory interpretations of the extant heritability data to permit unequivocal conclusions about the role of genes in intellectual development. However, a cautious conclusion is that both biology (manifested in the form of the various cognitive potentials) and ecology (e.g., schooling and social class) shape the ultimate faces of the intellect. In the bioecological theory, the cognitive potentials that are reflected on IQ test performance are but a subset of the totality of cognitive potentials. A less cautious conclusion, however, is also imaginable, and it, too, can be interpreted within the bioecological theory: The highest heritability estimates probably dramatically overestimate the importance of genes for *any* type of intelligent performance, including those that are relevant to IQ test performance, because the same cognitive operations involved in IQ performance are involved in many real-world forms of problem solving; and the correlation between these two domains is sometimes quite low. Undoubtedly, researchers like Plomin and DeFries (1985) and Bouchard and McGue (1981) have identified something real in the high heritability correlations they have obtained, and it does not seem worthwhile to gainsay or trivialize their results. (Indeed, the Colorado Adoption Project alone has produced over 50 reports in the past decade that clearly establish the fact that hereditary transmission exists.) But it is equally clear that no matter what degree of heritability has been unearthed in behavior-genetic studies of IQ, it is always a matter of restricting one's conclusions to particular populations, in particular historical and social epochs, on particular cognitive measures. And it is not at all clear exactly what meaning to attach to the results. (See Loehlin's, 1989, alternative path models that he fits to the same data, with equally good "fits" that are contradictory in the importance of shared environmental sources of varance.) It is not a trivialization of the issue to remember that if all environments were made 100% optimal, then 100% of the differences between individuals' IQs would be due to genes. Thus the size of h^2 could grow, even as the magnitude of observed differences on intelligence tests might shrink. To take a slightly different tack, if one viewed data from Germany collected in the 1960s, it would lead to a suggestion of a genetic basis for a meritocratic society because of a very high correlation between parents' and their children's level of education. That same correlation in the U.S., however, was less than half as large during the 1960s (Husén, 1967). Again, one must restrict conclusions to specific contexts.

The evidence for the influence of environment is equally clear (e.g., schooling, social class, parenting practices, national ideologies), despite claims by some that environmentalists have not yet specified in any but the crudest sense the mechanisms by which experiential variables affect intellectual development (McAskie and Clarke, 1976; Willerman, 1979). In this book, I have put forward a case that complexity of thinking is domain specific, the result of a combination of genetic and environmental factors, broadly speaking. These factors include biologically constrained cognitive potentials (e.g., size of working memory), schooling (both level of attainment and quality), family structure and disciplinary styles, macrolevel setting (agricultural vs. urban vs. mining), and motivation. Empirical support—in some cases, substantial—has been mustered for most of these (e.g., Schiff, et al., 1982).

The preceding suggests that a genetic account of the continuity of mental development frequently may be indistinguishable from a bioecological account, given the inclusion of the former in the latter. Thus, the seminal behavior-genetic evidence supportive of some current theories of intelligence is not detrimental to the framework being put forward here, as it is apparent that no single explanation (biological or ecological) is satisfactory. Along these same lines, Hoffman (1985) has recently examined the three most influential behavior-genetic analyses—the Texas Adoption Project (Horn, 1983), the Scarr and Weinberg (1983) study of adopted adolescents, and the Scarr and Weinberg (1976) transracial adoption study—and concluded that "despite the conclusions that have been drawn from the data, the behavioral geneticists have not shown that family environment fails to affect intelligence, nor even that it has less effect than genetics" (p. 140). (It should be emphasized that many behavior geneticists endorse a large role for environment (e.g., Plomin, 1985). Given, then, the ambiguity as to whether there is a single, isolable factor capable of accounting for large amounts of variance in intelligent behavior, it would seem prudent to view genetics as an important precondition for intelligent behavior (i.e., the ultimate limits of the cognitive potentials may be genetically determined), although inextricably dependent on the other sources mentioned (e.g., developmental contexts for the crystallization of particular cognitive abilities, or motivation). So a slightly reworded version of the well known aphorism that *biology proposes and ecology disposes* seems appropriate when discussing intellectual development, provided that the reader bear in mind that what is being proposed as *intellectual* is neither directly genetic nor a singular entity, but a galaxy of cognitive potentials that have their own developmental trajectories and are differentially susceptible to ecological differentiation.

CHAPTER 8

The Fallacy: Biology = IQ = Intelligence = Singularity of Mind = Real-World Success

In 1884, Sir Francis Galton had the idea that simple sensory and perceptual skills, because of their presumed link to biological processes, were related to intelligence and to real-world success (Eysenck, 1986; Sternberg and Powell, 1983a). To confirm this hypothesized link, Galton set up a pavilion at the South Kensington Science Museum in London where, until 1891, visitors could have their simple sensory and perceptual processes tested. He believed that one of the prime differences between the more and the less intelligent was a sensitivity to stimulation. For example, the distance needed between the placement of two pins on the arm in order to perceive that they are touching different spots (punctate sensitivity) was presumed to be greater for the less intelligent.

Unfortunately, Galton was unsuccessful in demonstrating that such simple (biological) measures were related to more global intellectual measures or to societal prominence. Recently, however, researchers have revived this idea and provided what appears on its face to be strong evidence for it, suggesting that Galton's failure was due to the low reliability of his measures and to the inefficiency of his statistical tests—many of which he developed (Kline, 1988). The fundamental finding in the recent research, however, is that the speeds with which individuals can perform very simple Galton-type tasks are correlated not only with each other, but with IQ. And when a group of cognitive tasks are administered, scores on them tend to be intercorrelated, yielding the so-called positive manifold of correlations discussed in Chapter 1. Because of the interrelatedness of these task performances, it is assumed that a singular processing force (cognitive, motivational, and/or biological) runs through them.

To some, it is perhaps paradoxical that so much is made by psychometricians of the single undercurrent that runs through different types of task performance. After all, one can point to the well-documented unevenness of aptitudes displayed by famous prodigies: If there exists a single biologically based undercurrent to all cognitive performance, then why are their cognitive performances so variable across domains, as argued in a recent biographical analysis?

> Most people underestimate the extent to which it is possible for an individual to possess particular skills in isolation from other abilities. . . . It is quite possible, for instance, for someone to be a brilliant chess player, or a superb musician, or an excellent mathematician, while having less than average ability at other intellectual skills, including ones that appear to be quite closely related to the field in which that person excels. To a surprising degree they (the different intellectual skills of prodigies) are quite autonomous. (Howe, 1989, p. 30)

But the unevenness of prodigies' abilities may not be a very good basis for arguing against a single underlying resource pool, that flows into all intellectual endeavors for several reasons. First, prodigies are, by definition, unlike the rest of us, and therefore what propels their performance may be qualitatively different than what propels the 99.9% of us who are not prodigies. Second, their variable abilities could be the result of focusing all of their energy into their main area of expertise and ignoring all others. And, third, no psychometric researcher has ever argued that a single resource pool (g) is the whole story. In fact, it accounts for only about 10% of the variation in performance, leaving much for specific talents to account for.

Having made these disclaimers, there is still a nagging insinuation, fueled by the validity studies mentioned throughout earlier chapters. Even if g is not the entire story, it is enormously important because of its usefulness for employment screening, educational placement, and sundry other matters. And it is claimed to be heritable to a nontrivial degree, suggesting that it may be the 10% that separates the haves from the have nots! A consideration of just these two assertions should clarify why g and the positive manifold deserve close scrutiny. First, Hunter (1983) has demonstrated that g is much more predictive of occupational success than are specific ability factors, suggesting that training someone with low g on specific skills is not as likely to be as effective as hiring someone with more g. And, second, Jensen (1980) reported that the average differences between blacks and whites on 13 cognitive tests were highly correlated with the extent to which these tests loaded on g—the larger a test is saturated with g, the greater the black's disadvantage.

So whatever its magnitude, g deserves attention, and an important task for any proposal that opposes it is to account for the well-established positive manifold of correlations among global tests like IQ and "basic" information-processing measures without postulating a singular source of processing variance or a direct linkage to real-world success (for excellent reviews, see Brody, 1985; Carroll, 1976; Humphreys, 1979; Hunt, 1980; 1984; Jensen, 1984; Larson, et al., 1988; McNemar, 1964; Thorndike, 1985).

CROSS-TASK COMMONALITY

What is the nature of this commonality across tasks? It is possible to conceptualize it (the underlying commonality among tasks) in terms of individual differences in cognitive processes (e.g., individuals may differ in their memory strategies), or of individual differences in motivation (e.g., general differences in how motivated people are on tasks). These nonbiological explanations have not been preferred by most researchers. Historically, many intelligence researchers assumed that the root of the common performance across tasks was a primordial form of mental energy, recruited during focused attentional states, which was relevant to performance across a wide range of tasks. This mental energy is seen as inherently biological in nature and provides the foundation upon which individual differences in societal attainment are based. This helps to explain why some tasks are correlated, even though they appear on the surface to have little in common. For example, the time it takes to decide whether two straight lines are the same length is correlated with vocabulary scores. The faster one can make these simple line judgments, the higher is his or her word knowledge score (e.g., Lubin and Fernandez, 1986). Similarly, the speed with which one can name letters is correlated with the ability to solve abstract visual mazes (e.g., Ford and Keating, 1981). And the time it takes educated adults to judge whether strings of letters are words (CAT, CAK) is moderately correlated with their accuracy on complex paragraph comprehension ($r = .3$ to $r = .4$; see Hunt, 1985). I think that most people who are confronted with these types of correlations for the first time are initially quite surprised, because the tasks that are correlated appear to have little in common by way of shared cognitive processes or likely motivational forces. After all, what do vocabulary knowledge and the judgment of a line's length have in common? The basis for their overlap, many researchers reason, is something more "basic" than either cognitive processes or motivation—something rather biological and singular in nature. Throughout the century, various writers have referred to this entity as g, processing capacity, resource pools, capacity to educe relations, finite pools of energizing forces, signal-to-noise ratio in the nervous system, span of apprehension (i.e., the capacity to hold "fundaments of knowledge" in consciousness while trying to solve a problem), the integrity of neural circuitry, or something else.

The Role of Task Complexity

Now it wouldn't necessarily convince a skeptic that just because IQ correlates with performance on simple tasks like the judgment of a line's length it is also an index of some singular biological energy that is common to both performance in IQ and simple judgment tasks. But what if the skeptic is told that the size of the correlation between the performance on these simple judgment tasks and the performance of more global tasks like IQ increases as these simple tasks are made to require more and more of this singular energy by making them more complex? If the simple tasks are made to require increased attentional capacity or memory load, then they tend to correlate even higher with IQ (and g) as well as with each other. For instance, the correlation between the difference between the IQ scores of gifted versus average students and the speed with which these students can push a button to indicate a simple sensory discrimination increases as the sensory discrimination task is made more complex by requiring additional memory. Cohn, et al. (1985), report a correlation of .94 between the complexity of a reaction-time task and the magnitude of the performance differences between these two groups of students . Moreover, the size of the correlation between g and a judgmental task increases as the complexity of the task increases (Marshalek, et al., 1983), as does the relationship between complexity and simple associative learning efficiency. For instance, the correlation between paired associate learning and intelligence increases as the pairs of associates are presented at faster rates, thus increasing the complexity of the task (Christal, et al., 1984.) But a skeptic reading this could still imagine alternative interpretations.

In order to complete the *IQ = intelligence = singularity of mind = real-world success* claim, consider this: the predictiveness of g increases as the real-world criterion is made more complex. For instance, g-loaded tests such as standard IQ tests predict occupational success in complex jobs better than they predict occupational success in less complex jobs (Gottfredson, 1986; Hunter, 1983, 1986; Jensen, 1980, 1981). In data provided by the U.S. Department of Labor, there was a substantial statistical relationship between g-based measures of general intellectual ability and job-related aptitudes such as manual dexterity, spatial ability, and so on. In these data, Jensen (1981) showed that the median validity of an optimally weighted composite of nine aptitudes, including a measure of g, and performance in over 400 different jobs was .36, and the median coefficient for the IQ measure alone was .27! In another study, the predictive validity of a measure of g increased as the job demands became more complex (Jensen, 1980). The average corrected validity of IQ (and related measures) for predicting occupational success is approximately .5.

g and Job Success

So, the positive manifold of correlations among cognitive tasks that gives rise to g is important to consider not only because of its support for the view that g is a single biologically based undercurrent of all forms of cognition, but also because of its relationship to real-world criteria such as job success and to the psychological construct of "complexity." I shall turn to these issues next, but as an aside, it should be noted that the predictive validity of intelligence tests for a range of occupations, although modest in magnitude, has prompted Hunter, Schmidt, Gottfredson, and many others to suggest that the use of such tests to assign workers to positions could save the government huge sums of money each year[1]. Gottfredson in particular has concluded from her analysis of the literature that

> General cognitive ability not only predicts job performance moderately well, but it also predicts performance better than does any other single worker attribute. It also predicts performance equally well for blacks, Hispanics, and whites. . . . and performance in training and on the job is linearly related to performance on. . . . general intelligence. On the average, increasingly higher intelligence levels are associated with increasingly better job performance. . . . It appears likely then that more extensive training or experience in relevant jobs skills can temporarily render less intelligent workers equally productive as more intelligent but less experienced workers, but that the latter will outperform the former within at least a few years, if not much sooner, depending on the complexity of the job. . . . There is no evidence that less g-loaded traits (e.g. motivation) compensate substantially for differences in intelligence, on the average. (Gottfredson, 1986, p. 395)

To recapitulate, the evidence, crudely put, states that performances on a host of seemingly basic information-processing tasks, such as recognizing letters, judging the length of lines, and rotating shapes, are related to each other because of their presumed mutual dependence on the efficiency of a single biological resource pool in the central nervous system of humans. And because of this, they are also related to IQ, as the latter itself is an indirect reflection of central nervous system functioning. (More efficient nervous systems allow individuals to process more information from their environments and thus to be able to perform better on IQ tests.) Moreover, the relationship between IQ and job success is further evidence that g is not simply an artifact of the laboratory or of some school-related experience, but an integral aspect of human functioning that permeates a wide array of endeavors, including schooling and work. In what follows, I shall try to dismantle this argument beginning with the positive manifold of correlations between information-processing tasks.

POSITIVE MANIFOLD

Although the magnitude of the intercorrelations observed between measures of g and measures of information-processing are usually quite modest (e.g., usually not able to account for more than 10% of the variance in the relationship), they do suggest that whatever it is that an IQ test tests, it is at least somewhat related to the size of one's underlying biological resource pool(s), which in turn affects the efficiency with which information can be received, transformed, rotated, and/or retrieved by the nervous system. While many so-called basic indicants of information-processing efficiency that appear to be involved in these intercorrelated tasks are clearly mediated by schooling, parenting, or some other aspect of the environment, and therefore do not pose problems for the position being advanced here (i.e., we can attribute their commonality to environmental, cognitive, or motivational variables at least as confidently as others have attributed their commonality to biological ones), this is not true of all of them. Some of the tasks that are correlated with each other (and with IQ) appear to be insensitive to all but the most extreme environmental conditions and therefore appear to present problems for the thesis being put forward.

Take, for example, Hunt's (1980) demonstration that individual differences in the speed of a same vs. different decision involving familiar alphanumeric symbols (Posner and Mitchel, 1967) are correlated with verbal intellectual ability. This is but one of many demonstrations of a moderately significant statistical relationship between microlevel information-processing mechanisms and IQ (see Brody, 1985; Jensen, 1982; Longstreth, 1984; Marr and Sternberg, 1987; Vernon, 1987, for reviews of this literature). In these studies, researchers have reported that subjects' performances on microlevel tasks like long-term memory retrieval, short-term memory scanning, mental rotation, and inspection-time/encoding are correlated with IQ—the higher the IQ, the faster the individual can scan, retrieve, rotate, and inspect such stimuli. This fundamental observation requires some commentary because it lies at the very heart of the argument made by Eysenck (1988), Vernon (1986, 1987), and others that IQ does index some basic aspect of biological (i.e., neural) efficiency. For example, based on his ongoing study of monozygotic and dizygotic twins, Vernon (1987) reported that the heritability of speed of access to long-term memory in a category-matching task was very high (h^2 = .71) and was correlated −.52 with verbal intelligence. (The sign is negative because the briefer one's scanning speed, the higher is his or her IQ.) Brody summarized these results, but, while he agreed that they supported the view that speed of information-processing is a fundamental component of intelligence, he remained skeptical as to whether they supported the biological foundation of IQ and the view that performance on these information-processing measures underpins individual differences in the ability to acquire complex knowledge:

> These results buttress our earlier conclusion that relatively simple measures of basic information-processing may correlate with general intelligence close to .70. Vernon suggested that differences in speed of information-processing may determine the ability to acquire more complex knowledge and skills that are assessed by omnibus measures of

intelligence. He implies that such measures stand in a causal relationship to other components of intelligence. However, given the absence of appropriate longitudinal research, such a causal account of these relationships remains conjectural. His research does support the notion that speed and efficiency of information-processing are components of intelligence. (Brody, 1985, p. 375)

In order to evaluate this argument, let us examine the nature of the correlations in two commonly employed research paradigms: the letter-matching task and the mental rotation task. I am omitting a discussion of the "choice reaction time" paradigm used by Jensen (e.g., 1982), as the problems underlying its measurement, logic, and theory have been extensively reviewed and the paradigm has been found wanting. The interested reader is referred to the thoughtful critique of this paradigm by Longstreth (1984) for details.

Same-Different Judgments

In one of the most popular information-processing tasks, subjects are asked simply to judge whether a pair of overlearned stimuli are physically identical (e.g., "a", "a") or have the same name ("a", "A".) It takes significantly longer to respond to pairs of stimuli that are similar in name only, and the difference between the time needed to make a physical identity match ("a", "a") from the time needed to match names ("a", "A") is assumed to reflect the time required to access and retrieve the overlearned lexical codes of these letters from long-term memory (e.g., Jackson, 1980). That all subjects are highly familiar with the numbers and letters they are asked to judge in this task but individuals with high verbal abilities or reading comprehension scores are faster at making these judgments (e.g., Hunt, 1985a; Jackson, 1980) has been taken by some researchers as evidence for the validity of IQ as an indicant of neural efficiency, or at least of some specific aspect of neural functioning associated with the retrieval of overlearned codes from long-term memory. What possible effect could differences in schooling or environment or motivation have on the performance of such simple but elegant tasks, given that familiarity with the symbols is equated across ability levels?

Before attempting to answer this question, it should be noted that similar analyses could be provided for other simple tasks, such as the detection of a line's length, judgments of a letter's mirror image as a function of its angle of rotation, and the speed of reacting to a colored pattern of lights. A vast literature is available that cannot be summarized here because of space constraints. This literature shows that, in general, relationships exist between measures of intelligence and performance on microlevel information-processing measures, although parameters derived from some of these are more highly correlated with intellectual measures in certain age and/or IQ groups than in others (e.g., Ford and Keating, 1981; Larson et al., 1988). These kinds of demonstrations have led some researchers to posit that while background knowledge can be extremely important in influencing some types of cognitive performance, in tasks such as the ones just described there exists a wide range of fundamental "mechanistic processes" that either are or can be made to be uninfluenced by knowledge, but which nevertheless contribute to differences in performance and these differences index differences in brain function. This argument has been made explicitly by Eysenck (1988), Jensen (1988), and others. Commenting on the correlation between various information processing tasks and IQ, Eysenck (1988) suggested that because the information processing tasks are so simple and the stimuli so familiar, their correlation with IQ suggests both tasks are measuring netural integrity: ". . . suffice it to say these tests depart significantly from Binet's suggestions for IQ tests in that they show no learning, do not involve problem solving, and are not dependent in any sense on previous learning. That they nevertheless correlate very highly with IQ tests must throw serious doubt on (environmental theories of intelligence)" (p. 92).

As alluring as the *IQ = Intelligence = Singularity of Mind* perspective may sound, there are several problems with it, the most serious of which has been the occasional failure to find the predicted relationship between IQ and these basic letter-naming and mental rotation tasks when stringent validity and reliability controls have been imposed. Two recent studies demonstrate just how fundamental this failure to replicate results has been. Both studies have employed the same-versus-different task and a letter rotation task, along with measures of both verbal and spatial aptitude. They represent rigorous studies because of the care taken to ensure the reliability and validity of the measures used.

Tetewsky (1988) reported a pair of studies from his doctoral dissertation in which subjects were required to make various kinds of stimulus classifications. The first experiment involved a letter-matching task, the second a mental rotation task. For both of these tasks, Tetewsky analyzed the various processes involved in making these classifications and examined how these processes were related to psychometric measures of intelligence. However, he also added the following twist to the experiments: Instead of using tasks with highly overlearned stimuli, he specifically structured the tasks so that they would produce different processing strategies, depending on his subjects' familiarity with the stimuli. The familiarity manipulation involved using characters from the Hebrew alphabet and recruiting subjects who either could or could not read Hebrew. By this design, Tetewsky was able to show that even though subjects could do the tasks regardless of their familiarity with the stimuli, the same task might measure a different kind of process for the two groups of subjects. In particular, he predicted that when the stimuli were relatively unfamiliar, the processing parameters from these tasks would be correlated with measures of fluid ability (e.g., abstract reason-

(a) whether the two letters were physically identical:

(b) or whether they had curved or straight physical forms:

(c) whether they were from the same alphabetic system:

| G M | M G | ٦ ם | ם ٦ |

or (d) whether they had the same name (sound):

| G ٦ | ٦ G | M ם | ם M |

ing), whereas when the stimuli were familiar, the same processing parameters would not be related to measures of fluid ability.

In the first experiment, subjects were shown a set of characters in which two letters were from the Hebrew alphabet and two letters were from the Roman alphabet. They were then instructed that these letters could be paired according to four distinct "rules of sameness." That is, they were asked to judge whether they were physically identical letters, whether they were curved, from the same alphabet, or had the same name.

In the second experiment, Tetewsky employed a mental rotation task in which the time taken to rotate two letters into congruence is usually found to be an increasing linear function of the angular distance separating them. Although the basic paradigm was very similar to the one that has been used extensively in previous research, Tetewsky also included a different form of this task that was intended to result in an alternative processing strategy for subjects who were familiar with Hebrew. Such a strategy was predicted to be different in that it would correspond more closely to the kind of strategy that subjects would use if they could make their comparisons without carrying out a rotational transformation.

Overall, the results of both experiments supported the predicted hypothesis. In the letter-matching task, subjects who knew Hebrew were able to make same name classifications in a way that was not directly related to the physical properties of the stimuli, while subjects who did not know Hebrew had a more difficult time with this task because they had to make such judgments by fully processing the physical properties of the stimuli. Similarly, in the mental rotation task, the subjects who knew Hebrew were able to use what might correspond to a nonrotational strategy in one form of the task, while subjects who did not know Hebrew clearly had to carry out rotational transformations in all of the various forms of the task. In addition, the two processing parameters for the subjects who did not know Hebrew were significantly correlated with measures of fluid abilities, while the same processing parameters for those who did know Hebrew were not correlated with such measures. Based on these findings, Tetewsky concluded that contextual variables can influence cognitive processes that are related to intelligence, and therefore the assumption that there are certain "knowledge-free mechanistic processes" needs to be re-thought:

> Overall, these results indicate that the familiarity of the stimuli had a direct influence on the relative difficulty of using various kinds of processing rules. When Hebrew characters were not in a highly overlearned state, it was easier to make classifications based on physical dimensions of form (curvedness) and system (language) whereas, when the Hebrew characters were in a highly overlearned state, it was easier to make the classifications based on the cognitive dimension of name (p. 20.)

> . . . . Stated differently, when subjects had to use the dimensions of system (language) and name to classify these physically identical stimuli, the physical properties of them became less important in determining how these stimuli were processed. . . .Thus, it appears that the idea that there are knowledge-free mechanistic processes that can contribute to human intelligence needs to be re-evaluated. (p. 64)

Interestingly, other investigators have also shown that knowledge influences processing outcomes. For example, Pellegrino and Kail (1982) reported that the mental rotation of well known letters has a different function over degree of angle than the rotation of abstract, unfamiliar figures does, indicating that subjects' preexisting knowledge influences their processing strategies. Schaie (1988) and Schaie and Willis (1988) have also found that knowledge (familiarity) exerts an effect on patterns of factors over adult years. The most familiar stimuli showed the largest age divergencies in these factor patterns. And in a related study, Keating, et al. (1985), reported the results of three experiments that also employed same-vs.-different and mental rotation tasks. Because their study is more accessible than Tetewsky's, I shall not spend as much time describing it, although their careful discussion of the validity issues that have plagued the former studies will repay those unfamiliar with it who decide to spend the time to read it. They reported that the evidence for a relationship between microlevel processing parameters such as those derived from encoding and rotation tasks was "bleak." Of 30 correlations for the adults, only 8 were statistically reliable (and for eighth graders, only 5 of 30 were statistically reliable). Moreover, the average inter-correlations for the rotation measures were only .17. Interestingly, these researchers found similar magnitudes when correlating measures from di-verse domains ($r = .17$). As if this were not bad enough, they found that for the measures of retrieval from memory, the average intercorrelation was actually negative (−.13)! After summarizing the results, the authors went on to examine the relationship between the microlevel measures and measures of intelligence and concluded that:

> The evidence here is more strikingly negative. Of the six theoretically relevant correlations for eighth graders, only one is significant. . . . The average verbal ability/retrieval parameter correlation is −.14; the average verbal ability/rotation parameter correlation is −.01. The theoretically less related ability/process correlations average is −.09, which is indistin-guishable from the predicted correlations. . . . The virtual absence of sig-nificant correlations in the predicted directions, together with the equal likelihood of across-domain compared to within-domain correla-tions. . . . provides clear disconfirmation of the. . . . theoretically critical processing/ability relationships. . . . The findings from this construct va-lidity investigation of the connection between cognitive processing vari-ance, as assessed in a series of standard experiments, and mental ability, as assessed by both group and individual ability tests, are unequivocally disconfirming. We examined a number of possible artifacts to explain these negative results but none were compelling. (Keating et al., 1985, p. 168)

In a related study, Keating and colleagues have also reported that parameters derived from individual differences in long-term memory re-trieval were uncorrelated with intellectual ability measures (List, et al., 1985). In that study, they cautioned researchers against accepting previous claims of a relationship between retrieval parameters and IQ, since their study had various methodological advantages over those purporting to find such rela-tionships: "Our study. . . .provides strong evidence that this (traditional) method of parameter estimation confounds other cognitive processes. In our investigation, LTM retrieval efficiency correlated with verbal abil-ity. . . .when the more confounded estimate was used but was not related to individual differences in complex task performance when the less con-founded estimate was used" (p. 149.) Other studies that bear on this conclu-sion will be reviewed in other sections of this and the following chapter, as they are most relevant to other issues (e.g. Larson, et al., 1988). But for now, suffice it to say that the first chink is visible in the argument that some underlying biologically driven resource pool directly influences performance across a range of tasks.

The Role of Practice

As if the failures to find the expected pattern of intercorrelations is not enough to give the supporters of the direct biological view pause, other problems exist to magnify the interpretive difficulty. These concern the role of "homogeneity of action" and the role of practice effects on basic information-processing parameters.

Homogeneity of Action. During the 85 years that researchers have studied the phenomenon of positive manifold it has been assumed that this singular entity (g) operates uniformly across all ability levels, i.e., there existed a *homogeneity of action*. Thus, if g is .32 in a study, we assume that this means that about 10 percent of the linear variance among subjects' test scores in that study can be accounted for by the first principal component. So, the g that underlies 10 percent of the variance of high IQ subjects' test scores is supposedly more efficient than the g that underlies 10 percent of the variance of low IQ subjects' test scores. Clearly, comparisons like this assume that both groups have about 10 percent of their variance explained by g. But what if it were the case that IQ and microlevel cognitive tests have dramatically lower values of g at the high end of the IQ range than at the low end? This could require serious rethinking of a host of interpretations having to do with g because it essentially would mean that the IQ test is more g-saturated for high IQ individuals than for average or low IQ ones. If true, then the heritability findings that were reported in the previous chapter might need to be qualified (is h^2 higher for low IQ persons?) as would our thinking related to the use of IQ scores for practical decisions (is there more intra-subject variability among higher IQ individuals and if so does this lessen their use for employment prediction?)

In a recent study, Detterman and Daniel (in press) reported two independent analyses of the positive manifold of correlations of test scores. They divided their subjects into various IQ groups—in the first analysis they were divided into high (above 125) and low (below 68), and in the second analysis they were divided into five gradients, from low IQ to high IQ. They reported that the intercorrelations among the cognitive scores (e.g., memory, encoding, perception) were approximately twice as large for low IQ persons as for high IQ persons, and the correlation between g and IQ was also much higher for low IQ subjects. Furthermore, the average correlation among the different Wechsler IQ subtests were nearly three times larger for low IQ individuals (below 76) than for those with IQs greater than 111. Thus, by virtue of discovering greater intercorrelations among microlevel and macrolevel (IQ) measures at the lower end of the IQ scale, Detterman and Daniel may have forced us to rethink the meaning of g. One can ask if it even represents the same entity (i.e., is due to homogeneity of action) across all IQ levels. And even if it is due to the same action (cognitive, biological, motivational, environmental, personality, etc.) across all IQ levels, the fact that g is only half as large for persons with IQs above 111 than for those with IQs below 76 raises confusion about the practical utility of the general factor (or of IQ, for that matter) for persons with IQs above 111. IQ tests are more g-loaded at the low end of the IQ distribution and at the high end there exists far greater intertest scatter or variability. In fact, Detterman and Daniel report that their highest correlation between microlevel cognitive measures and IQ only reached .26 for persons with IQs above 115, whereas for persons with IQs below 69 it reached .6. This raises questions about the meaning of g and its practical usefulness at the higher end of the scale.

Practice Effects. In addition to conceptual confusion about the meaning of g and its implications for related issues (e.g., is the meaning of h^2 qualified by the apparent heterogeneity of action described above), additional problems exist that complicate the traditional meaning attached to g. The best documented of these has to do with the possibility that a person's performance on tasks used to derive g can be substantially altered by practice, i.e., they can be improved by increased practice with the procedures or familiarity with the stimuli used. It is known that the minimum trace duration, followed by a mask, that is needed to identify an alphanumeric stimulus is moderately correlated with IQ (Walker and Ceci, 1985). Subjects with high IQs are faster at identifying numbers and letters than those with low IQs, and similarly, subjects with high verbal SATs are faster than those with relatively low verbal SATs (Hunt, 1985.) This could be taken as support for the view that IQ directly reflects the speed and efficiency with which the central nervous system can encode material because the material to be encoded is so simple and well known that no environmental variable would be expected to influence its encoding. Thus, individual differences on IQ tests reflect basic biological processes because they mimic individual differences on something as simple as the speed of identifying familiar words and numbers, which might seem to provide an uncontaminated index of brain functioning. This is precisely the argument made by Eysenck (1982), Razz, et al. (1983), Jensen (1986), Itzkoff (1989), and many others (see the contributors to Eysenck's 1982 volume for further support). For example,

> Let us now consider what is the major import of the work here discussed. The major finding is that, along several independent lines,. . . . IQ correlates very highly (.8 and above without correction for attenuation) with tests which are essentially so simple, or even directly physiological, that they can hardly be considered cognitive in the accepted sense. . . . Thus we arrive at the astonishing conclusion that the best tests of individual

differences in cognitive ability are non-cognitive in nature! (Eysenck, 1982, p. 9)

Since these [reaction time] types of tests require at most a third grade education, it is the view of the researchers, given the high correlations of reaction time with IQ, that we are now assessing a deeper stratum of neurological, physiological reality in our research for the variable *g* of biological intelligence. . . . At this rate we may yet be able to discover the underlying neurological and morphological dimensions of intelligence. Eventually we may be able to predict educational, social, and cultural potential at quite early stages in a human being's development. (Itzkoff, 1989, p. 82)

However plausible such a straightforward biological view may be, there is an even simpler explanation: The encoding speed–IQ correlation is, at least to some extent, inadequate to demonstrate that individual differences in IQ are due to differences in central nervous system functioning (reflected in differences in encoding speed variance) because of the unproven assumption that individual differences in encoding speed are themselves due to CNS differences. Even though it would appear that the stimuli used in these studies are well known to all subjects, there can still be enormous differences in how they are mentally represented across various ability groups and how this representation may interact with specific procedures and practice. In the only studies of which I am familiar that have examined the effect of practice on encoding speed, the researchers reported reliable benefits due to practice in their experiments. Subjects become significantly faster encoding stimuli as they go through the lists. That is, even after the usual warm-ups, subjects substantially improve their performance with practice (Larson,. et al., in press; Razz, et al., 1983; Longstreth, 1984; Regian, et al., 1985; Tetewsky, 1988); and the critical interstimulus interval required to identify a word between its removal from the screen and the onset of a mask is highly sensitive to one's familiarity with the word, including its meaning (Rawling, et al., 1988). Therefore, encoding speed ought not, it seems, be accepted as *prima facie* evidence of central nervous system efficiency, because other variables such as familiarity with the procedures and stimuli can significantly affect encoding speed. Little is known about the nature of this improvement over trials and whether a rank ordering of individuals' encoding speeds after the usual period of warm-ups is similar to that found at the middle or end of the study, that is, whether there is heterogeneity of regression in the processing improvement due to item (or procedure) familiarity as a function of IQ level. Regian, et al. (1985) did find that all ability groups became faster at encoding, comparison, and rotation with practice and that these improvements generalized to paper-and-pencil ability tests. Unfortunately, the presence of practice effects or carry-over effects is seldom directly tested by those who construct aptitude tests or do basic information-processing research, although during the past several years Longstreth (1984), Tetewsky (1988), and Larson, et al. (1988) reported practice effects in their subjects' reaction times to judge whether two stimuli were the same or different or to make a choice. To the extent that these effects occur, they compromise the construct validity of a test (e.g., the extent to which an encoding measure is really measuring encoding) because they permit the intrusion of extraneous variables to account for subjects' correlated performances across items, even after their aptitude on the construct in question has been partialed out of their total performance. This problem of local dependence has been known for some time:

> The content of preceding items may influence responses on later items. Stimulus sets which have carry over or learning effects between the items are not locally independent. Unfortunately, this violation of local independence is rarely tested directly, although research has shown that context can influence estimates even for aptitude test items that have been developed in latent trait models (Whitely and Davis, 1975.) Furthermore, order of item presentation has also been shown to influence item difficulty (Sax and Karr, 1962), and similarly influence item parameter estimation in latent trait models. (Whitely, 1980, p. 107)

Additionally, Rabbitt (1986) has warned against drawing conclusions based on the interaction of processing speed and IQ. He rightly notes the nonlinear nature of the processing speed function, even when making within-group comparisons. The interaction that results from low-IQ groups requiring disproportionately more time to process difficult or degraded stimuli than they take to process simpler ones assumes that the scale is linear, e.g., that the difference between 100 msec and 250 msec is the same amount of processing time as the difference between 500 msec and 650 msec. But even though the time scale is arithmetic, the psychological function is logarithmic or exponential. A definitive test of the relationship would seem to require, among other things, a determination of the upper limit of an individual's encoding speed under high-incentive conditions, following ample practice. Otherwise, speed as an indicant of biological capacity is hard to separate from speed as a matter of personal style—i.e., some people prefer to process more slowly than they need to. Yet, to determine upper limits is itself problematic because it requires tailoring the presentation of the stimulus for each subject, using different durations and interstimulus intervals to control for baseline differences. This can have unforeseen consequences in a processing task because timing is all important. For example, the nature of visual pattern masking undergoes a change around 12 milliseconds after its onset, thus enabling quite different processes to be measured when differing interstimulus intervals are employed.

Finally, Stanovich and his colleagues have outlined a number of technical problems that are involved in isolating processing deficits in low-IQ persons (Stanovich, 1978; Stanovich and Purcell, 1981). They point out that high- and low-IQ groups differ on many variables that are extraneous to those being measured by processing tasks, but which can nevertheless influence their performance on such tasks. These variables include attentional states, report strategies, and differential familiarity with letter and number stimuli. They argue (persuasively, I think) that encoding differences cannot be inferred from group differences in the duration of exposure for a stimulus or from differences in the interval between offset of the stimulus and onset of the mask, when the latter is employed as a *dependent* variable. Rather, encoding parameters (such as duration of exposure and interstimulus interval) must be manipulated as *independent* variables, and an interaction between IQ and these parameters must be demonstrated to ensure that extraneous variables are not promoting group differences. To illustrate the force of this argument, Stanovich and Purcell (1981) point out research by others showing that the slower encoding of alphanumeric stimuli by low-IQ individuals is the result of reduced familiarity with these letters and numbers, despite their ability to readily name and recognize them when they were presented without a mask. When alphanumeric stimuli are made equally unfamiliar for both high- and low-IQ individuals by using Chinese characters, the so-called encoding deficit of low-IQ persons disappears. So much for a "hard-wired" biological basis for differential encoding by high- and low-IQ persons!

In short, it is not unreasonable to suspect that the speed-of-information-processing–IQ correlation is due to individual differences in familiarity with written letters, numbers, and words and correlated differences in identifying them. Two exceptions to this are, first, the aforementioned study by Walker and Ceci, 1985, in which the speed-IQ correlation remained significant after partialing out the influence of education. One hesitates to press the results of this exception too far, however, in view of the fact that the subjects were hospitalized psychiatric patients who were receiving drug therapy. Second, Jackson (1980) reported that undergraduates who were classified on the basis of their speed and comprehension of reading stories differed not only at encoding highly familiar alphanumeric stimuli but also newly learned characters. Because good readers could encode these novel stimuli faster than the poorer readers—even though all subjects had learned these stimuli to the same criterion (about 10 minutes of practice), Jackson was led to speculate that the differential "familiarity" with stimuli was an implausible basis for explaining individual differences in encoding. Although Jackson's research is highly creative and well-designed, it was conducted to address a different issue from that being discussed in this chapter. Hence, there are no data on his subjects' IQs. An extreme group design probably led to larger correlations, and there was no auditory analog task. (He did not find a reliable relationship between visual encoding of the novel stimuli and subjects' listening comprehension scores, suggesting that if he had used an auditory encoding analog, differences between high and low IQ may have disappeared.) Finally, Jackson's task cannot rule out certain forms of verbal mediation wherein good readers, who presumably have larger vocabularies, code the novel shapes more effectively and thus remember them better.

Consistent with the differential familiarity–practice hypothesis are several other studies. Before reviewing these, however, it may be informative to put together two lines of research that have not been combined previously, namely the "age-of-word-acquisition" and the "word-memory" literature. The former is replete with demonstrations that the earlier in one's life a word has been acquired, the faster it can be named later in childhood (see Baumeister, 1985; Done and Miles, 1988, for relevant citations). Moreover, this literature indicates that some of the poor performances of children with learning problems on naming tasks can be tied indirectly to their later age of acquisition of the words used in the tasks. If age of acquisition is viewed as a substitute for the amount of practice or familiarity children have had with words or with pictures that depict the word's meaning, then such effects would seem to suggest that nonneurological variables can significantly affect basic encoding speed. This inference does not imply that there is no neurological basis responsible for learning disabled children's later acquisition of certain words, but it does imply that it is the length of time one has known a word, not some neurological insult, that determines how fast the word can be named. Otherwise, why would children with learning problems perform as fast as nondisabled children when naming words for which they have equivalent familiarity?

Age of Acquisition and Memory

Now, let us add the age-of-acquisition findings to the word-memory literature. In the latter, we find that the physical properties of words, such as their length and the speed with which they can be pronounced, directly influence both children's and adults' ability to remember them (Hoosain and Salili, 1988; Ellis and Henelly, 1980; Lorsbach and Gray, 1986). There is a substantial correlation between a nation's speed of naming numbers and its' citizens digit-span memory for those numbers. For example, compared to their English-speaking peers, Welsh-speaking children appear to exhibit inferior digit-span memory and arithmetic performance, and this has been linked to the Welsh language's increased length of number terms (Ellis and Hennelly, 1980). It takes a lot longer to count in Welsh than in English, and this added time interferes with Welsh children's memory for numbers and also may be responsible for their poorer arithmetic performance. (To convince yourself of this, try to memorize a list of numbers, and then try to memorize the same numbers multiplied by one hundred. The longer time needed to verbalize the hundreds results in some decay and, therefore, in less accurate memory.) Similarly, the Chinese (Cantonese) language contains briefer number terms than does English, and so the time it takes native Chinese speakers to read multiplication tables is shorter than the time it takes someone to read these tables in English. As a result, Chinese digit-span memory is superior to that found among speakers of English (Hoosain and Salili, 1988). The latter researchers suggest further that national differences in duration of pronunciation may influence not only digit-span memory, but also the capacity for mental manipulation of numbers. If true, this would constitute an overlooked source of cultural influence on cognition.[2]

Earlier, I mentioned that researchers working with children who have learning problems have suggested that part of their inferior performance on word memory tasks can be tied to their age of acquisition of these words. If the mechanism underlying this poor performance is an effect of practice, then if poor learners are given sufficient practice with words until their naming speed is equivalent to that of their normally functioning peers, we can ask whether their memory for those words will then be equivalent, too. Lorsbach and Gray (1986) addressed this question, at least in part. They identified a list of words that was named much more slowly by poor learners than by their normally functioning peers and a list of words from the same taxonomic category (e.g., utensils) that was named as fast by both ability groups, although the specific words differed for the two groups. Lorsbach and Gray demonstrated that the poor learners' memory for the former list of words was significantly inferior to that of their normal peers, but their memory for the latter list of words was equivalent. A similar result has been reported by Bjorklund and his colleagues in a series of studies in which there was an attempt to equate the mental representations for typical and atypical category knowledge (see Bjorklund and Muir, 1989, for a review of this work). In general, it has been found that IQ-related differences in memory and organization disappeared when knowledge was equated across IQ groups.

By putting together these two lines of research, we can begin to imagine how a lack of both practice and familiarity with words and numbers could lead to slower encoding speeds for these words and numbers (and perhaps even lower IQ scores on the two most popular tests, since both digit span and arithmetic are part of what is tested on the Wechsler and Stanford-Binet IQ tests). If true, then part of the IQ-encoding relationship could be explained without recourse to the sorts of direct neurophysiological mechanisms that have been put forward by some, e.g., the signal-to-noise ratio in nervous system transmission or the integrity of neural circuits (e.g., Eysenck, 1988; Hendrickson and Hendrickson, 1982; Schafer, 1987; Razz, et al., 1983.) A number of recent studies have reported that familiarity continues to influence processing speed well into adulthood. Words encountered more frequently by adults in written and spoken language can be named faster and recognized and/or recalled better than less frequently encountered ones

(Jackson and Morton, 1984; Johnson, et al., 1985; Lorch, 1986). And it appears that a similar familiarity mechanism accounts for the superior accuracy of expert musicians at detecting musical structures and chess experts at detecting chess moves (Beal, 1985; Chi and Ceci, 1987). Finally, lest one forget, the words, numbers, and letters in these studies were familiar to all subjects, regardless of their age or ability; what was not the same was their *degree* of familiarity. For this, we need to go beyond simply assuring ourselves that all children can name them or in the case of Jackson's study, that all subjects can learn nonsense shapes well enough to recite them twice in a row. Finer grained measures of children's familiarity and representation of these seemingly simple stimuli are needed before we assume that they directly tap brain functions and this is the basis of their correlation with IQ.

With the merger of the literature on age of acquisition and that on familiarity and practice effects as a backdrop, let us turn to a few studies that illustrate the labile nature of the information-processing–IQ relationship and, in so doing, caution against a simplistic assumption that because they are related, the variance in both tasks directly reflects variance in neural functioning. Bosco (1972) has reported that middle-class children's encoding superiority over their lower class peers in the early grades (using a backward masking procedure similar to the ones used with adults) disappears by the time the children reach sixth grade. If the faster encoding speed of middle-class children is taken as evidence of some neurological superiority, then how do we account for its disappearance with increased schooling? Granted that one could claim that the ability to encode is genetically programmed to unfold in a time-locked manner, and those persons with better genetic endowments (middle-class persons) develop it earlier. But such explanations are unparsimonious, if not downright farfetched. Similarly, Bakker (1972) has reported that differences between good and poor readers' visual integration speeds (another microlevel measure of information-processing) at the early grades vanished by the later grades. Again, one could suggest that the basis for this narrowing of the gap between ability groups is due to biological or environmental causes, but my goal here is just to demonstrate the *possibility* of the latter, as many researchers reflexively assume that these microlevel information-processing parameters are impervious to environmental factors, and therefore, any link between them and general intellectual ability must lead inevitably to the conclusion that the latter is an index of neurological status because the former appears to be. These studies document the possibility that environment—specifically practice, familiarity, and schooling—can influence the encoding of words and numbers that are nameable by all children. Along these same lines, Ceci and Tishman (1982) demonstrated that the minimum interstimulus interval (ISI) needed to identify simple words decreased from age six to nine whereas the ISI needed to identify pictures of objects represented by these words did not. Thus, because verbal encoding speed appears to be susceptible to practice effects (including schooling, familiarity of stimulus, and modality of presentation), it is a problematic indicant of neural functioning, and therefore its relationship to IQ should not be the sole source of evidence for the physiological basis of IQ.

One final word in this regard. Presently, my colleagues Steve Cornelius and Narina Nightingale and I are in the throes of a lengthy examination of both inter- and intra-individual differences in the encoding of a variety of stimulus types (letters, numbers, static and dynamic shapes, words, and auditory pitch and duration). Subjects are asked to identify target stimuli, each of which is followed by a mask (a patterned mask in the case of visually presented targets, white noise in the case of auditorily presented targets). We have discovered one subject who performs at ceiling on the visually (but not the auditorily) presented targets. He is able to identify these briefly exposed targets with almost perfect accuracy, even when onset of the mask is at the minimum interstimulus interval (16 2/3 msec). In contrast, my colleagues and I required far longer ISIs (by a factor of two to three!) to achieve levels of accuracy comparable to that of this young man. Perhaps he is genuinely faster at encoding than the rest of us because of a more efficient nervous system. But one thing more that the reader needs to know before making this assumption is that the young man was the technician who programmed the visual (but not the auditory) experiments. He spent countless hours creating the graphics for each target and experienced thousands of informal "practice" trials in troubleshooting the program. Although the targets looked much like any other letters, numbers, and shapes, he quickly could detect their identity from subtle shape and contour information that had escaped our attention at brief exposures.

Since my initial experience with this task, I have personally experienced hundreds of practice trials in an attempt to make revisions to the program before the inception of data collection. Following this lengthy practice, I, too, became consistently able to encode at the fastest ISIs, a remarkable change from my original encoding performance levels several thousand trials earlier (which my assistants at that time only half jokingly took to be the confirmation they had long sought for my cognitive turbidity!). Although the study is as yet incomplete, it appears from our first set of analyses that our findings will mimic those of Keating, et al. (1985), discussed earlier in the chapter, and those of Detterman (1986) and Larson et al. (1988) that will be discussed in the next chapter, in that they do not indicate a positive manifold of correlations among the various encoding measures (Ceci and Cornelius, 1989.) In addition, we have replicated what Tetewsky (1988) reported in his doctoral dissertation regarding the lack of a relationship between psychometric measures and information-processing measures for highly familiar materials. As more studies like these are carried out, it may be that the arguments that have been made concerning the presumed singularity of mental ability (e.g., Brody, 1985) will require rethinking.

Taken together, the preceding studies seem to suggest that the speed of encoding stimuli, including even well known symbols, may be a direct measure, not of CNS functioning, but rather of an aspect of behavior that is shaped by processes associated with practice, familiarity, and motivation. A corollary of this line of reasoning, for which I know of no data to test, is that the correlation between encoding speed and IQ would be greatly reduced (if not disappear completely) if the stimuli used in experiments were themselves uncorrelated with schooling or other avenues leading to differential practice, e.g., if instead of using alphanumeric symbols and words, one is asked to encode the angle of rotation of a baseball thrown 90 mph and followed by a mask. If both encoding speed and IQ performance were indicants of CNS functioning, we would expect the correlation between them to be observed for all types of stimuli (not just alphanumeric ones) and that any practice effects would not diminish the size of the correlations. We are pursuing just such an experiment at present, and our hunch, based on as yet incomplete data, is that encoding—as that term has been traditionally used by psychologists—will be shown to be a narrowly conceived concept that is unstable across stimulus types and paradigms. Elsewhere, Whitely (1980) has described her own research demonstrating that the relative contributions of cognitive processes to test performance "are not similar across item types." Also, in her work with latent trace modeling with Schneider, she has shown that underlying cognitive processes may be highly specific to particular test items (Whitely and Schneider, 1981).

No matter how interesting one may find some of the foregoing speculations, another argument for the correlation between IQ and encoding efficiency can be refuted on the basis of already completed data. From developmental analyses, it has become clear that the underlying representation of stimuli used in many information-processing tasks, even stimuli as simple as numbers and letters, undergoes transformation with development (Chi and Ceci, 1987). These studies demonstrate that it is not sufficient for stimuli to be familiar to children of various ages; the stimuli must be shown to be represented in a similar fashion, too. In the absence of differences in the underlying mental representations of stimuli, developmental differences in basic cognitive skills are much rarer than had once been thought (Roth, 1983). For example, it is known that digits, although highly familiar to young children, are represented differently by them than by adults. While an older person may represent digits in terms of their ordinality, oddness or evenness, root properties, cardinality, etc., young children may possess only the dimension of ordinality. In addition, reference points may change with development (schooling), so that while an adult may assent to "97 is nearly 100, but 103 is not nearly 100" because it is past 100; for young children the reference points of the number system are different (Rosch, Mervis, Gray, Johnson, and Boys-Braem, 1976). It is plausible to conceptualize the underlying representation of stimuli as a process forged through experience with the stimuli; and schooling is obviously a significant part of this experience. The speed with which microlevel processes such as encoding can access such stimuli depend on the elaborateness of their representation.

Thus, on the basis of both the changes in encoding speed as a function of age and practice (Bakker, 1972; Bosco, 1972; Ceci and Tishman, 1982; Rawlings, et al., in press) and the developmental data showing changes in the underlying mental representation of even well known stimuli with development (Chi and Ceci, 1987), as well as on speculation based on work already under way, we can at least begin to doubt that the positive manifold finding is convincing evidence that individual differences in IQ tests are the result of individual differences in neural processes, unrelated to background familiarity and school-induced knowledge.

To close the chapter, while we have focused exclusively on the attempt to invalidate IQ as a direct indicant of neural functioning through its association with microlevel information-processing measures such as encoding and inspection time that some have argued are a direct reflection of central nervous system efficiency, there are other candidates that, although more global than encoding, have also been used to validate and interpret IQ scores. For example, Sternberg (1977) has provided a systematic and thorough analysis of verbal analogies, in terms of the underlying cognitive processes needed for their solution, and their individual relationships to IQ, and others have similarly analyzed deduction, spatial reasoning, word meaning, and reading processes. But so far, none of this highly interesting work has tried to make the same case that has been attempted in the literature dealing with microlevel information-processing measures, and none has shown exactly the processes that contribute to consistent individual differences in aptitude (Whitely, 1980).

CHAPTER 9

How Abstract Is Intelligence?

The ability to engage in abstract thinking has been heralded as one of the hallmarks of intelligence since the very beginnings of the testing movement and continues to this day to be seen as the centerpiece by researchers and laypersons alike (e.g., Cattell, 1971; Sternberg, et al., 1981; Terman, 1921). In fact, Jensen (1969), commenting on the difficulty of defining intelligence, stated, "If we must define it in words, it is probably best thought of as a capacity for abstract reasoning and problem solving" (p. 19). Moreover, in the child development literature, abstract thinking has been viewed as the pinnacle of intellectual development (e.g., Case, 1985; Piaget, 1976; Vygotsky, 1962). In short, it has been suggested that one can be knowledgeable but not intelligent, but one cannot be intelligent without also being abstract. But exactly what does it mean to be abstract?

THE RELATIONSHIP BETWEEN INTELLIGENCE AND ABSTRACTION

When someone with a low-IQ functions on what appears to be an abstract level, an argument is frequently made that such complexity can only occur for such persons within a highly elaborated domain of knowledge and never across the board. Thus, when low-IQ persons exhibit complex behavior, it is thought to lack abstractness. By abstractness, I assume it is meant that these persons do not detect and/or rely on principles or rules that underlie problems they are attempting to solve, because if they did, they would be better able to generalize from the solution at hand to similar cases in other domains.

That they do not generalize implies that they do not solve problems in terms of underlying rules or principles.

According to this line of reasoning, instances of complex behavior among individuals who possess low IQs are the result of their hypertrophied fixation on the contents of a single highly elaborated domain of knowledge (e.g., chess, work, sports, gambling, love). Fixation on a single domain allows one to accrue elaborate information which is often sufficient to solve even very complex problems in that domain. Accordingly, low-IQ individuals do not solve complex problems from the vantage point of having abstracted an underlying rule or principle, but from the vantage point of having learned a great deal of highly specific material that is yoked to highly specific solutions. They cannot transfer their skill to related problems in domains that are not well known to them, and therefore, their complex behavior is not thought to be "intelligent." (Incidentally, I find this line of reasoning somewhat ironic in that the cart has been placed before the horse; after all, achievement scores should serve as validators of IQ scores—not the reverse!)

The preceding argument begs the question of abstractness in two ways. First, it presupposes that high-IQ individuals *are* complex across different domains—something that bears challenge, I believe, since it has not adequately been demonstrated that they regularly do so in nonacademic or nonverbal domains. And second, the argument assumes that even if high-IQ persons *were* found to function on a complex level in a variety of domains, it is because of their greater invocation of abstract rules and not because of their greater possession of elaborate but highly specific information in each domain. Thus, some have argued that low-IQ persons' complex performances at the racetrack or at the town manager's task are qualitatively different from the complex performances of high-IQ persons, presumably because the latter's complex performances are borne out of deeply analytic, abstract principles and rules that can be applied across many domains, while the former's complex performances are due to the tremendous amount of rather inert, domain-specific information they have acquired through years of commerce in that domain. Low-IQ individuals who excel at problem solving in a given domain are, in effect, depicted as *virtuosi* of single things in life, much like the classic *idiot savant* is alleged to be.

I am troubled more by this abstractness argument than by any other in the field of intelligence research. There is very little empirical and theoretical work on this topic, and what work there is happens to be more relevant to "local" transfer tasks (Brown and Kane, 1989; Bossok and Holyoak, 1989) or has provided more complicated results (see, for example, Nisbett, et al., 1983, for a description of attempts to teach general inductive reasoning skills) than pertains to the preceding assumptions. Later, I shall discuss the transfer literature and its relevance to the issue of intelligence, but first I want to make a semantic point.

Transfer Within and Across Domains

Transfer refers to a recognition that various elements of one set can be mapped onto those of another set. Transfer research is a complicated field and there are numerous variations for mapping elements from one set onto those of another set, as well as numerous variations in constraints that can be imposed on these mappings (see Gholson, et al., 1988).[1] Transfer between two tasks may occur within the very same domain, for instance, from an appreciation of the scheduling sequence that is required to solve the classic "missionaries and cannibals" problem, wherein a single missionary cannot be left alone with two or more cannibals, to an appreciation of the routes that may be traversed to solve an isomorphic problem involving "jealous husbands and wives." (In the latter, it is never made clear whether a single wife, left in the presence of two men who are not her husband, also gets eaten!)

Transfer can also take place across two completely different domains. Although a mapping *can* be guided by the abstraction of a general rule, it need not be, and it often appears to those who do this kind of research that it is not. Brown and Kane (1989), for example, report the case of a child who learns to transfer without even being presented the problem simply because he had come to expect that the solution to all problems of this type required the same procedures, thus allowing him to answer before the question was even posed! Moreover, Gholson, et al. (1988) have shown that transfer from the problem of the three jealous husbands to the isomorphic problem of the three missionaries is superior than vice versa, whereas transfer from a problem dealing with two missionaires to an isomorphic one dealing with two jealous husbands is better than the reverse!

Clearly, the specifics of the stimuli have a large bearing on whether transfer occurs. This has been known for at least 20 years, since Birch and Bortner (1970) got children to transfer from a task using wooden dolls when they appeared unable to do so when geometric shapes had been employed (also see Cole, 1976 for a similar demonstration). So transfer can occur even when a subject does not appreciate an underlying rule, and this explains the less than pervasive nature of transfer when it does occur. So even if one can transfer without being abstract, one cannot, some have alleged, be abstract and yet fail to transfer, except under very unusual and artificial circumstances. I intend to probe this claim in some detail in the following pages, and, as will be seen, definitional issues become problematic.

Abstractness Defined

Throughout this century, researchers from such diverse fields as psychology, philosophy, and neurology have provided definitions of abstractness, but these definitions have been very different (e.g., Beaumont, 1983;

Jaques, 1978; Pikas, 1966). The first comprehensive empirical treatment that I am aware of was provided by Goldstein and Scheerer in their 1941 monograph entitled *Abstract and Concrete Behavior: An Experimental Study with Special Tests*. Based on observations of returning brain-injured veterans from World War I, these researchers began to assemble a picture of abstractness that was quite global. They termed it the "abstract attitude," to convey their belief that abstractness was something more than just cognitive; it pervaded the entire personality of the organism—hence the term "attitude." Veterans with frontal lobe damage who had difficulty with tasks that were thought to require abstract thinking (e.g., assembling a cube out of four smaller ones) convinced Goldstein and Scheerer that the frontal lobe was the seat of their deficit. This touched off a half century of debate and the ultimate refutation of their neurological explanation, as it was shown that patients with posterior lobe damage had similar difficulties on the Goldstein-Scheerer battery (Beaumont, 1983).

Despite the problems with their neurological model, Goldstein and Scheerer's description of abstraction continues to be influential. The abstract individual, according to them, is one who can shift perspectives flexibly, plan ahead, hold in mind various actions simultaneously, and analyze separate parts in relation to a whole. The concrete person, on the other hand, is someone who is unreflective, rigid, and yoked to an immediate perceptual experience. This last point gets at the heart of the distinction: Concrete individuals' performances are confined to their immediate perception of experience, whereas abstract individuals can go beneath the surface of perceptual appearances and infer an underlying rule. Goldstein and Scheerer (1941) regarded this as an essential distinction of the two modes of behavior and commented on the pronounced demarcation between abstract and concrete thought:

> The greater difficulty connected with the abstract approach is not simply one of greater complexity, measured by the number of separate, subservient functions involved. It demands rather, the behavior of the new emergent quality, generically different from the concrete. (p. 22)

But it should already be apparent from the examples given earlier that there are many ways to solve problems successfully, including ones that, despite their tremendous complexity, do not appear to satisfy the foregoing notion of abstractness. Recall Von Neumann's inability to perceive the simplest solution to the problem involving the two cyclists. He solved this problem using a very complex mental computation of mathematical sets because he didn't perceive the simple structure (i.e., was *not* flexible in shifting). And then there was the mathematical logician, Alan Turing's, seeming slowness to fathom the simple structural matter involving his bicycle's broken chain which he solved in a much more complex manner than was necessary. According to the preceding definition, neither man's behavior on these problems could be described as abstract, as they were quite inflexible in their solution path and ended up sacrificing time and/or accuracy.

The Role of Knowledge in Abstraction

I think that what often happens in this area is that there is a tendency to rate as "abstract" those solutions which those of us who are making the rating find to be personally difficult due to our own lack of relevant domain-specific knowledge—even though the actual solution path employed by the problem solver may be quite concrete (inflexible, yoked to domain-specific knowledge, etc.).[2] This can be seen by asking raters who are themselves mathematically unsophisticated to judge the degree of abstractness of students' solutions to problems involving exponential expressions. Most students are taught to solve such expressions by recourse to a plausibility argument of the form that, say, $3^4/3^8$ ought to equal 3^{-4}. As Jere Confrey (1988) has noted, what students seem to learn is not the underlying abstract rule but a positional argument that allows the transformation of 3^{-4} to $3\frac{1}{4}$. Because of this, they might impress raters who are themselves ignorant of this heuristic as being abstract when they deploy this notational change— that is, until they encounter examples that actually require a deeper understanding of exponential expressions. At such times, they can get into great difficulty even when they are trying to transfer within the same exponential domain. Yet some might imagine their solutions as reflecting abstract rules because they themselves do not have sufficient knowledge of the underlying rules to appreciate that these students are in actuality behaving rather concretely. A *bona fide* abstract approach to exponential problems would require the recognition that the rule underlying exponents is one that relates subtraction to the construction of additive inverses (negative numbers) and at the same time links division to the construction of multiplicative inverses (rational numbers). It would seem that few students ever approach true abstraction on such problems, despite what raters who lack knowledge may think. And still fewer students ever approach true abstraction that goes beyond the domain in question, i.e., abstraction that is *transdomainal*.

The more I have looked at this problem of defining abstractness in terms of the eduction of underlying rules and principles that are not tied to perceptual information contained in the problem itself, the more I have tilted toward a view of abstractness that is highly domain-specific and relies on elaborate factual knowledge as much, if not more so, than on underlying rule eduction. My current view holds that being abstract means having lots of knowledge relevant to solving a specific class of problems. This view is bound to displease many of my more cognitively oriented colleagues.

RACE AND ABSTRACTION: THE JENSEN STUDY

Perhaps one can criticize my jaundiced impression of the difficulties involved in judging abstractness and argue that most observers would have no difficulty judging solutions that require abstract reasoning versus those that do not. One way to see whether this is in fact the case is to ask individuals to rate problems according to how much abstract reasoning is involved in solving them. It was Jensen who fueled this debate on abstractness by arguing that "level 1 ability" involved untransformed, rote, associative thinking that lacked abstractness whereas "level 2 ability" was more abstract (and had higher heritabilities).[3] Accordingly, Jensen (1980) has claimed that racial differences in intelligence are the result of a lack of abstract ability on the part of African-Americans and it is level 2 performances that saturate g. While the genotype for rote, associative thinking is said to be distributed across all racial groups equally (indicated by similar performance on tasks that supposedly do not benefit from abstraction, like memorizing strings of digits), the genotype for abstract thinking is said to favor whites, according to Jensen. In his book, *Bias in Mental Testing*, Jensen presented what has become the standard evidence for this position, including a demonstration that African-Americans do relatively better on IQ items (and entire subtests) that do not require abstraction. For example, Jensen cited a study by Richard Figueroa and himself (1975) which found an item × race interaction that resulted from larger racial differences memorizing a string of digits backwards than memorizing them forwards, presumably because memorizing them backwards requires more transformation and mental effort.

Yet both Jensen and those who have supported his view rely on an untested assumption about abstraction, namely, that scholars looking at the items in question will agree which ones require abstract reasoning and which do not. Several of my Cornell colleagues and I have put this assumption to a test, and the results ought to give Jensen pause, as will be demonstrated.

The first test of Jensen's position occurred here at Cornell in 1981. Two colleagues asked 20 of their students to rank-order a problem set that included many of the IQ items found to favor whites, as well as those found to favor neither race. There was no consensus as to the degree of abstractness among the students' ratings of these problems (Boyce and Darlington, 1981).

Items favoring whites were not rated as being more abstract than those that did not favor them. Now, it might be argued that college students—even those familiar with psychological writings on the concept of abstractness—are not the best people to do the type of rating required in this test. Therefore, Narina Nightingale and I recently asked a group of truly eminent scholars to do this kind of ranking. We solicited the ratings of individuals who had as much right to opine about the nature of abstractness as any psychologist I know. These individuals, all eminent scholars who can claim to be working on deeply abstract levels in their own work, included a group of Nobel prize winners (mostly from physics, literature, and chemistry), as well as a group of other individuals who were acknowledged by those working in their disciplines to be truly outstanding. Mathematicians in the survey were recipients of the highly coveted Sloan Award, and linguists, and philosophers/logicians were members of the prestigious American Academy of Arts and Sciences; their peers had repeatedly nominated them in our survey as among the most brilliant scholars in their disciplines. We also asked past and present Poet Laureates, winners of Pulitzer prizes for fiction, and a number of eminent psychologists. A complete list of questions and respondents can be found in our recently prepared paper that describes the survey (Ceci and Nightingale, 1990).

We sent all of those surveyed a list of 32 questions and problems and asked them to rate them as to how abstract their solution processes were. The particular questions and problems were selected by us for a reason. They include those that are associated with the largest black-white differences, as well as those that are associated with the smallest differences on the most commonly used IQ test (Wechsler). They were chosen on the basis of their psychometric properties, including overall pass rates and canonical variates with Full Scale IQ (Sandoval, 1979, 1982; Jensen, 1977).

The survey results showed that not only did scholars disagree from one field with those from another field about the abstractness of the 32 items, but even within a given field of inquiry there was often a lack of consensus. For example, some respondents rated the question *What does "brave" mean?* as more abstract than the mathematical problem *A jacket that usually sells for $32 was on sale for $\frac{1}{4}$ less. When no one bought it, the store owner reduced the sale price by $\frac{1}{2}$. How much did the jacket sell for after the second price reduction?* Their judgment was that the latter problem was algorithmic and did not therefore require deep abstraction, whereas the meaning of "brave" did. Others, however, rated these two problems' abstractness levels exactly opposite. And such lack of consensus was found for most of the items in the survey. Interestingly, if the items are divided into those that are associated with the largest racial differences and those that are not, and then the mean ratings for each group are computed, the result is no difference in abstractness of the two groups.

I think the moral of the abstractness rating exercise is that if there is no agreement as to whether the level of abstractness involved in items associated with the largest racial differences is greater than the level of abstractness associated with the smallest differences, then how can one assert that abstractness is the basis for the reported racial differences? The results should serve as a caution to those who would glibly dismiss the complex performance of low-IQ persons (when it occurs) as somehow less abstract than it seems and simultaneously elevate the complex performance of high-IQ persons (again, when it occurs) on the grounds that the latter are more abstract in general. What *is* abstractness anyway?

From this discussion, it should be obvious that it is difficult to construe abstractness in a way that leads to agreement in its identification when it is manifested. When individuals exhibit what, for a given observer, is a high level of complexity, the most parsimonious explanation is that they possess sufficient domain-relevant knowledge to behave complexly, not that they are endowed with a greater capacity for being abstract. But does this mean that the cognitive skills involved in complex performance in one domain are unimportant in others? No, it means only that the relevant contextual elicitors need to be available in these other domains for them to be used. And these may be especially important for persons who are most lacking in knowledge, such as those with low IQs. The existence of high levels of complexity among low-IQ persons suggests that these forms of cognition are relevant to successful performance on a wide range of cognitive tasks (not only those involved in the specific domain of expertise), including certain subtests of IQ tests. That persons low on IQ can be highly complex on such tasks, but not on others that require the same forms of cognition, suggests the enormity of the contextual constraints operating on cognition rather than the qualitative differences between individuals' thought processes. No current theory, including information-processing, has paid sufficient attention to this issue, other than recognizing in a rather offhand way the importance of how one "codes" or "recodes" a problem.

The other side of this issue is equally interesting: That persons with high IQs can perform so much more poorly than low-IQ experts on certain tasks, even though they are equated with the latter group on relevant experience, also implicates the role of contextual elicitors. As stated earlier, the presence of a cognitive potential is, by itself, insufficient to guarantee its exercise. Kunda and Nisbett (1986) showed that statistically sophisticated individuals were unable to detect various statistical phenomena when they were couched in unfamiliar domains. Similarly, it has been reported that individuals reason more consistently across dissimilar problems that involve the same content (i.e., knowledge) than they do on similar problems that cut across different contents domains (Pellegrino and Goldman, 1983).

Findings such as these suggest that knowledge sets limits on problem-solving efficiency and is the basis for consistency of performance—when consistency is observed. Thus, not only is an appropriate rearing context needed for a biologically influenced cognitive potential to crystallize, and a certain level of motivation required to benefit from the context, but it is important at the time of testing to understand better than we presently do the impact of the testing context and knowledge domain on consistency of performance. Slight physical, semantic, and motivational changes from circles to butterflies or shapes to wooden dolls may lead to psychologically large differences.

GENERALITY

Whatever one's notion of intelligence entails, *generality* is apt to be a part of it. It is the concept of generality that separates the intelligent individual from the idiot savant who happens to have acquired some hypertrophied skill within a single domain. In turn, it is the concept of abstractness that is invoked to account for generality: To be able to generalize implies, to some, a certain degree of abstractness. *Generality-cum-abstractness* is offered as an explanation of the frequently reported psychometric finding that when one administers a battery of tests, an individual's relative rank is likely to remain fairly stable from test to test. And, generality-cum-abstractness is invoked to explain the allure of information-processing models of intelligence because fundamental cognitive processes are presumed to operate across domains— i.e., memory is memory, whether in the service of recalling digits or remembering dance steps.

These beliefs regarding generality and abstractness are dangerously close to becoming ideology in some research quarters, despite the fact that the evidence to support them does not appear to be sufficiently stout to withstand empirical scrutiny. Even if one forgets for the moment the results of the survey of eminent scholars mentioned earlier, there are still several additional bases for this conclusion.

There have been few attempts to assess the stability of basic cognitive processes such as memory or induction across widely differing domains and in those attempts that have been undertaken, the results are often inconsistent with the notion of a "transdomainal" deployment of processes. Keating, et al. (1985), Keating (1984), Herrmann and Gruenich (1985), Coles (1978), Lorenz and Neisser (1986), and Tetwesky (1988) have examined this issue by administering a battery of memory tasks and intercorrelating subjects' performances on them. Without exception, the evidence for the "memory is memory" position was not supported. In most of the studies, the correlations were practically zero. For example, Wilding and Valentine (1988) tested 10 adults who professed to be superior at remembering, on a battery of memory tasks, including facial recognition, digit span, free recall of words, and sentence recall. They found that the ability of these adults was quite variable, both in terms of the battery administered and in terms of their relative performances in relation to memory norms collected by others for these same tasks. That is to say, being superior (or inferior) on one type of memory task was not a good predictor of being superior (or inferior) on another.

The Effect of Training on Transfer

Such lack of correlation across tasks that tap the same ability or a highly similar one is not limited to memory. Mirriam Bossok and Keith Holyoak

(1989) have shown that there is considerable difficulty involved in training college students to transfer certain statistical rules (like the Poisson distribution principle), even when a great deal of support and practice is provided, and indeed, even when the transfer is "local," that is, within the same domain, albeit one that is probably not elaborately structured for these subjects. Their work calls into question the presumption that abstraction is a "transdomainal" process that is freely invoked by those with high IQs. In a similar failure to obtain generalization, Barry Leshowitz (1989) gave a broad cross-section of college students everyday problems that contained principles taught in introductory social science courses (e.g., need for comparison groups, need to control third variables): "Of the several hundred students tested, many of whom had taken more than 6 years of laboratory science in high school and college and advanced mathematics through calculus, almost none demonstrated even a semblance of acceptable methodological reasoning" (p. 1160). I should point out, however, that Leshowitz (1989) and Nisbett, et al. (1988) have found limited evidence that other kinds of statistical reasoning can be taught to university students and that some of them are able to transfer this knowledge to new domains immediately following training. In the Nisbett et al. study, however, after two weeks much of the transfer appears lost. So, given the difficulty of achieving transfer within some domains, it should come as no surprise to find that the generality of training across knowledge domains, although it can occur, frequently does not occur—which has prompted Carmi Schooler to comment that "the question for which we do have some empirical answers has to do with how generalizable cognitive training is from one subject area to another. As of now, the answer is not very much" (Schooler, 1989, p. 11).

Transfer and Cross-Task Correlations

Keating and his colleagues (1985) have also reported no significant correlations between three tasks that ostensibly involve the same mental spatial rotation operation (rotation of faces, rotation of geometric figures, and rotation of letters). If the same cognitive operation (e.g., mental rotation) is not correlated across such highly similar tasks, then there is little reason to expect it to be correlated with other cognitive operations like memory and inferring. And this is precisely what some recent research has shown (Herrmann and Gruenich, 1985; Keating, et al., 1985; Laufer, 1985; Lorenz and Neisser, 1986; Pellegrino, 1986).

It is hard to overstate the importance of this point. It drives home the need for researchers to expand their test batteries and settings to better sample the diversity of real-world contexts in which a skill may be exhibited. It may even be that when this occurs, we will discover a great deal more diversity in people's performance and begin to define closer "fits" between people and real-world work contexts than is possible with current *g*-loaded tests. (See Hunter et al., 1984, for a review of the state of the art.) That intelligence test scores fail to predict occupational success or social status when educational attainment is statistically controlled (Duncan, et al., 1972; Lewontin, 1982), together with what was learned about the role of education in fostering IQ scores in Chapter 5, leads to the conclusion that the psychometric findings are plagued with interpretative doubt:

> Given these considerations (i.e., failure to find a positive manifold), *could* we design a series of studies that would yield a pattern of results conforming more closely to the theoretical assumptions of componential analysis? By selecting experimental and psychometric tasks which are more similar, by systematically varying such factors as difficulty level and meaningfulness, by altering sample variability, and by considerable fine-tuning, one could almost certainly generate such a pattern of findings. But would they indicate the purer, almost physiological indices of ability for which some strive? Consider the theoretical value of such results. Their interpretability is compromised for precisely the same reason that factor analysis is an unreliable guide to mental structures: the tautology of presumed internal entities and the carefully structured and selected tasks used to discover and then "prove" their existence. We suggest that the more productive path lies in the direction of using. . . . real-world complex cognitive tasks: to study cognitive processing developmentally and contextually. (Keating, et al., 1985, p. 169)

As a final example of this argument, Detterman (1986) reports the results of several correlational analyses in which he demonstrated that the multiple correlation of a set of ten microlevel cognitive processes with Full Scale IQ is in excess of .65 and approaches .95 when the appropriate corrections are made for reliability and range size (median value = .81; with correction it jumps to .95). Yet the first principal component among these measures accounts for little variance—only about a quarter of the magnitude that is found when batteries of IQ tests are factor analyzed. Now lots of researchers have reported first principal components around .3–.4 (10–16% of the common linear variance among the performances), but Detterman went further by showing that these cognitive measures, although only moderately correlated with each other, could be combined to account for most of the diversity on IQ tests. If intelligence is a singular entity, then why do these microlevel processes (perception, encoding, memory, etc.), which together predict IQ better than some IQ tests predict other IQ tests, not intercorrelate higher? This and other analyses have led Detterman to the position that *g* (or the positive manifold that has been observed by others) has arisen out of the use of tests that are themselves quite heterogeneous in terms of the underlying microlevel processes. Each test item could be comprised of a variety of these microlevel processes, and they would tend to overlap with microlevel processes contained in other items. The first principal component would then merely reflect the average extent of this overlap. Because few of the microlevel measures are even modestly correlated with each other, despite being highly predictive of IQ when taken together, Detterman's findings add force to the idea of multiple statistically independent cognitive abilities as opposed to a single underlying mental force or resource pool (g).

In a similar vein, Larson and his colleagues have reported an interesting analysis of the relationship between simple and complex measures of information processing and their relationship to intellectual aptitude. A factor analysis of a battery of simple and complex reaction time tasks yielded principal component scores. The scores were entered into a stepwise regression analysis to predict scores on the hierarchical g factor derived from 11 psychometric tests. According to Larson et al. (1988), "This analysis appears to indicate that g, rather than being unitary, is a composite of more than one factor in these subjects". In other words, these researchers found multiple independent contributions to the size of g, and the simple reaction-time tasks that had been assumed to index biological efficiency were poorer measures of intelligence than were the more complex measures. This view is becoming widely accepted among psychometric researchers.

The literature reviewed earlier describes several cases in which high-IQ subjects were actually inferior at times to low-IQ subjects in solving complex problems, despite similar levels of experience and motivation (Ceci and Liker, 1986a; Dörner, et al., 1983). Now, if intelligence is general and abstract, then why would persons with presumably high intelligence (i.e., high-IQs) not be *at least* as quick to acquire a complex model that facilitates their chances of picking winners at the racetrack? Along these same lines, economists have long complained about the irrational fiscal behavior of seemingly intelligent people who persist at pouring good money after bad money, the so-called sunk principle. These persons reason that they have already sunk too much money into an economically decelerating investment to abandon it, even though a rational analysis would be to do just that.

Numerous other examples of seemingly intelligent people engaging in irrationale and even self-destructive behavior abound as was seen earlier. A favorite example of this is to mention to colleagues that you were browsing through a baseball record book (one of those almanacs that contain all of the truly trivial information that only a baseball buff would wish to know about), and you noticed that the New York Yankees lost over 70% of the games they played during daytime last year, whereas they won over 70% of their nighttime games. Then, you mention that you noticed that the Minnesota Twins also had lost about 70% of their daytime games last year. When I have done this, my colleagues have leaped in with grand theories to explain why daytime games are harder on *all* teams, what makes it harder for *all* teams to win games under natural illumination, etc. They appear to have lost sight of the fact that the percentage of daytime wins must be identical to that of nighttime wins, i.e., 50% and have fallen into a simple case of fallacious reasoning based on small numbers! For a different type of fallacious reasoning, I asked several colleagues at a professional meeting to estimate what they believed to be the likelihood of boarding a plane in which one of the passengers was a terrorist who was hiding a bomb. My colleagues' estimates ranged from one in ten thousand to one in a million. Next, I informed them that I had discovered a means of reducing the chances of boarding a plane in which a terrorist bomb might be hidden. Merely, always carry a bomb yourself. After all, what are the odds that two people are carrying a bomb on the same flight? Surprisingly, only one of my colleagues appreciated the fallacy implied by my solution. (One actually thought that my suggestion had tremendous merit!)

The Problem of Problem Isomorphs

Nisbett and Ross (1980), Leshowitz (1989), and Kahneman, et al. (1982) have amply documented the biased and simplistic modes of reasoning about social situations adopted by Stanford, Arizona, and Ann Arbor undergraduates, individuals known to become very competent at complex reasoning in at least some settings. Two of the notable "reductionist" biases that these students have been documented to exhibit are the tendency to accept single-cause attributions and their rejection of information that is inconsistent with a personally held causal theory. Recently, several researchers have provided a partial explanation of this paradox (Johnson-Laird, 1983; Olson, 1987). Olson demonstrated that the types of tasks and instructions that are typically used have led researchers to a needlessly ungenerous view of their subjects' capacity for reasoning, and by experimentally manipulating the instructions the researchers can often get their subjects to perform at a higher level. Related to this work, Johnson-Laird (1983) and Chi and Glaser (1985) have researched the basis of their subjects' differential performances on "problem isomorphs." Problem isomorphs are two analogous forms of the same problem that are cast in different contexts. A simple example of a problem isomorph is the following set of two formally identical problems:

1. $5 - 3.70 = ?$
2. How much change will the cashier at McDonalds return to you if your burgers, cokes, and fries cost $3.70, and you pay with a five-dollar bill?

The surprising result with problem isomorphs is that slight shifts in wording or experiential context frequently change the subject's performance considerably. In Johnson-Laird's study, a problem was first framed in the context of a playing card decision in which one must decide whether a rule is true; for example, if a card has a vowel on one side, it will have an even number on the other. Then they are shown four cards (e.g., A, B, 2, 5) and asked to turn over only those cards that are critical to verifying the rule. In this example, the best decision is to turn up the 5 and the A. College subjects have great difficulty with such tasks. Then the problem is framed in the context of a travel game in which one must decide whether to turn up a card with the name of a British town on it or the name of a mode of transport on it. The latter isomorph proved to be far easier for the subjects to deal with, even though it was logically identical to the former. Moreover, subjects who had greater experience riding trains (in England) or who had more experience with the towns in question performed even better. This result is obviously due to a combination of the greater concreteness and motivational value of the latter context for the subjects.

Recently, Nisbett, et al. (1988) studied over 1,000 students on various problem solving tasks similar to the one used by Johnson Laird. In one task, students were presented four cards with *disembark, not disembark, cholera,* and *diphtheria* written on the sides facing up. The goal was to turn over as few cards as possible to discover whether it was true that cards with *disembark* on one side had *cholera* on the other. Most students could not figure out they had to turn over both the *disembark* and *diphtheria* cards because, if the latter card had *disembark* written on its other side, they would have disconfirmed the rule. However, when students were told that the cards were airplane passengers who were permitted to disembark in foreign countries only if they had been inoculated against cholera, they easily figured out the optimal rule for turning over the fewest cards. Students who take statistics courses improve their reasoning, as do law students after taking contract law, vis-a-vis chemistry students who get no training in probability or permission theory. Again, it is an illustration of the importance of background knowledge in one's ability to reason.

The foregoing findings are not specific to problem isomorphs, but can be found in virtually every domain. A nice demonstration was recently provided by Logie and Wright (1988), who asked subjects to remember information that would be important to a house burglar. In addition to a group of "normal" adults, Logie and Wright asked another group to remember this information "from the perspective of the burglar." If you think their results indicate the superiority of the "burglar perspective" condition, you are wrong: Both groups remembered the information equally well (or poorly). But Wright and Logie also recruited a group of convicted house burglars to do the same task. Although an ordinary convict would probably not be expected to excel on memory or any other cognitive task, these house burglars certainly did. They recalled more significant details from the scene than did the two nonoffender groups. Similar effects for "experts" remembering more about things relevant to their expertise occur throughout the literature (Coltheart and Walsh, 1988). Again, the same conclusion reached in the problem isomorph example is warranted in these cases: Observation of subjects in a limited range of contexts leads to underestimating their mnemonic and reasoning skills.

Another piece of evidence against the generality-cum-abstractness assumption is that individuals who perform very well on academic tasks frequently do poorly on similar tasks if their personal belief systems get in the way of solving a problem. An example comes from Nisbett and Ross (1980). They cite a case in which individuals who know how to calculate the probability of two independent events co-occurring (by multiplying their individual probabilities) sometimes actually add the probabilities when the task is embedded in a context that permits their personal biases and stereotypes to come into play. It appears that these individuals are not very general in the application of their knowledge. In fact, as already mentioned, the subjects in much of the research reported by Nisbett and Ross were students at Stanford and the University of Michigan—groups who are known to score well above average on IQ tests. Yet the authors repeatedly document instances in which these students commit some rather basic reasoning errors (e.g., a failure to regress to the mean in their estimates, or discounting multicausal explanations), despite their ability to reason correctly along these lines in other contexts (e.g., in their courses). In fact, a mystery among some social psychologists who study inferential processes is how persons with such seemingly poorly developed problem solving abilities can function so well in most contexts outside the laboratory.

Finally, there is research showing the importance of specific social-contextual variables in evoking strategies (e.g., Ceci and Bronfenbrenner, 1985). Children and adults from middle-class backgrounds and with presumably at least average IQs often behave more or less complexly as a function of the context in which their cognition is assessed. Numerous cross-cultural researchers have made similar observations (e.g., Cole, 1975; Labov, 1970; Scribner, 1976; in press). Take, for instance, the logic behind the "law of large numbers" (LLN). According to the LLN, generalization should proceed in accordance with the rule that holds that large samples are necessary to generalize about the groups that are more variable on a given attribute than to generalize about groups that are less variable (Nisbett, et al., 1988). Adults seem to appreciate this rule in some contexts quite well (appreciating that a small sampling of a slot machine's behavior or a rookie's batting performance is inadequate for generalizing about their performance over the long haul). Yet, in other contexts, adults are quite unlikely to appreciate this rule (e.g., realizing that empathy or altruism expressed by someone at a party is a poor basis for generalizing about their manifestation of such traits over the long haul of a social relationship). Thus, notwithstanding the pervasive and often-stated assumption that cognitive tests measure important underlying abilities that generalize beyond the problems actually presented, the empirical picture is far less clear. This has prompted Rogoff (1981) to ask critically, after reviewing the cross-cultural literature on generality, "What conclusions about generality can be drawn from successful performance on a syllogism problem? That the individual (a) will do well on the next syllogism? (b) will do well on other kinds of logic problems? (c) will be logical in many situations? or (d) is smart?" (p. 271).

If the preceding arguments put forward to challenge genetic, psychometric, and information-processing theories of intelligence have lessened their explanatory power in ways not true of the bioecological framework, then this leaves only the task of explaining the fact that mental test performances are moderately intercorrelated, as well as frequently correlated with measures of everyday success (occupational prestige, income). In other words, a person who scores high on one mental test is likely to score high on all of them and, in general will be rated more successful in jobs and earnings. There are two subclaims to the claim being made here: (1) There exist substantial intercorrelations among the mental tests themselves, and (2) a (causal) relationship exists between mental test performance and everyday success. But the intercorrelations among mental tests can be explained as a result of the restricted range of these tests and the contexts in which they are typically administered. By now, it should be apparent that what one almost always means when discussing this phenomenon is that performance is consistent across a homogeneous group of academic-type tasks, assessed in a disembedded context with no attempt to "test the limits" of cognitive processes or manipulate the incentive structure. When the ecological context of the tests (as well as their content and administration) changes, so may the stability of performance. Earlier it was argued that schooling acts as a moderating variable: Perhaps the reason verbal and numerical performances tend to be correlated in this culture is because both of these abilities are nurtured in school. If our culture valued some other ability, for example, the ability to read upside down, then this would be correlated with verbal and numerical scores, too, because the schools would begin to teach it. Even when tests are readministered, but in different combinations, their "g-loading" can change dramatically, as Jensen (1980) himself has noted. As if this were not sufficient to dispell the myth of the unchangeable nature of g, it should be noted that covariation among mental tests has been shown to increase (and so does g, according to its proponents) as a function of age (Willis and Schaie, 1986).

To these conclusions regarding the disparate results of studies that have assessed cognitive abilities in a variety of contexts and the findings from studies that demonstrate multiple cognitive abilities, one can add the work of Detterman (1989), Keating, et al. (1985), Keating (1984), Herrmann (1985), Lorenz (1987), and others who have directly confronted the "positive manifold" phenomenon and found it wanting. With this last bit of evidence discounted, it would appear that no theory is capable of handling the diversity of findings reviewed earlier, unless it consists of the three prongs of

biology, environment, and motivation. An important feature of the bioecological framework has been to suggest mechanisms by which these three
factors combine to produce contextually tied performance (e.g., the elaborateness of the representation of knowledge and its role in producing cognitive
and developmental differences), as well as to provide criticism of the main
arguments put forward by information-processing, genetic, and psychometric theories.

CHAPTER 10

Taking Stock of the Options

In this final chapter, it is appropriate to ask how the bioecological account of intellectual development put forward here differs from other accounts. To begin with, it is inherently "developmental," unlike 7 of the 9 classes of theories with which it will be contrasted. A developmental approach permits one to assess the causal links, if any, between the onset of a given microlevel cognitive operation (e.g., encoding) and its later role in some macrolevel performance (e.g., IQ, induction, thinking, and reasoning tasks). Because of the importance of such linkages, it is necessary to study the. temporal course of the microlevel and macrolevel developments in information-processing. Another reason for the bioecological theory's developmental approach is that a central tenet of the theory is that age-related alterations in the organization of "local knowledge structures" are seen as an important constraint on intellectual functioning (Chi and Ceci, 1987; Keil, 1984). A developmental approach allows one to chart the interrelationships between structural changes in knowledge representations and concomitant changes in processing efficiency. The "ecological" part of the bioecological perspective requires a developmental orientation to explain why "period" effects and cohort effects are vital to explaining developmental outcomes. Independent of the microlevel-macrolevel relationship at any given time, it is crucial to know something about the larger contexts in which the organism grows up. This is why it was so difficult to be in one's 20s during the 1930s, as seen in the poorer outcomes of the older cohort of the Terman men (Elder, et al., 1989). Finally, the "bio" part of the bioecological perspective requires a developmental orientation to explain why the onsets and rates of arborization of neurons create enabling conditions for various cognitive competencies at various moments during the course of one's life.

Two of the nine classes of theories that will be contrasted with the bioecological theory are also developmental: structural theories (e.g., Piaget's and Case's) and knowledge-based theories (e.g., Chi's and Keil's). The bioecological theory is quite unlike these, however, in its "person × process × context" orientation, as well as in the range of its evidentiary base. As will be

TABLE 10.1 A Comparison of the Nature and Scope of the Bioecological Theory with Nine Other Theoretical Approaches to Understanding Intelligence

Nature of assumptions	Psychometric	Knowledge Based	Information Processing	Genetic/ Biological	Contextualist	Triarchic	Multiple Intelligence	Modular Theories	Structural Piaget/Case	Bioecological
Existence of a Substantial "g"	Mixed	No	Mixed	Mixed	No	Yes	No	No	Yes	No
Existence of Specific Abilities	Yes	Yes	Mixed	Yes	Yes	Yes	Yes	Yes	—	Yes
Biological Bases	Mixed	—	—	Yes	—	Yes	Yes	Yes	Yes	Yes
"Transdomainal" Processes	Mixed	No	Yes	Yes	—	Yes	No	—	Yes	No
Context Sensitive	—	—	No	No	Yes	Yes	Yes	No	No	Yes
Process vs. Product Oriented	Product	Both	Process	Product	Both	Both	Product	Process	Process	Both
Role of Motivation	—	—	—	—	Yes	Yes	Yes	No	—	Yes
Developmental	Mixed	Yes	Mixed	Mixed	No	No	Yes	No	Yes	Yes
Inductive/Deductive Balance	I	D	D	—	—	D	D	D	D	—
Scope of Evidence										
Sociological	—	—	—	Mixed	Yes	Yes	—	—	—	Yes
Anthropological	—	Mixed	—	Mixed	Yes	Yes	Yes	—	—	Yes
Historical	Mixed	—	Mixed	—	Mixed	—	Yes	Yes	—	Yes
Genetic	—	—	—	Yes	Yes	Yes	Yes	Yes	—	Yes
Educational	—	Mixed	Mixed	—	—	—	—	—	Yes	Yes
Microlevel Processing	—	—	Yes	—	—	Yes	—	Yes	Yes	—
Macrolevel Processing	Yes	Yes	Mixed	Yes	Yes	Yes	Yes	Yes	Yes	Yes
Role of Knowledge	—	Yes	Mixed	—	Yes	Yes	—	—	—	Yes
Reaction Time	Yes	—	Yes	—	—	Yes	—	—	—	Yes
Positive Manifold	Yes	—	—	—	—	Yes	—	—	—	Yes
Life-Course	Mixed	—	—	Mixed	—	—	—	—	Yes	Yes

Notes: "Mixed" indicates variability within an approach; for instance, some psychometrically derived theories postulate the existence of a larger g than others. A dash indicates that the theory either does not address the indicated assumption or evidence, or alternatively, that different versions of the theory make different claims.

seen, the other developmental theories focus almost exclusively on *process* (i.e., mental structures) and have relatively little or nothing to say about the role of context or individual differences. Table 10.1 depicts the various types of evidence that each theory attempts to address, as well as the assumptions it makes. My original goal in promulgating the bioecological theory was to attempt to address a much broader corpus of data than past psychological theories have done. Readers can judge for themselves whether that goal has been met. In the course of considering data and arguments from the disciplines of philosophy, anthropology, sociology, biology, and education, it became necessary to amend many of the assumptions and tenets of strictly psychological theories of intelligence; therefore, the bioecological theory has unfolded in a way that leads to a different pattern of assumptions and draws on different types of evidence than do most psychological theories.

CONTRASTING THE BIOECOLOGICAL FRAMEWORK WITH EXISTING THEORIES

In this chapter, I shall compare the bioecological framework with Piaget's (1952) theory of cognitive development, Case's (1985) neo-Piagetian theory of intellectual development, Keil and Chi's theories of ontological (knowledge) development, Gardner's (1983) theory of multiple intelligences, Fodor's modularity theory, and Sternberg' triarchic theory. In addition, Table 10.1 lists generic theories related to the psychometric, information-processing, and contextualist traditions for comparison. Although the bioecological theory is derivative of many of these, especially some aspects of Sternberg's and Keil's theories, it has aspects that are unique as well as features that are contrary to aspects of these other theories. Because some of the entries in the table are classes of theories (e.g., psychometric theories) rather than a single theory, it is hard to precisely compare the bioecological theory with them without delving into each variant at some length. Still, the scheme in the table should serve to point out broad differences between these accounts.

Contextualist Theories

Several related theories of intelligence assert the importance of the context in which intelligent behavior unfolds. As pointed out by Sternberg (1985), these theories can be arranged from the "radical cultural relativist" view of Berry (1980), which states that indigenous conceptions of intelligent behavior are the only valid basis of assessments of cognitive complexity across groups, to the least radical contextualist views that combine psychometric approaches within a contextualist orientation (e.g., Baltes and Willis, 1979). In between these two positions are a variety of others, including

Sternberg's own, which we shall discuss separately in view of its scope and importance.

The bioecological framework shares features with all of the contextualist positions, especially Charlesworth's (1979) view of adaptation and Cole's (1975) cognitive-anthropological approach. Like all of these theories, the bioecological framework proposes that cognitive competence is honed by, and sharpened through, efforts to surmount one's most important environmental challenges. Hurdles that recur and are of substantial interest or survival value to the organism are seen as fostering intellectual development through the recruitment of strategies and the integration of structures that were initially independent. The bioecological theory leaves open the possibility of assessing intelligence across various cultures, provided that the domains of knowledge required to behave intelligently in such a cross-cultural comparison are equivalent in their structure and meaning, have similar attainment value across the cultures, and are experientially similar in regard to the operations required to access them. Such equivalences are possible, at least in principle, because cultural groups often interact and certain values and experiences are presumably shared. For example, it might be possible to compare two culturally different persons on the same task, say, solving a soccer problem, if each of their cultures provided experience both in this domain and the operations required by the task, as well as if the values associated with performance on the task were similar.

Where the bioecological theory differs from these contextualist theories is in its scope, developmental orientation (described previously), and motivation. With the exception of Sternberg's (1985) theory, the contextualist theories are not well developed beyond their emphasis on the role of context. They provide interesting insights into the ways in which cultures organize the experiences of their members and consequently affect the manner in which they perceive and solve problems of importance. But because these theories were formulated in response to a prior era's uncritical acceptance of deficit models that emanated out of research comparing non-Western individuals on Western-style tasks, they seldom have gone beyond explication of the external world's influence. They do not explicitly describe either the "inner cognitive worlds" of people or the transition mechanisms that lead to progressively higher levels of cognitive functioning. That is to say, within every culture that has been studied, there have emerged individual differences among its peoples in solving various types of everyday problems. Researchers who have studied even the most traditional hunting and gathering societies have remarked on the presence of differences within the group—differences sufficiently salient that the members of the group use labels like "stupid," "wise," and the like to describe its members (Edgerton, 1981; Reuning, 1988). It would seem that contextualist accounts would need to go beyond the external world to explain these intracultural individual differences. Specifically, these theories must provide an analysis in terms of individual differences of the mechanisms involved in transforming sensations into cognitions, that is, the interaction between knowledge and the information-processing components described earlier. Therefore, although the bioecological account is a contextualist account of human intelligence, it is more than that: It is also an account of the cognitive mechanisms involved in intelligent behavior, the role of context in their crystallization, and their biological underpinnings. These differences originate out of the focus of the bioecological framework on individual and developmental differences within similar cultures.

Information Processing Theory

Because information processing is more a methodology (derived from task analysis) than a theory, it has been helpful in identifying isolated mechanisms by which individuals solve problems, sometimes in different ways. We now know that a great many competencies that Piaget thought were absent from young children's thoughts are in fact present. And much of this work was undertaken by researchers within the information processing camp. But a shortcoming of information processing approaches to intellectual development is their incompleteness and "disembeddedness": There is no agreed-upon nexus between the information-processing parameters and their biological substrate, nor is there a description of the contextual mechanisms by which these parameters are initially shaped and later elicited. Finally, there is no currently agreed-upon structural representation of the interrelatedness of various microlevel and macrolevel processes (see Rabbitt, 1988, for a few alternative models). Some of these shortcomings are the result of information processing's status as a methodology rather than a theory. Certainly theorists like Sternberg (1985), Keating (1984), and Hunt (1985) have been able to incorporate information-processing parameters into their accounts of intelligence. But by itself, information-processing should not be expected to provide a self-contained theory of intelligence. The bioecological framework is concerned with information processing, although it stipulates that the efficiency of such processing is a function of context, both during intial development and later at the time of testing, as well as the structure of knowledge that gets processed. It is possible to infer that an information processing mechanism is intact if it is required in two different tasks, even though it may be manifested in only one of them. The bioecological framework attempts to account for these contextual and knowledge-based determinants of information processing efficiency.

Several information-processing approaches to intelligence have already been discussed in some detail in Chapters 1 and 7, and several more (Sternberg's, Fodor's, and the bioecological) will be dealt with individually in this chapter .

Structural Theories: Piaget's Theory and Case's Theory

Piaget's theory has long been praised and criticized for the same reason: It depicts intellectual growth through a series of stages and resultant structures that are genetically programmed to unfold in response to what are normally thought to be universal experiences, in invariant order, although not invariant rate. Piaget saw development as a "structured whole," with a singular, integrated intellect always orchestrating all of one's cognizing, regardless of the specific domain involved (moral, spatial, linguistic, esthetic, mathematical, or whatever). Piaget's theory assumes that the structures that comprise the intellect are powerful "transdomainal" algorithms that are "blind" to content (cf. Chi and Ceci, 1986; Gardner, 1983; Keil, 1984): Once a logico-mathematical structure is in place, it can be recruited by any task that requires it.

In response to Piaget's claims, it has frequently been pointed out that there are problems having to do with *horizontal décalage*, wherein variation in the deployment of a structure across different domains of knowledge is a frequent occurrence (e.g., the attainment of number conservation precedes that of liquid conservation by as much as several years, despite the ostensible similarity of the operations; and furthermore, conservation of number may appear earlier with some stimuli than with others—see Gelman, 1979).

Missing from Piaget's conceptualization of intellectual development is the acknowledgment of the role of context in the manifestation of independent biologically constrained abilities. Although Piaget does not explicitly dismiss the role of domain-specific experience in accounting for age differences in causal reasoning, his writings imply that specific experience with objects used in his tasks should have minimal impact on performance (Piaget, 1953). Yet Berzonsky (1971), Nass (1956), and others have called into question this assumption. They have shown that six- to eight-year-olds reason more "animistically" or "magically" when the problem is unfamiliar to them than when it is familiar (e.g., being asked why clouds move rather than why cars move). When objects are familiar, children's reasoning is more sophisticated. In her interesting monograph on conceptual change, Susan Carey (1985) has made a similar finding, viz., that children's reasoning about animistic concepts is confused by their lack of biological knowledge. (See also Glaser, 1984, for a review of failures to find cross-situational reasoning skills.)

This lack of knowledge on the part of children, including that of linguistic conventions, can be seen dramatically in a study by Aebli (1987), who found that six-year-olds did not transfer their faulty knowledge on a liquid conservation task to a familiar practical situation that required them to add water to get a drink. Over 80% of the children who made a conservation error by claiming that water in a short, wide container had increased in quantity when it was transferred to a tall, thin beaker, did not assert that transferring the water to the tall, thin beaker fulfills the need to add more water in the practical task of getting a drink of water. Aebli suggests that it is the artificial nature of the task that prompts the erroneous judgments; the error is invented by the children *ad hoc* and does not reflect underlying cognitive capacities. He claims that the standard conservation tasks are so disembedded from children's ecologies that they provide little insight into their thinking.

Such demonstrations have apparently persuaded former co-workers of Piaget to acknowledge explicitly the crucial role that context plays in reasoning. Recently, Barbel Inhelder, Piaget's long-time collaborator, wrote of the new contextual orientation in Piagetian research:

> The new approach was meant to help in getting closer to a real subject, including his intentions, plans and means for controlling his action. . . . This research is in the field of pragmatics, i.e., the subject's construction of cognitive tools that are adapted to a specific situation. . . . The idea here is that discoveries take place in particular contexts. (Inhelder and DeCaprona, 1987, p. 11)

In addition to its failure to consider the impact of context on cognitive attainment, Piaget's theory dealt so much with logico-mathematical structures that it had little to say about the full range of abilities known today to exist independently of these, such as inferences about social situations, everyday problem solving, and causal beliefs. As Gardner (1983) has noted: "It is poignant that Piaget, a biologist by training who believed he was studying the biology of cognition, should have been so insensitive to diverse biological proclivities in the cognitive realm" (p. 322).

The bioecological framework shares with Piagetian approaches the emphasis on the individual's construction of structures as part of his or her efforts to adapt to important environmental challenges and its endorsement of the view that some developmental epochs are more likely to be associated with the crystallization of certain cognitive potentials than others. Where it differs, of course, is in its explicit sensitivity to the role of context and specific knowledge domains that constrain one's use of reasoning algorithms. Whereas Piaget's theory is concerned with broad (some might even say "overarching") normative descriptions of individuals at a given point in their development, the bioecological framework is a response to concerns about individual differences at any point in development. So without minimizing the importance of normative mechanisms, such as disequilibrium, for explaining cognitive growth, the bioecological framework asserts that these types of mechanisms are invoked by individuals in response to changes in their highly personalized knowledge domains and in response to their environmental challenges. Contrary to Piaget, social forces do more than simply pace intellectual development: They are capable of altering its inherent structure and outcome (see Chapter 3).

Case's Theory of Intellectual Development

Robbie Case has provided the most ambitious revision and extension of Piaget's theory to date (Case, 1985). Comprehensive coverage of Case's theory would carry us beyond the scope of this monograph; hence, I shall describe only those aspects that are most directly relevant for a comparison with the bioecological theory.

Like Piaget (1952) and Baldwin (1894) before him, Case finds the evidence for stage-like developments in children's thinking quite compelling. Children of each age group are seen as deploying different, but characteristic, forms of thinking for that age. Case reviews research by his students and others that suggest that age-related differences in children's thought processes reflect the development of a sequence of increasingly sophisticated mental structures. These structures are *somewhat* similar to the developmental epochs (each with its own substages) described by Piaget and others (5 months, 20 months, 5 years, and 11 years). Case describes transition mechanisms that permit movement from stage to stage, ending up with an account of intellectual development that has a structure governing both within-stage and between-stage developments:

> Children go through the same sequence of substages, across a wide variety of content domains, and they do so at the same rate, and during the same age range. In short, it has been suggested that there is a great deal of both vertical and horizontal structure in children's development. (Case, 1985, p. 231)

The transition mechanism most central to Case's theory is the size of what he terms "short term storage space" (STSS). Case distinguishes between a biologically constrained capacity to store information and the availability of strategies that facilitate such storage. He argues that the former (the biological basis of STSS) constrains the amount of information that can be stored in any domain, while the latter is domain sensitive. Thus, he nicely establishes an apparatus that will permit both vertical and horizontal change. Case sees *total* capacity as a combination of the amount of "operating space" and STSS. The former is that portion of storage space required to carry out mental operations (e.g., counting). As children increase the efficiency of these mental operations, use of the operations requires less and less storage space, thus freeing more and more of it for STSS of the material being operated upon. If an operation or algorithm requires a lot of storage capacity, then little residual space is available for short-term storage of material. In a series of five experiments, Case and his students demonstrated that when there are no developmental differences in the amount of storage space required by an operation, STSS is equivalent across a wide age range, and hence, there are no age-related differences in memory.

Like Piaget's theory, Case's was not crafted as an account of individual differences, although several of his tenets are important for just that. The aspect of Case's theory that is most relevant for us to consider has to do with the notion of cognitive asymmetry across domains, or what some developmentalists refer to as "horizontal décalage." Because Case, like Piaget, endorses a view of development characterized by age-related logical acquisitions, a concept such as conservation, once acquired, should be applicable across a variety of content domains. In the past, many developmentalists had attacked this aspect of Piaget's theory because of evidence that children acquired the same logical concept in some domains long before they acquired it in others. One already mentioned example of horizontal décalage is the finding that children can apply the operative knowledge needed to conserve numbers several years earlier (and theoretically, even before they possessed the requisite "reciprocal" logic) than they can to conserve liquid or volume. Similar instances of décalage have been demonstrated for the ability to deploy decentering, reciprocal compensation, etc. (Fischer, 1980; Gelman, 1979; Kuhn, 1983).

Developmentalists responded differently to these purported counterexamples. Some Piagetians and neo-Piagetians maintained that the basis for horizontal décalage lay at the level of measurement; i.e., the tasks being used to assess conservation, decentering, etc., were flawed in some way. Non-Piagetians like Fischer (1980), Gelman (1979), Klahr and Wallace (1976), Deutsche (1937), and Grigsby (1932) responded by challenging the notion of horizontal structure and by entertaining the idea of domain-specific cognition. (Presumably, a child's interests, experiences, and aptitudes combine to determine his or her particular age of conceptual acquisition within each domain.) Case makes the asymmetry across domains work to his advantage by analyzing each situation for the complexity of the executive control mechanisms involved. On the basis of an *a priori* analysis of the executive processes involved in various tasks, Case is able to predict the age at which each will be successfully completed. By emphasizing the growth in STSS that results from more efficient use of operations and algorithms, and the domain-specific nature of the operative and algorithmic knowledge, he avoids the aforementioned criticisms of Piaget while continuing to portray cognitive development in an age-normative manner, with cognitive changes coming in a sequence of stage-like developments.

In terms of a theory of individual differences, Case's major contribution is in departing from Piaget's description of mental structures as domain independent. Instead, Case explicitly characterizes the structures associated with each stage as "domain-specific devices for achieving executive control" (Case, 1985, p. 259). He argues that the different ages of acquisition of knowledge about the various conservation tasks are due to the fact that these tasks are not equally representative of the conservation concept. Instead of reflecting horizontal décalage, the conservation dilemma really demonstrates vertical development, with conservation of number (the earliest acquisition) requiring only an appreciation of variation in one variable as a function of variation in another one, i.e., $X = f(Y)$. On the other hand, conservation of substances that are continuous, such as liquid, require a more complex logic of the form $X = f(Y \bullet Z)$, in which the two variables X and Y can be traded off against each other. According to Case's analysis, children can acquire concepts at any age, but their level of understanding those concepts is tied to their form of logical structure, which itself is age related. Thus, developments of logical structures will result in a better understanding of conservation *per se*, not in the ability to tackle all types of conservation problems. So, almost by sleight of hand, Case has finessed the thorny problem of décalage by arguing that Piaget was mistaken in his belief that logical concepts were closely tied to the nature of their underlying representations. Rather, Case suggests that logical concepts such as conservation (as well as abilities like decentering) can be acquired at any age, but that within a given concept or ability, one can identify a range of levels that are of increasing sophistication.

Carried to its extreme, Case's position is similar to that of the bioecological theory. But it must be emphasized that carrying it to its extreme would obviate the need to posit logical structures that are stage related; the determining factor would be the nature of the knowledge representation, rather than *age* or *logical structures*. Consider, for example, the conservation of continuous substances like area or liquid. Within plane and solid geometry, it is valid to say that "whatever liquid goes into a container must either come out of it or be absorbed"; that is, conservation is operative. But can anything comparable be said about conservation of other continuous substances? Consider the conservation of "area" as a counterexample. If one forges a 22 inch piece of wire into a perfect circle with a 22-inch circumference, it will encompass 38.5 square inches. (Try it.) But what will happen if the same 22 inch piece of wire is reconfigured into another shape (e.g., a triangle, rectangle, or trapezoid)? Will area be conserved? The answer is no. (For instance, a 10" × 1" rectangle will encompass only ten square inches even though its perimeter is 22 inches.) Obviously, there are two parameters being covaried (length and area), only one of which is conserved. Area is conserved whenever the number of square inches contained in a given shape stays the same when the shape is reconfigured into other shapes. The fact that this insight is often lacking in adults is evidence for the domain-specific nature of concepts.

Whether stage-related logical structures do indeed exist, and if they do, whether they are "transdomainal," is a matter for further empirical testing. Regardless of the outcome of such testing, the bioecological framework is still clearly different from Case's (or Piaget's) theory. It is more "componential" than either theory in its attempt to discover the kernels that underpin everyday macrolevel cognitive performance. Thus, it views logico-deductive reasoning as neither the epitome, nor even the most important state, of everyday thinking. Deduction and axiomatics, although able to accurately model knowledge for which one is given valid premises, is not seen as the means by which knowledge is constructed by individual children or even by scientists. (See DiSessa, 1984, for a detailed discussion of this point of view.) Moreover, the bioecological framework tends to look in different places than does either of these theories in accounting for developmental or individual changes. It is less process oriented, emphasizing as it does the importance of the knowledge representation. In the accounts of Piaget and Case, the mental blueprints that children develop (i.e., habitual ways of representing a problem) are largely acontextual and knowledge independent, except for rudimentary perceptual knowledge.

While Case has avoided the problem inherent in Piaget's account of horizontal décalage, the bioecological framework, because of its knowledge-driven nature, would suggest that horizontal structure exists for some children and not others, and the reason is only partly a function of the structural developments discussed by Case. An important omission is that knowledge becomes increasingly elaborated to allow new insights that earlier had been opaque. For example, perhaps the problem of low intertask correlation (conservation of area being acquired years later than other types of conservation) is due to the nature knowledge of areas. As the previous wire example shows, area does not appear to many adults to be conserved in all contexts. On a more esoteric level, neither mass nor distance is conserved across all contexts either; they depend on certain physical assumptions about linear surface geometry. A "law" of conservation of mass or distance can be disconfirmed by those who are knowledgeable about spheres within moving planes. Again, we return to the domain-specific nature of one's knowledge.

Knowledge-based Theories of Intellectual Development

Despite the similarity with Case's and Piaget's theories in some respects, a knowledge-based theory of intelligence differs from theirs in several important respects. Since the foci of these two classes of approaches are sufficiently different, they end up with opposite expectations in some instances. In knowledge-based theories, *a priori* constraints on the way knowledge is represented influence the way in which new learning proceeds. Many psychologists hold such views, of course (e.g., Brown, 1975), and in the past they have attempted to explain intellectual and developmental differences on the basis of general developments that are assumed to hold for each domain. But Keil (1981, 1984), Chi (1978), Chi and Ceci (1986), and others go further than this by recognizing that there exist differences in the way knowledge is represented in "local knowledge structures" or domains. For these researchers, the notion of knowledge-independent intellectual processes holds for very few developments. Of course, the bioecological framework incorporates this assumption also, as has been seen.

As an example of what I mean by the phrase "knowledge-independent structures," consider that third-grade children will usually correctly insist that the answer to the following deductive statement cannot be known: "All football players are strong/ This man is strong/ Is he a football player?" Yet adults will insist on an affirmative answer to a similarly structured deductive statement: "All oak trees have acorns/ This tree has acorns/ Is it an oak?" (Nisbett, et al., 1988). To assume that there exists a disembedded cognitive skill called deductive reasoning, which operates with uniform efficiency across knowledge domains, does not accord with such data. Interestingly, if one's knowledge of oak trees includes a belief that only oaks have acorns, then the deduction "invites" such implicit premises, which is another way that knowledge can influence reasoning.

According to the knowledge-based view, the actual dynamics of change (the mechanisms that lead from one cognitive state to the next) result from the way in which knowledge is structured, with a premium placed on its degree of differentiation. Such an account is at the opposite end of the process-knowledge spectrum from Piaget's and Case's, as well as from Sternberg's, in that it argues that cognitive operations and algorithms are often present and operable in some domains of knowledge but inoperable in other domains. For example, what explains development for Keil and Chi are the advances in the representation of knowledge that make possible the detection of new relations and, hence, the deployment of existing operations. The "process bias" of both Piaget and Case (general developmental growth occurs as a result of attempts to accommodate the cognitive system in the face of disequilibrium) is for Keil and Chi often an instance of "local" knowledge changes, thus making it possible to notice the inconsistency that resulted in disequilibrium in the first place. Chi bases her account on the gradual alterations to the knowledge base that unfold in the course of gaining expertise while Keil makes his argument on the basis of his analysis of the developmental changes observed in four areas: comprehension of metaphor, detection of inconsistency, a characteristic-to-defining shift, and the influence of boundary conditions. For instance, in the case of metaphor comprehension, Keil suggests that children are able to comprehend some metaphors long before others. The controlling variable is the degree of elaboration of the specific domains of knowledge involved in comprehending the metaphor:

> When two semantic fields become sufficiently differentiated that a common set of relations between them can be perceived, a whole class of metaphors becomes comprehensible. Since the child is able to juxtapose some semantic fields to perceive common relations much earlier than other fields, the development can not be described as the emergence of a

general metaphorical skill. Thus, young children are usually able to perceive metaphorical relations between animal terms and automobiles ("the car is thirsty") but no relations between human eating terms and ways of reading a book ("he gobbled up the book"). . . . It would be a mistake to infer that most metaphor development is a consequence of general shifts in processing ability. . . . highly intelligent adults also fail to comprehend metaphors if one of the domains is unfamiliar to them. (Keil, 1984, pp. 84–85)

Both Keil and Chi, despite some important differences in their theories, agree that traditional accounts of cognitive development that are predicated upon the occurrence of "transdomainal" changes in representation or ability (e.g., the acquisition of a new reasoning algorithm) do not withstand empirical scrutiny. Although many others have commented on the similarity of children's and adults' thinking—children's at first being more tied to specific domains of knowledge (Bullock, 1981; Fodor, 1973)—Chi and Keil have provided a fine-grained analysis of purported instances of "transdomainal" development. They almost always reject that position in favor of the view that the ability in question was already present, but could be expressed only in a limited range of conceptual contexts. In describing the results of his analysis of one such area of presumed "transdomainal" development, that of the shift from exemplar-based representations (representations based on highly *characteristic* features of a concept) to algorithmic representations (those based on a concept's *defining* rules), Keil notes that

The shift occurs at different times for different concepts, suggesting that it is determined primarily by the structures of the concepts themselves rather than by a general transition from instance bound knowledge to more rule-governed knowledge. . . . The shift occurs at widely differing ages for the different conceptual domains, but at roughly the same time for the terms within each domain. Clearly then, a primary determinant of when the shift occurs is the degree of knowledge the child has of a particular domain. . . . (thus) one cannot speak of learning mechanisms in isolation, independent of the content of the knowledge they operate on. This point of view follows from the assumption that the complexity and richness of human cognition lie much more in prestored knowledge structures than in general-purpose computational routines or procedures. (Keil, 1984, pp. 89–90)

The bioecological framework is predicated on some of the same knowledge-based assumptions described by Keil (1984) and Chi and Ceci (1987), although it goes further than these approaches in a number of respects. Importantly, the bioecological framework that has been put forward is a complete *person × process × context* model. It attends not only to characteristics of the stimulus domains (i.e., the context, including knowledge), but to characteristics of the person (i.e., biology) attempting to construct these domains and process information within them. Because of this broader function, it addresses a far wider body of phenomena, such as biological developments and sociological influences. And, rather than asserting that a representational shift occurs simultaneously for all terms within an entire class of concepts (e.g., as Keil seems to do with "nominal kind" terms from a given domain, such as kinship or morality, which possess highly salient characteristic features and clear definitional features), the bioecological framework anticipates a more gradual incremental change. According to the bioecological framework, it is an unproven assumption that the characteristic-to-defining shift will occur around the same time for all terms within a domain (even though it may not be tied to shifts in other domains). It would seem an open question as to whether this claim really captures the richness of the emergent knowledge base. It may be that even terms *within* the same domain will shift at different times whenever their definitional features are complex enough to be acquired gradually over a long period of time. For example, children may treat land-water relations as defining of islands, but fail to appreciate that climatic features can also be defining. The point is simply that knowledge can accrue gradually, and it is possible that we may discover sizable asymmetries in a representational shift within a given domain. If this were so, it would call into question the very usefulness of the idea of a representational shift, because an explanation couched solely in terms of the gradual accumulation of semantic features would be more parsimonious. Since the bioecological framework maintains that the structure of the knowledge base sets important constraints on the efficiency of existing cognitive processes, it is possible to find processing differences even within the same domain if terms within it are elaborated to different degrees.

Theories of Multiple Intelligences

Long before the time of Brentano and the "act psychologists," it had been popular to posit discrete mental faculties or factors of the mind to account for specific types of cognizing (Boring, 1953). Certainly, Gall and the phrenologists were on firm footing when they postulated the existence of isolable neurological systems that subserved different types of processing, even if they were discredited for some other aspects of their theory. Today, neurological research has confirmed the importance of different neuroanatomical structures for various types of cognition. Modern replacements of faculty psychology and phrenology can be found in the highly refined factor analytic treatments of tests as well as in the more "subjective factor analyses" conducted by scholars as diverse as Gardner (1983) and Fodor (1983). Both schools of thought share a view of the human mind as a modular computational device that has separate and qualitatively different analytic procedures for processing different kinds of sensory information. Differences exist between various modular accounts, of course, but their exposition would carry us beyond the bounds of this monograph. For example, Fodor and Gardner appear to differ in the importance they attach to sensory representations themselves. For Gardner, they play a less salient role, if any, in the actual nature of the intelligences, while for Fodor they represent the most unambiguously modular systems. In this section I shall focus on Gardner's theory; in the next, I discuss Fodor's.

All multiple-intelligence (MI) approaches share a macrolevel orientation (e.g., spatial reasoning skills and verbal fluency) that can be distinguished from the microlevel processing orientation (e.g., encoding) found in the structural-cognitive theories and in the information processing theories. Factorists may arrive at different *group* and *specific* factors, depending on such things as the number and placement of axes, the number and nature of the knowledge domains considered, and the availability of comprehensive task analyses. But whatever the nature and number of their factors, they usually are summarizations of macrolevel processes. For example, suppose that a "verbal" factor emerges (or, in Gardner's case, a "musical" factor). What this means is that a set of performances on macrolevel linguistic or musical tests were found to covary in certain determinate ways. The nature of the linguistic or musical processes that support these performances (e.g., fluency, reauditorization) may or may not be revealed. Their discovery is a matter of chance, as is the discovery of the subprocesses that underpin these processes (e.g., pitch encoding). Unlike other approaches that have the discovery of such subprocesses as their primary aim, in the MI approach the level of reduction is a function of the type of tasks administered. Thus the discovery of subprocesses is chancy. Rabbitt has made the point most cogently:

> The observed degree of transfer from one such learned production to another will, likewise not depend on any intrinsic structure of the human cognitive system, such as psychometric models putatively describe, but rather on the structure of the particular tasks that we compare to obtain our data. . . . Thus, in all cases, the actual structure of the tasks that we compare will determine both the nature of the inferences we draw about the structure of the human cognitive system and the particular structures with which our factor analyses are consistent. The moral is that without detailed functional models for all the tasks we use, psychometric models based on factor analytic techniques may be very misleading ways of drawing inferences from experiments. (Rabbitt, 1988, p. 182)

For ease of exposition, I shall focus here on Gardner's approach to MI because (1) it is the one that is most adaptable to the comparative analysis to be made, (2) it has had the most influence in both psychology and education, and (3) factorial approaches are covered in some detail elsewhere.

Gardner's theory of multiple intelligences is similar to a traditional factorial account in the sense that he reports evidence for the existence of what he terms "factors of mind" (e.g., spatial, logico-mathematical, musical, motor, and two types of personal factors) from his analysis of the literature in different areas (e.g., developmental psychology, neurology, and cross-cultural analyses). Where he differs from traditional psychometric approaches, of course, is that traditional tests tap only the first three "factors of mind." He also explicitly rejects the notion of a central, pervasive processing mechanism (g).

Although Gardner has revised his theory several times since its publication in 1983 and has increased the number of factors of mind, it still retains many of its original features. Probably the most defining feature of his theory of multiple intelligences, besides the fact that he maintains that there are many, and not one, types of intelligence, is his world view. Gardner shares with most intelligence researchers the belief that not all individuals are "at promise" for equally high levels of functioning in all areas. He does not believe that if all children were exposed to a comparable environment, they would excel to the same degree. Having said this, Gardner does believe that most children have strengths that are not reflected in their IQ scores.

Project Spectrum, which Gardner has been conducting with David Feldman at Harvard, is a project aimed at chronicling children's development in the various arenas that represent his multiple intelligences. He reports that some children catch on to some things faster than others, and their mastery would seem to be undetermined by their background experiences. For example, one child may figure out how to disassemble a meat grinder faster than others, and this propensity may be related to later intellectual development in a logico-deductive domain. Others may spontaneously contrive stories that integrate disparate events and this skill could be related to their later linguistic intelligence.

Gardner argues that the "hard" side of his theory of intelligence postulates these individual differences, while the "soft" side of the theory holds out the promise that all children can excel in one or more of the intelligences, relative to the others. Even a child who is below average on one type of intelligence may be above average on another. Because he has described many types of intelligence, Gardner regards his theory as a more egalitarian theory than, say, psychometrically based accounts. This egalitarian world view is an important driving feature in all of Gardner's work: It is less important to him that one child outperforms another on some skill; what is more important is to identify each child's relative strengths.

Of all of the theorists in this area, Gardner is the one most clearly outside the "guild" of traditional intelligence researchers. He carries no membership in any of the traditional theoretical camps and indeed appears to have come to his view in part out of an objection to what he saw as the domination of thinking about intelligence by a few well-entrenched theorists.

As a result, he sometimes is criticized for failing to mesh his theory with empirical realities like the positive manifold phenomenon discussed earlier. At times he has stated that he takes no position on it, seeing it as educationally irrelevant, while at other times he has sought to debunk the notion of *g* on the basis of neurological data, without debunking the psychometric findings that appear to run counter to it. At least, this is my impression, because he does not present direct statistical evidence for the separateness of his multiple intelligences, nor does he examine the possibility that various cognitive abilities exist *within* each type of intelligence.

Gardner's theory of multiple intelligences is astonishingly disjunctive to both Piaget's and psychometric views. Unlike their assumption that all of intelligence is a structured whole and performance on one task is correlated with that on others, Gardner maintains that his everyday observation of preschoolers shows this to be untrue. Many preschool teachers and parents, he claims, realize that one child may have a propensity for interpersonal skills (one type of intelligence), while another may have a propensity for numerical reasoning (another type of intelligence). These propensities do not come about merely because their parents expose the children to situations and activities involving these things. Gardner suggests that if Piaget and others who endorse the singularity of mind viewpoint stepped outside their "pet" areas (e.g., logico-mathematical reasoning) and observed children's responses to music, art, space, bodily kinesthetics, social relations, and so on, they would be persuaded to abandon such a view. This is because children have "jagged cognitive profiles," much like the classic learning disabled child who may have trouble reading but may be able to do arithmetic satisfactorily.

One of the supporting observations for Gardner's theory comes from the work of David Feldman on prodigies. Feldman (1980) has found that most of these individuals excel in only one domain and perform average or slightly above average in the others. (Exceptions like John Stuart Mill do exist, but they appear to be rather atypical.) This "jagged profile" is compatible with the idea that intelligence is manifold, not singular, or else why wouldn't these persons be prodigies across the board, and why wouldn't a child who had difficulty in one area not have difficulty in others?

It is easy to criticize something as ambitious as Gardner's undertaking, and some have done so. The concerns that have emerged have ranged from a lack of precision to a failure to confront empirical findings that form the basis of one or another of the competing theories of intelligence. Gardner's very definition of intelligence, "solving problems or shaping products that are valued by a culture," has been criticized by theorists like Sternberg on the ground that it does not distinguish between behaviors that a honeybee exhibits from what an Einstein does. Moreover, his MI theory has been denied theoretical status by many who argue that it is not a set of formally interrelated hypotheses that can be subjected to adequate testing, despite claims in certain educational quarters that massive gains in children's intelligence have been observed in response to the use of a curriculum based on Gardner's theory. (To the best of my knowledge, Gardner himself has had absolutely nothing to do with the so-called integrated mastery based learning approaches that tout his theory for their rationale; some of these have made extravagant claims about their effectiveness and have suggested that their success validates Gardner's theory.)

A more telling criticism could be leveled, however. A careful reading of Gardner's work indicates that the "subjective factor analysis" that formed the basis for his seven intelligences was itself subjective, a claim I am sure he would be the first to acknowledge. (In his book, he described it as only a "first cut" and subject to future revision.) But the best evidence he mustered for the existence of multiple independent intelligences was the neurological findings that showed that if one area of the brain was ablated, a particular cognitive function was lost. For Gardner, this above all else refuted the notion of *g*. As was pointed out in an earlier chapter in which an analogy with athletic prowess was made, however, this neurological evidence is really not convincing and should *not* be seen as contradictory to the idea of *g*. For that, one must go beyond Gardner's arguments, while acknowledging the importance of his insights.

Sternberg's Triarchic Theory

By far the most extensive and impressive theory of intelligence is Sternberg's (1985) triarchic theory. This theory makes specific allowances for context, information processing mechanisms, experience, and the interaction between these "outer" and "inner" ingredients of cognizing. Setting out a detailed description of the interaction between knowledge representation and information acquisition, the theory pays attention to the nature of task demands (degree of novelty versus automatization), the type of contextual adaptation (shape the context to fit the organism, adapt the organism to fit better into the context, or "de-select" the context), and the type of information-processing components that individually or in combination are implicated in task performance. It is probably the single most influential theory today among cognitive researchers.

In several ways, the bioecological theory is derivative from the triarchic theory but there are important differences, too. Among the similarities is an explicit consideration of contextual factors, experiential factors, and componential factors. Both accounts are sufficient to address the mechanics of information processing, I think, and both accounts have mechanisms by which development takes place. Foremost among the differences, however, is a dramatically different view of the nature of context and, no doubt because of the theoretically different goals, a dramatically different view of the role that knowledge plays in producing individual differences. The role of context in the bioecological framework is far more pervasive than in the triarchic theory. Context shapes not just the solution to intellectual challenges, but the perceptions of them as well. And knowledge, a type of context, is elaborated and given a more fundamental role in the bioecological framework than in the triarchic theory.

In addition to these differences, the two accounts differ in the status they assign to the componential subtheory, particularly the metacomponential factor. Metacom-ponents are elevated to the status of "transdomainal" processing algorithms in the triarchic theory, but are relegated to domain-specific processes in the bioecological framework—at least when they are initially acquired. Recall that I have asserted that many individuals' cognitive processes probably never achieve truly "transdomainal" status. Among these are Sternberg's metacomponents (defining the nature of a task, selecting microlevel processes to solve it, forming a mental representation for processes to access, etc.). Although Sternberg does endorse the view that these components will vary for persons from different cultures, depending on the tasks employed to measure them, he appears to hold the view that such components can be used to explain the basis of individual differences and serve, in some fundamental way, to underpin general intelligence (g). For Sternberg, what separates two individuals who possess similar knowledge is the efficacy of their information-processing components in interaction with the elaborateness of their knowledge base. He suggests that because these components (especially the metacomponents) can be seen to play a role in a wide range of tasks, they serve as the basis for general intelligence or g:

> Individual differences in general intelligence are attributed to the effectiveness with which general components are used. Since these components are common to all tasks in a given task universe, factor analysis will tend to lump these general sources of individual difference variance into a single general factor. As it happens, the metacomponents have a much higher proportion of general components among them than do any of the other kinds of components, presumably because the executive routines needed to plan, monitor, and possibly replan performance are highly overlapping across widely differing tasks. Thus, individual differences in metacomponential functioning are largely responsible for the persistent appearance of a general factor. (Sternberg, 1985, p. 119)

As is plain in this quote, Sternberg believes that the existence of g is due, not so much to the persistence of knowledge acquisition components (e.g., encoding and high-speed memory scanning) across a wide range of tasks, but to the persistence of metacomponents (e.g., planning and introspective awareness of one's cognitive state). According to the triarchic theory, executive components such as planning and revising are important in virtually all meaningful endeavors, and the existence of g is a testimonial to their widespread use in all domains. Sternberg departs from Eysenck, Jensen, and others in this view because he does not agree with them that the basis of variation in individual difference is at the level of microprocesses such as encoding and memory scanning. So, while they are inherently g-based theorists, he is not. Indeed, for Sternberg, the source of the variation in individual differences is at a higher level of analysis, namely at the executive level or metalevel. This is probably why his theory is so attractive to educators: Intelligence, according to his account, is congenial to remediation in a way that theories based on microlevel differences (e.g., encoding efficiency) are not. It is conceivable that someone may be trained to plan, revise, and replan; it is not clear that a person may be trained to use his or her microlevel processes more efficiently.

The empirical basis for the preceding claims is not strong, at least not yet. According to the research that forms the basis of the bioecological framework, evidence suggests that the metacomponents are not "highly overlapping across widely differing tasks," as Sternberg proposes. Rather, the metacomponents are seen as originating in a particular domain of knowledge (e.g., one begins by being planful and self-monitoring in response to a specific environmental challenge), and they only achieve the "transdomainal" status after prolonged development, if at all. The developmental literature indicates that children begin appreciating only certain aspects of metacomponents— for instance, that some potential distractors (e.g., cafeteria noise in the background when one is trying to study) are indeed distractors while others are not. They do not seem to apply this insight, however, across a wide range of settings, at least from what I can discern (Miller, 1983). Thus, noise in other study settings may not be appreciated as distracting, even though its consequences can be shown to be. In Figure 7.2, metacomponents are depicted as being tied to specific domains. By the processes of induction, analogy, and simple comparison, they may be enlisted in other domains. An earlier criticism of the psychometric approach is relevant here: To the extent that one steps outside of academic types of tasks, the generality of metacomponential functioning may diminish, and with it, the size of g. Thus, the "transdomainal" nature of metacomponential functioning may prove to be more illusory than real, and the search for g will need to be fastened to specific types of knowledge differences that exist among individuals.

Sternberg's conceptualization of the "transdomainal" nature of performance and metacomponents prompts him to shift the emphasis in explaining individual differences from the *products* of one's knowledge base, including its elaborateness and differentiation (i.e., cognitive complexity) and the contexts that shaped those products and their subsequent elicitation, to the *processes* of acquiring the products. (That is, the movement is away from knowledge products and in the direction of cognitive processes.) Sternberg rightly notes that individual differences in knowledge can emerge from individual differences in the efficiency of components, but he appears to attach a priority to the latter on the basis of the claim that sheer differences in experience are not perfectly correlated with differences in performance: "Many individuals play the piano for many years, but do not become concert-level pianists; chess buffs do not all become grandmasters, no matter what the frequency of their play. And simply reading a lot does not guarantee a high vocabulary" (Sternberg, 1984, p. 58).

Here, Sternberg comes close to falling prey to the tautology that has ensnarled Jensen's (1980) theory. That is, verbal skills like vocabulary are the best predictors of verbal intelligence ($r = .85$, Matarazzo, 1970) and, for that matter, are the best predictors of Full Scale IQ. The reason for this predictiveness, it appears, is that the acquisition of vocabulary is governed by the same processes that influence IQ—for example, the skillfulness with which one can "decontextualize" the linguistic environment. The evidence for such a view is not strong, chiefly comprising studies like those reviewed by Sternberg and Powell (1983b) in which children of various IQ levels attempted to define nonwords or low frequency words from their context (e.g., "John fell into a *contavish* in the road—What is a *contavish?*"). Yet the bioecological framework would argue that those children who appear to be poor "decontextualizers" on these linguistic tasks could be quite stunning in their use of context in other domains, including verbal ones, depending on the nature and importance of the environmental challenges in their lives (e.g., detecting a nuance in one's mother's voice to decide when would be the appropriate time to ask for a favor).

Despite its differences with traditional theories of intelligence, Sternberg's triarchic theory describes a continuum along which individuals can be compared. Because one of his three subtheories, the componential one, is subject to practice effects, this leads to uncertain outcomes as far as individual differences go, a point made by Rabbitt (1988) in his review of the theory. Specifically, it is unclear which factors determine an individual's level of expertise within a domain, especially if most individuals can achieve expertise with practice. Since speed of learning has not been demonstrated to be the underlying factor, the question arises: Can intelligence, according to Sternberg's model, be anything more than a manifestation of expertise in a lot of domains, such expertise in any one of which is also attainable by persons of average and below average IQ? Finally, the triarchic theory has not yet been able to specify the relationship that exists between microlevel and macrolevel processes. As Rabbitt notes, it is in need of careful thought in regard to the functional model it posits in which the nature and interactions between microlevel and macrolevel processes are specified. And, as already mentioned in the discussion of factor analysis, the characteristics of the tasks that are employed to gather data will be critically important in determining whether metacomponents control lower level processes.

Modularity Theories

Modularity theories comprise a class of theories whose proponents range from Howard Gardner to Noam Chomsky, and from Franz Joseph Gall to Thomas Reid of the Scottish school. Although I have already discussed several different modular views elsewhere, some historical and some contemporary (Doris and Ceci, 1988), I shall focus here on the theory put forward by Jerry Fodor, the well-known philosopher of mind. In a provocative treatise entitled *Modularity of Mind*, Fodor (1983) revisited the assumptions implicitly held by proponents of one type or another of "faculty psychology" over the past two centuries. According to this view, mental life is best seen as a concatenation of many isolable cognitive processes, possessing various organizations.

For Franz Joseph Gall, the father of modern "vertical" theories of modularity, claiming that one has an aptitude is equivalent to saying that one possesses competence in its use within a given domain. Unlike nonfaculty theorists and "horizontal" modular theorists, both of whom distinguish between one or a few processes that operate uniformly across all knowledge domains, Gall believed that *all* aptitudes were content specific. To make his 18th-century terminology accord with modern psychometric terms, we can describe Gall as a non-*g* theorist in the restricted sense that he believed that all of the faculties were manifested quite differently across different knowledge domains. As Fodor states,

> Gall's major argument against horizontal faculties turns on the idea that if there is only one faculty of (say) memory, then if somebody is good at remembering *any* sort of thing, he ought to be good at remembering *every* sort of thing. That is, Gall thinks the existence of a unitary horizontal faculty of memory would imply that an individual's capacity for recalling things ought to be highly correlated across kinds of tasks. Similarly, *mutatis mutandis*, for judgment, imagination, attention, and the rest. (Fodor, 1983, p. 13)

It is hard to emerge from reading Fodor's description of Gall's views without the impression that the latter suffered from some of the worst "press" in modern psychology. Had Gall not speculated about each aptitude's location near the site of cranial bumps, he might be seen by today's cognitive researchers as thoroughly modern in his views about mental organization, and not as the nut many unfortunately think him to be. Gall noted, for example, that there existed a relationship between the shape of some classmates' heads and faces and their mental abilities: Boys with widely spaced eyes tended to be dull, while those with prominent eyes tended to have good memories. Throughout the 19th century, neurologists turned up brain centers for language, sensory and motor areas, etc., but they remained unable to provide convincing evidence for the location of the great silent "association areas" in which resided the presumably higher cognitive operations of reasoning, thought, and beliefs.

For Fodor, it makes a great deal of sense to posit the existence of two sets of faculties, one that is bound by (or, "encapsulated within," to use his jargon) specific knowledge domains and the other which is able to range across information from a host of sources. A good example of encapsulated faculties are the linguistic and perceptual processes that can be shown to operate only in response to specific linguistic or perceptual inputs. (I believe that an argument could be made for the case that motor faculties are encapsulated as well, using Fodor's rationale.) Encapsulated faculties, at a certain stage of processing, do not have access to information that other faculties know about. (This feature resembles Pylyshin's (1980) depiction of "cognitively impenetrable" processes.) They are comprised of a small set of modules that behave as though they are domain-specific computational systems that undertake their work at high speeds. Also, they have very limited access to other sources of information and have definable neural structures. Important for Fodor's argument is the idea that the output of these linguistic and perceptual (encapsulated) processes is not directly available to other faculties, nor are the outputs of these (nonencapsulated) processes available to the encapsulated ones. For instance, the auditory mechanisms involved in the segmentation of phonemes differ from those involved in the segmentation of nonspeech sounds, and there is no communality of resources between the two tasks. Similarly, the structure of the sentence recognition system is triggered by specific language-related features and can therefore operate efficiently only in domains that share some of these features. It is on this basis that Fodor argues that only a narrow family of stimuli can throw the cognitive switch that turns a particular process on. Hence, the class of encapsulated systems is small and pretuned to be uniquely sensitive to the eccentricities of information in their domains, rendering their output unavailable to other domains.

While Fodor's view is not avowedly developmental in orientation, it suggests that modular systems are endogenously determined to unfold under the impact of the environment. We can contrast such bounded faculties with ones that are unbounded or unencapsulated. The latter, such as problem solving, are thought by "horizontal modularists" to be applied to knowledge *in all domains*. After all, problem solving is problem solving: No matter whether it is exercised on legal, scientific, esthetic, or moral issues, it is still the same problem-solving process. (Gall, however, maintained more separation between processes that operated on different knowledge bases than does Fodor or the horizontalists.) So, for horizontalists the mind is comprised of functionally separable cognitive operations that cut across knowledge domains, while for vertical theorists the faculties that are associated with various types of cogitation (e.g., identifying objects and appreciating music) not only are different with respect to the psychological mechanisms that subserve them (something most modern information-processing theorists would assume without argument), but are further distinguished by reference to the domains of knowledge they operate upon.

For Fodor in particular, mental life is best characterized as an amalgam of encapsulated and nonencapsulated faculties. The former refer to such current cognitive themes as "modality-specific pathways to a higher order cognitive system." The latter are part of the higher order system. But Fodor takes exception to Gall and others who would maintain the existence of separate faculties for all cognitive feats, including those thought to be most central. For example, one can argue for the existence of separate memories on the basis of the fact that people are better at remembering some things (e.g., chess positions) than other things. Gall believed that the memory involved in facial recognition shares no cognitive resources with the memory involved in music, and perceiving music shares no resources with perceiving speech—all of these are functionally independent. For Fodor, on the other hand, the issue is not simply one of functional separation of processing systems, but of "hard-wired" differences. He believes that the experimental psychological research and the neurological research converge to support the view that there exists a characteristic neural architecture for each of the input systems to the central cognitive system. But that central system is itself not associated with a specific neural architecture and does not appear to be informationally encapsulated. Here Fodor acknowledges the need for central, higher order cognizing that has access to information from a multitude of domains. He feels that this must exist because at some level of analysis it becomes necessary to integrate inputs from the modular systems, and such integration requires wide-ranging access:

> For these and other similar reasons, I assume that there must be relatively nondenominational (i.e., domain *in*specific) psychological systems which operate, *inter alia*, to exploit the information that input systems provide. Following the tradition, I shall call these "central" systems, and I will assume that it is the operation of these sorts of systems that people have in mind when they talk, pretheoretically, of such mental processes as thought and problem solving. Central systems may be domain specific in *some* sense—but at least they aren't domain specific in the way that input systems are. (Fodor, 1983, p. 103)

Fodor's most important contribution may very well be his cogent argument against a modularity of the sort espoused by many cognitivists. He rightly argues against the arbitrary division of so-called higher cognitive processes, noting that they interact with one another in ways that belie their individuation. And while there is much in Fodor's speculations that is provocative and relevant for cognitive scientists interested in individual differences in intellectual development, there is in an important sense a mismatch in discourse between his thesis and the bioecological view being proffered here. I am grappling not so much with formalisms about the extent to which some central system is able to "look at" information from a nonrestricted class of inputs in arriving at the solution to a problem, but rather with the extent to which such a look-up process functions equivalently across various domains of data. In other words, even if I agreed with Fodor's depiction of a "bottom-up" model in which the unencapsulated (central) systems have access to all of the input data beneath them, this does not imply anything, in point of logic, about the uniformity or efficiency with which "consultation" with the diverse data is undertaken in various content domains. My thesis is that the central system is more or less able to solve some problem, depending on the way the information that it considers is represented. One might imagine information that is poorly suited to the representation of the problem preventing the discovery of its solution because it required too many resources from memory, attention, etc., to reconfigure it appropriately. This does not mean that the central system is not unencapsulated and nonmodular—only that its efficiency depends to an important extent on the nature of the information it accesses and that individual differences in its functioning can be tied to domain specific sources. Thus, the question asked is not one regarding the formal delimitation of the informational resources affected by central, unencapsulated systems like problem solving, but whether the system performs equally efficiently using different types of informational resources. Therefore, a formalism that specifies a mix between modular inputs and nonmodular, or at least nonencapsulated, central processes still leaves open the issue of how efficiently the latter consult the former—are there domain-specific constraints in this regard? Part of the problem encountered in addressing Fodor's argument is that he argues by analogy (to the process of disconfirmation by expert scientists) for the unencapsulated nature of central systems, and this analogical reasoning forms an important evidentiary basis for his view. In this sense, Fodor's theory of intelligent systems is a *competence* model of rationality as opposed to a *performance* model, although one would expect some overlap between the two. The bioecological view addresses the competence question without displacing the fascade of Fodor's argument, at least as I understand it, because I regard it as an imaginative and quite possibly correct position.

CAVEAT LECTOR

As far as accounts of individual differences in intellectual development go, the bioecological framework is a "new kid on the block," and it will require substantially more definition before it can be elevated to the formal status of a theory. There are lacunae, to be sure, and these will need to be filled-in in the coming years if the framework is to progress beyond the current *hypothesis-generation* phase to a *hypothesis-testing* phase. For example, there is a need to specify how many biologically constrained cognitive potentials exist and exactly what are the processes by which they are crystallized. As mentioned earlier, there is a need to specify the manner in which cognitive potentials are organized and deployed during problem solving, for example, the degree of subordination among the structures that control processing. (See, for instance, Rabbitt's, 1988, description of four classes of theories.) Key constructs like "context" are somewhat circular and used promiscuously. For example, if context is stipulated to be "any physical or social feature of an activity that channels behavior" (Mistry and Rogoff, 1986, p. 2) or a "set of control mechanisms. . . .for the governing of behavior" (Geertz, 1973, p. 44), then asking whether or not intelligent behavior arises from its context is begging the question, as these sorts of definitions are based on the assumption that context influences behavior. Some of the theories criticized in this monograph may eventually prove to contain features that are inherently unfalsifiable as, for example, the presence of a central processor. The bioecological framework is on shaky ground to the extent that it attempts to gain support by disproving unfalsifiable constructs.

Yet some gaps in the framework are only "apparent"; they simply require a more detailed treatment for their clarification than is possible here. Finally, the bioecological framework has as yet failed to specify the mechanism for generalizing a process across domains, other than to suggest the rather obvious ones like induction, analogy, and comparison. Like Keil (1984), Chi and Ceci (1987), and others, it predicts that reconfigurations in the knowledge base often generate a search for new processes. As this problem of how psychological processes become "transdomainal" has stymied both stage and nonstage theories, it may prove to be the most difficult support to muster. Related to this problem, the bioecological framework will need to be much more precise about the distinction between processes that initially can operate in only restricted domains (but which, with development, may spread to others) and processes that can operate transdomainally from the start but at deeply different levels of efficiency (e.g., encoding).

Neisser (1985) once asserted that cognitive psychology is still a long way from proffering an adequate theory of the environment. Many of the lacunae I mentioned are a reflection of this lack of theory, I think. With further conceptual advances of the sort provided by Bronfenbrenner (1989) and others recently, it may be that some of these holes can begin to be filled.

Notwithstanding these real limitations, the bioecological framework does appear to provide an inductively driven basis for escaping the often circular reasoning that surrounds much of the psychometric research. And it appears to provide an alternative basis for explaining many of the central findings from both information-processing and behavior-genetic theory. There is much that may seem "fuzzy" about the bioecological framework—in fact, there are places in this monograph where I debated using alternative wordings because the phenomenon under consideration appeared to "slide off the page" as it was being dissected. There were even several instances where the argument I was making seemed to become less precise even while I was struggling to make it more so. And there were cases in which some rule or principle was supported in general but refuted in the particular. Such problems could be due simply to imprecision on my part, but they also could be due to the indeterminate nature of some of the phenomena under discussion. At times like these, I found some small degree of solace by reminding myself of the philosopher John Searle's reply to critics of literary criticism:

> Many theorists fail to see, for example, that it is not an objection to a theory of fiction that it does not sharply divide fiction from non-fiction, or an objection to a theory of metaphor that it does not sharply divide the metaphorical from the nonmetaphorical. On the contrary, it is a condition of the adequacy of a precise theory of an indeterminate phenomenon that it should precisely characterize that phenomenon as indeterminate; and a distinction is no less a distinction for allowing a family of related, marginal, diverging cases. (Searle, 1983, p. 79)

In sum, I believe that I have raised a number of empirical and logical concerns about existing theoretical approaches and provided a "first cut" for a new way of thinking about intellectual development that avoids many of the problems I have raised about these other approaches. I have tried to broaden the debate by explicitly incorporating literatures from outside of psychology (anthropology, sociology, and education) as I have blended developmental, biological, and information-processing strands into a single account. Of course, there were places where I wish I could have been less tentative in refuting someone's claim, and there were times I wished I could have presented an argument that was "beyond a reasonable doubt" against some theory, instead of merely presenting a "preponderance of evidence" case against it. Yet I never found myself thinking that the inability to *unequivocally* refute any single strand in the argument required me to accept the opposite argument—any more than my inability to explain certain so-called "miracles" requires me to accept the existence of a supernatural deity. The theory that *IQ = intelligence = generality = everyday success = heredity* involves an intricately interwoven argument, and I think that I have dismantled some parts of it more effectively than others. Whether the bioecological framework holds up to the ultimate canons of scientific acceptability remains to be seen, as the necessary research to fill the gaps is completed. In the meantime, I look forward to the reactions of readers with much excitement (and modest hope).

Epilogue

In the first edition of *On Intelligence* (1990), I described a "smuggling operation" in which social scientists had collected massive amounts of new data that, taken together, validated their belief that America's woes can be directly traced to the genetic inferiority of some groups' intelligence. I called this a "smuggling operation" because though the constituent data were well known to members of the scientific community, they were gradually but deftly assembled by a group of IQ researchers (the New Interpreters) into an argument for a genetically based meritocracy. This meritocracy argument was unknown to those who worked outside the psychometric community. Quietly, steadily, and impressively, the research accumulated until it could be put together into a five-part syllogism which essentially claimed that:

i)   individual differences in general intelligence exist;

ii) they are well documented by standard IQ tests;

iii) they are primarily genetic in origin;

iv) they are associated with differences in real-world attainments (e.g., schooling and earnings);

and consequently

v)   real-world attainments depend, at least in part, on genetic differences among individuals.

Simply put, the New Interpreters' argument claims that individual differences as well as group differences (racial, ethnic, and social class) in important life-course outcomes (e.g., school success, supervisor ratings of work efficiency, lifetime earnings, criminality-proneness, welfare dependency) are the result of genetic differences in intelligence. Individuals are said to inherit nervous systems that differ in the speed and accuracy of their information processing, and it is this difference that is claimed to underlie their subsequent differences in both IQ *and* life-course outcomes. Sluggish nervous systems impede the ability of some to perform well both on IQ tests and in the world of school and work. For lack of a better phrase, I refer to this set of tenets and the conclusion drawn from it as the "genetically based meritocracy syllogism."

Some of my colleagues criticized me six years ago when I first described this smuggling operation because they felt that I was erecting a "straw man"; they politely suggested that no one but the most outrageous extremists would attempt to argue that racial and ethnic differences in native (i.e., inherited) intelligence were the cause of differential educational success and social class differences in society. Six years later, the publication of a spate of books, most conspicuously *The Bell Curve* (Herrnstein and Murray, 1994), regressed them to reality: prominent members of the scientific community did indeed take this syllogism, or at least most of its tenets, seriously. Fifty-two self-described "experts in intelligence and allied fields" signed a full-page op ed in the *Wall Street Journal* endorsing twenty-five conclusions reached by Herrnstein and Murray (*Wall Street Journal*, 1994).

My colleagues' misunderstanding six years ago resulted from the fact that few who worked outside the area of intelligence research could appreciate just how comprehensive the validation of the genetically based meritocracy syllogism had become. Even in those rare instances where scholars outside the psychometric area were aware of the constituent pieces of this validation enterprise, they seldom understood how these pieces fit together into an argument in favor of a genetically based meritocracy. For example, they had no idea that the speed with which one makes very simple sensory decisions (e.g., saying which of two lines is longer) is modestly correlated with IQ, is not linked to schooling experience (Deary, 1993, 1995), and is somewhat heritable. Moreover, they had no idea how such pieces of information related to biological measures of nervous system functioning, such as central nerve conductance velocity and ocilation (Reed and Jensen, 1992, 1993) or heritability coefficients; nor did they realize how these biological measures were related to school success.

Finally, my colleagues outside the intelligence arena had little awareness of a burgeoning literature that demonstrates correlations among IQ, nerve conductance ocilation, cranial capacity, glucose metabolic uptake rate, and group differences in elementary information-processing speed (e.g., speed at naming digits and letters). These pieces fit together into an argument that was unknown to outsiders, an argument that seemingly validated the genetic meritocracy syllogism.

In the six years that have passed since the first edition of *On Intelligence* appeared, my colleagues have stopped suggesting that I erected a straw man. As mentioned in the Preface, there has been a spate of books and articles by New Interpreters that has made explicit the genetic-meritocratic argument as well as its presumed validation (e.g., Herrnstein and Murray, 1994; Lynn, 1991; Rushton, 1995; Seligman, 1992). If anything, when I sounded the clarion call in 1990 I underestimated the vehemence of the genetic meritocracy argument. In the final pages of this edition of *On Intelligence*, I will critique the New Interpreters' argument and provide what I take to be a scientifically more compelling alternative, a bioecological framework. I will focus on the syllogism laid out in Herrnstein and Murray's *Bell Curve*, as it is easily the most prodigious and influential of the new interpretations, though it is by no means unique.

Seldom in the history of science has an 840-page tome, packed with tables and formulae, had as great an impact as *The Bell Curve*. It has been the cover story in major print media (*Newsweek, U.S. News and World Report, New Republic, Chronicle of Higher Education*), the source of editorials in leading newspapers (*New York Times, Wall Street Journal, Washington Post*), the target of numerous discussions in the electronic media (National Public Radio's *All Things Considered*, NBC's *Good Morning America*, ABC's *Meet the Press*), and the object of lively and heated debates (e.g., *The Bell Curve Wars*, Fraser, 1995; *The Bell Curve Debate*, Jacoby and Glauberman, 1995). It is not surprising in view of the frenzy surrounding it that *The Bell Curve* sold more than a half million copies in its first eight months, and reached number two on the *New York Times* Best Sellers List for nonfiction.

I begin my critique of the New Interpreters' argument by challenging the evidence that provides its very foundation; namely, their three-pronged thesis that: 1) variation in both individual and group intelligence is primarily determined by heredity; 2) intelligence is best depicted as a singular or general construct (*g*) that permeates all intellectual activity and is well captured by a standard IQ test; and 3) racial, ethnic, and socioeconomic groups differ on important life-course outcomes (e.g., school success, criminality, illegitimacy rates, earnings) primarily because they differ in their genetically derived capacity to be intelligent.

In dealing with these claims, I will tackle several subsidiary claims that Herrnstein and Murray tout as accepted wisdom among members of the scientific community; namely, that a divergence is building in America between the cognitive elite and the cognitive have-nots; that this divergence is the result of a dysgenic trend whereby poor, primarily black and Hispanic, mothers with low IQs begin having babies earlier than white, middle-class women, and therefore end up producing more babies over time (i.e., even if the number of offspring is the same between racial groups, the group that begins child-bearing earlier will produce more generations per unit time).

The New Interpreters arrive at this remarkable set of conclusions through a synthesis of "heritability" research showing that IQ scores tend to run along blood lines, and that such scores are somewhat predictive of a range of developmental outcomes, such as poverty rates, job performance, earnings, and school success. Reviewing the New Interpreters' use of the concept of heritability gives me an opportunity not only to critique the logic of their argument, but also to provide a conclusion to the second edition of *On Intelligence*—for in refuting their genetic meritocracy claim, I hope to make the validity of an alternative model of human development, "the bioecological theory of intellectual development," clearer.

THE HERITABILITY OF INTELLIGENCE OR THE INTELLIGENCE OF HERITABILITY?

The heart of the empirical evidence cited by the New Interpreters is the commonly reported differences between social-class and ethnic- and racial-group IQ scores. Poor children score about 10 IQ points below middle-class children, and blacks score below whites, on average, by 1.08 standard deviations (around 16 points on the most common IQ test, the WAIS). Although there are many variations by geographic region, there is no disagreement among scholars on the basic thrust of these points, for the racial difference has remained stable since the first Stanford Binet test was normed in the early 1930s.

The conflict among scholars begins with the realization that the New Interpreters fuse the common observation of group differences on IQ with two others; namely, the heritability for IQ is quite high (approximately 60 percent of the *differences* among individuals within a group or race is accounted for by heredity), and the magnitude of the commonly reported differences (16 points) is too high to be reasonably explained solely in terms of the environmental disadvantage of blacks (Herrnstein and Murray, 1994). I will explore these claims and their implications more fully below.

Researchers have only one scientific strategy for estimating the relative contributions of heredity and environment to the formation of an IQ score, and even this method, which produces a *heritability coefficient*, is limited in its scope and generalizability (for a full exposition of these limitations see Cavalli-Sforza and Bodmer, 1971; Falconer, 1989; Jacquard, 1973, 1983). The heritability coefficient measures the extent to which individual differences in intelligence are attributable to differences in genetic endowment, and is formally defined as "the proportion of the total phenotypic variance that is due to additive genetic variation" (Cavalli-Sforza and Bodmer, 1971, p. 536). There are two general strategies for estimating this proportion. The first and most widely used involves comparing persons of contrasting genetic similarity (typically monozygotic and dizygotic twins) who live in the same environment. The second strategy examines the degree of dissimilarity between first-degree blood relatives (often identical twins) who are separated and living in different environments. Herrnstein and Murray correctly note that the average heritability coefficient is around .6 (and it is actually slightly higher if we restrict ourselves to data based on monozygotic and dizygotic twins). A heritability coefficient of .6 means that 60 percent of the *differences* among the IQ scores of a group of individuals is a reflection of genetic differences among them; for example, if the average difference was 15 points, then 9 points would be due to genetic differences (including gene-environment covariation) and the remaining 6 points would be due to environment (plus measurement error). There are a series of problems, however, with basing this conclusion on either of these strategies for computing heritability.

Generalizations based on either strategy are limited to the extent that the sample on which heritability is calculated encompasses the range of contrasting environments experienced by persons in the population to which one wishes to generalize. To the best of my knowledge, this condition has not been met by any of the New Interpreters. If I am correct, heritability coefficients that have been reported cannot be used to estimate the contribution of genetics to average differences between groups if these groups live in vastly different environments. As one straightforward criticism, one can ask: Is the heritability of blacks the same as that of whites? The answer to this surprisingly simple question is not at all simple. Very few studies have even bothered to examine this issue, and we know very little about heritability within specific racial and ethnic groups, or even among poor white families. What we do know about heritability comes largely from data contrasting members of middle-class white families. Later I will describe features of the environments of poor individuals that are known to have baleful consequences for the development of IQ but that are systematically underestimated in heritability studies because this group is undersampled in published studies. But I am getting ahead of my story line.

All of the New Interpreters realize that one cannot infer anything about the source of between-group differences on the basis of within-group differences. Hence, they know that just because heritability among white families is .6 (i.e., 60 percent of the differences among the IQs of middle-class white individuals is associated with systematic genetic variation among them), does not imply that 60 percent of the differences among the IQs of members of other racial or ethnic groups is also the result of genetics or, more important, that 60 percent of the differences among the IQs of various racial or ethnic groups is the result of genetic differences between them. All of this is copybook maxim to modern researchers, and there is no disagreement. Despite this awareness, however, most New Interpreters are inconsistent in acknowledging the implications of this truism for crucial steps in their syllogism. For instance, in *The Bell Curve*, Herrnstein and

Murray explicitly state their awareness of this principle, emphasize that it is often misunderstood by others, and then, to bring home their awareness of its validity, provide a cogent example of why one cannot infer the source of between-group heritabilities based on within-group heritabilities:

> Scholars accept that IQ is substantially heritable, somewhere between 40 and 80 percent, meaning that much of the observed variation in IQ is genetic. And yet this information tells us nothing for sure about the origin of the differences between races in measured intelligence. The point is so basic, and so commonly misunderstood, that it deserves emphasis. *That a trait is genetically transmitted in individuals does not mean that group differences in that trait are also genetic in origin.* Anyone who doubts this assertion may take two handfuls of genetically identical seed corn and plant one handful in Iowa, and the other in the Mojave Desert, and let nature (i.e. the environment) take its course. The seeds will grow in Iowa, not in the Mojave, and the result will have nothing to do with genetic differences. (Herrnstein and Murray, 1994, p. 298)

Even though their next paragraph begins with the acknowledgment that "the environment for American blacks has been closer to the Mojave and the environment for American whites has been closer to Iowa," Herrnstein and Murray reaffirm their claim of the primacy of genetic factors in producing differences in IQ between the races, and proceed to base this conclusion precisely on differences between heritability coefficients (see also Rushton, 1995, for a similar argument). Their evidence and argument run as follows:

> We . . . stipulate that one standard deviation (fifteen IQ points) separates American blacks and whites and that a fifth of a standard deviation (three IQ points) separates East Asians and whites. Finally, we assume that IQ is 60 percent heritable (a middle ground estimate) . . . The observed ethnic differences in IQ could be explained by the environment if the mean environment of whites is 1.58 standard deviations better than the mean environment of blacks and .32 standard deviation worse than the mean environment for East Asians, when environments are measured along the continuum of their capacity to nurture intelligence. Let's state these conclusions in percentile terms: The average environment of blacks would have to be at the 6th percentile of the distribution of environments among whites . . . for the racial differences to be entirely environmental. Environmental differences of this magnitude and pattern are implausible. (Herrnstein and Murray, 1994, pp. 298–299)

There are two major problems with this line of reasoning. First, it assumes that the magnitude of the heritability coefficients computed within-groups is a valid basis for inferring the contribution of genetics to between-group differences in intelligence—an assumption that directly violates the general principle the authors had so earnestly endorsed in their earlier quotation. The second major problem is even more fundamental, and ample empirical evidence exists to refute it.

Herrnstein and Murray assert that environmental differences of the magnitude required to outweigh genetic effects between groups are "implausible." This claim has been tested in the sphere in which heritability coefficients do have some validity; namely, as estimates of the extent to which genetic endowment accounts for within-group individual differences in intelligence, with the residual variance due to environmental factors (plus measurement error). These same studies also speak directly to Herrnstein and Murray's claim that between-group differences in intelligence are principally the result of genetic selection rather than environmental influences.

Two alternative models can be tested. According to the first model, environments are viewed as forces that affect the extent to which individual differences in genetic potential for intelligence are actualized. This model, called the "environmental disadvantage hypothesis," has been evoked to explain how an "unspecified complex of environmental factors associated with poverty may prevent an organism from achieving its optimum development" (Scarr-Salapatek, 1971, p. 1286). According to the second model, the one implicitly endorsed by Herrnstein and Murray, environmental niches are themselves viewed as the result of genetic selection rather than as an independent influence on its expression.

If the first alternative is correct, then the contribution of genetics to within-group individual differences in IQ should be significantly greater in environments that provide the resources and conditions that foster intellectual growth, and it should be smaller in settings lacking such developmentally fostering features. This is because it is only in environments that are good enough to bring to fruition genetic potential that such potential gets actualized. In environments that are chaotic, disorganized, and violent, much genetic potential may go unrealized—much like the child who has the genetic potential to acquire foreign languages easily but attends a school that does not teach foreign languages. Although endowed with the genetic potential to be superior at foreign languages, such a child will never become as superior to her peers as she might have if she attended schools that emphasized foreign languages.

By contrast, if the New Interpreters are correct in viewing environments as the result of genetic selection (with environmental differences merely reflecting such genetic differences), then variations in the average IQ of persons living in different environments will be the result of genetic selection, and therefore heritability coefficients would be expected to be consistently high across all types of environments.

These two models have in fact been assessed in two studies of differences in heritability as a function of social class (Fischbein, 1980; Scarr-

Salapatek, 1971). In her pioneering investigation, Scarr-Salapatek provided a simultaneous test of the two competing hypotheses separately for black and white samples of twins. The results from both samples supported the environmental disadvantage hypothesis, not only with regard to social class, but also with regard to race. In each case, the within-group individual differences attributable to genetics were significantly greater in more advantaged settings than in poorer environments. In her words: "The major finding of the analysis is that advantaged and disadvantaged children differ primarily in what proportion of the variance can be attributed to environmental sources" (p. 1292). The major shortcoming of this study was that Scarr-Salapatek had to estimate the proportion of monozygotic (identical) twins in her sample based on their frequency in the general population, because no birth-zygocity information was available to her. Nevertheless, very similar results were obtained in an investigation of Swedish twins in which zygocity was based on laboratory procedures (Fischbein, 1980). This finding supports the environmental disadvantage hypothesis even more decisively than Scarr-Salapatek's. Heritability coefficients calculated across three social-class levels (high, middle, low) for a verbal test of IQ were .78, .48, and .30, whereas for an inductive test of IQ they were .96, .48, and .22.[1] Not only are these findings ignored by all of the New Interpreters, but other results in support of them (Fischbein, 1980; Sundet, Tambs, Magnus, and Berg, 1988) are also ignored, despite the often extensive bibliographies in the New Interpreters' books.

This pattern of findings is hardly in accord with the New Interpreters' claims that group differences in intelligence are the result of genes. First, it is clear from the results of such studies that the heritability for a particular population varies significantly as a function of the environmental resources available to that population. Not only do these results violate a genetic-distribution explanation of mean differences in IQ between groups, but they simultaneously provide strong evidence for the influence of environmental factors on individual differences in IQ within groups.

Second, given the additive nature of heritability, its complement, termed *environmentality* (Fuller and Thompson, 1978), provides an estimate of the extent to which within-group individual differences in IQ are a function of environmental factors. Herrnstein and Murray, like many New Interpreters, argue that the environment hardly matters, except in the most abusive circumstances (see also Scarr, 1992, for a similar argument). Contrary to the claims of the New Interpreters, however, the contributions of the environment to within-group individual differences in IQ can be quite substantial. In fact, if one computes the differences between pairs of environmentality coefficients for each of the six comparisons mentioned above (i.e., verbal and inductive sections of the test used by Fischbein for each of three social-class levels), then half of them exceed the difference of 44 percentile points (50 percentiles minus 6) set by Herrnstein and Murray (1994) as the standard that must be met if environmental influences are to compensate for the genetically-based difference in intelligence between blacks and whites (see their text for a description of how they arrived at this standard).[2]

The preceding analysis brings us back to an issue mentioned briefly above; namely, the samples employed in the studies on which all of the New Interpreters' conclusions about heritability (and corresponding environmentality) coefficients are based are not representative of the populations to which they wish to generalize. For generalization to occur, the samples must be representative, especially with regard to those features of the environment that research has demonstrated to exert an influence on intelligence. A substantial body of research has shown that three types of environmental contexts have deleterious consequences for the development of IQ: 1) growing up in a single-parent family; 2) being born to a teenage mother; or 3) being raised in a family living in poverty. The debilitating effects of these factors are greatly magnified when they co-occur (see Duncan, Brooks-Gunn, and Klebanov, 1994; Haveman and Wolfe, 1994; McLanahan and Sandefur, 1994). Herrnstein and Murray devote an entire chapter to these same three domains, but treat them as *destinations* of genetic selection arising from biologically driven lowered IQs, rather than as environmental engines that drive cognitive development independent of genetics.

Accordingly, the question arises: To what extent are these three developmentally vulnerable environmental contexts adequately represented in the samples the New Interpreters used for assessing heritability? Although it is true that the samples Herrnstein and Murray and other New Interpreters relied on heavily for some of their arguments do include these three deleterious contexts (The NLSY survey), they were not used to compute the heritability coefficients they reported. This point is important because the omission of these substantial sources and effects of environmental influence from the samples used to compute heritability coefficients risks a serious underestimate of the proportion of variation in IQ that is attributable to the environment, and thus leads to a corresponding overestimate of the influence of genetics.

The samples that Herrnstein and Murray do use for calculating heritability raise several concerns that they were not representative with respect to the three environmental domains under consideration. First, in none of the relevant investigations of heritability cited by the New Interpreters was an effort made to ensure systematic representation of children living in these developmentally disadvantaged environments. Second, most of the studies that do report heritability coefficients were conducted with middle-class samples, in which poverty, single-parenthood, and teenage pregnancy are considerably less frequent than in society as a whole. Third, with few exceptions, the actual data from which the heritability coefficients were computed were collected between ten and thirty years ago, when rates for such phenomena as unmarried teenage motherhood and childhood poverty were much lower than they are today.

The most striking feature of these social changes is the rapidity with which they occurred over comparatively brief periods of time. Within the space of a decade or two, there have been dramatic increases in the number of young children living with a single, unmarried, separated, or divorced parent; teenage birth rates have first fallen and then risen markedly; and many families with young children have experienced an economic turnabout, first moving out of and then back into poverty (for relevant data to support each of these claims, see Bronfenbrenner, McClelland, Wethington, Moen, and Ceci, 1996). All these phenomena call into question the New Interpreters' claim that such changes are mainly the by-products or destinations of genetic selection, a process that must take place across, and not within, generations. Although one cannot rule out the possibility that the deleterious patterns are actually the result of genetic selection in the preceding generation, and then serve as precursors of a continuing genetically driven trend in the next, such an explanation appears highly unlikely in view of the absence of any support in the studies by Scarr-Salapatek and Fischbein already cited. Clearly, something other than genes is responsible for changes in these secular trends in teenage pregnancy, poverty, and divorce over such brief periods of time. After all, what genetic trend can explain why individuals move into and then out of poverty within a single generation? The answer is, quite simply, none.

The Nature of Intelligence: One vs. Many?

A central issue in contemporary cognitive science is also central to the New Interpreters' overall thesis: How should intelligence be conceptualized and analyzed—as a single, integrated general mental capacity (i.e., g) best measured by an IQ score, or as a complex of relatively independent factors, each assessed separately? The New Interpreters take for granted that the *sine qua non* of intelligence is singular (g), inherited, and accounts for more of the differences among individuals' developmental outcomes (e.g., school and work success) than any specific cognitive factor. If intelligence was not singular, this would open the possibility for a different depiction, one in which individuals could be strong at some things and weak at others, and these various strengths and weaknesses could have different levels of heritability, some being much lower than others, and each of these individual abilities and strengths could lead to important predictions that are independent. In other words, moving away from g-based depictions of intelligence dents the facade of the genetic meritocracy syllogism because it allows for social mobility along some cognitive paths even when other paths are ineffective though equally heritable.

Especially over the past decade, the non-*g* view of intelligence has taken on a new scientific life, mainly made possible by the development of theoretical and analytic models, the use of advanced statistical techniques for data analysis, and the design of controlled experiments for testing alternative hypotheses. This work represents a serious theoretical and empirical challenge to proponents of a single-factor *(g)*, underlying capacity for general intelligence (e.g., Ceci and Liker, 1986; Detterman, Mayer, Caruso, Legree, Connors, and Taylor, 1992; Dörner and Kreuzig, 1983; Keating, List, and Merriman, 1985; Rosnow, Skleder, Jaeger, and Rind, 1994; Sternberg, 1985; Sternberg, Wagner, Williams, and Horvath, 1995; Streufert and Nogami, 1989). As pointed out in Chapter 9, a frequently reported finding among cognitive scientists is the low cross-task correlation between performance on tasks that are essentially identical, except for their context. For instance, inductive reasoning is highly contextualized, with individuals being able to reason more complexly in some domains than in others even when the formal requirements of the reasoning tasks are the same in both (Johnson-Laird, 1983). This calls into question the presumption that what is measured in a single context (e.g., an IQ test) captures the essence of cognitive ability. There are many examples of such low cross-task correlations, and they have provided the empirical basis of multifactor conceptions of intelligence (see Ceci, 1990).

The New Interpreters' treatment of this new research follows a curious and somewhat contradictory course (see Itzkoff, 1989; Seligman, 1992). For example, Herrnstein and Murray's (1994) introductory chapter is devoted to this subject. In two sections of that chapter, the authors discuss the work of researchers who base their analyses and experiments on multifactor conceptions of intelligence. Referring to members of this latter group as "the Revisionists" and "the Radicals," Herrnstein and Murray ally themselves with "the Classicists"; that is, those who subscribe to a single general-factor theory of mental ability, i.e., *g*. With respect to competing research traditions, Herrnstein and Murray describe their own orientation as follows:

> Given these different ways of understanding intelligence, you will naturally ask where our sympathies lie . . . We will be drawing most heavily on the classical tradition. That body of scholarship represents an immense and rigorously analyzed body of knowledge. By accepted standards of what constitutes scientific evidence and scientific proof, the classical tradition has in our view given the world a treasure of information that has been largely ignored in trying to understand contemporary policy issues. (1994, p. 19)

Writing in the *New Republic* two months after the publication of *The Bell Curve*, Murray and Herrnstein (1994) elevated what was once a scientific preference to the status of a scientific fact. In an extended invited article they wrote:

> First, the evidence, beginning with this furiously denied fact: intelligence is a useful construct. Among the experts it is by now much beyond technical dispute that there is such a thing as a general factor of cognitive ability on which human beings differ and this general factor is measured reasonably well by a variety of standardized tests, best of all by I.Q. tests. These points are no longer the topic of much new work in technical journals because most of the questions about them have been answered. (Murray and Herrnstein, 1994, p. 27)

As I have argued throughout this book, this is simply not the case. New research is constantly being reported that is in accord with the multi-factor view of intelligence (e.g., see Detterman et al., 1992; Neisser et al., 1996; Rosnow et al., 1994; Sternberg et al., 1995). While this debate continues in the scientific journals without a clear winner, it is premature to imply, as the New Interpreters do, that the issue has been conclusively resolved in favor of the general factor, *g*. It has not (see Horn, 1989).[3] This is yet another instance of the New Interpreters' presupposing what ought to have been the object of serious testing and verification.

LINGERING PARADOXES

In the remaining half of this Epilogue I want to focus on several seemingly paradoxical claims made by the New Interpreters. For the sake of brevity, I will focus on the paradoxes as they are laid out by Herrnstein and Murray (1994), though similar *non sequiturs* can often be found in the writings of other New Interpreters.

Herrnstein and Murray's Conundrum

Twice Herrnstein and Murray lay down a pointed challenge to their critics (Herrnstein and Murray, 1994; Murray and Herrnstein, 1994), the second time with an explicit call for a response:

> Why, if the black-white difference is entirely environmental, should the advantage of the "white" environment compared to the "black" be greater among the better-off and better-educated blacks and whites? We have not been able to think of a plausible reason. Can you? (Murray and Herrnstein, 1994, p. 32)

I can think of not one but several reasons. And the answer to one of them does not lie far from Herrnstein and Murray's analogy of growing corn in the rich Iowa farmland (the white environment) versus the depleted soil of the Mojave Desert (the black environment). The authors correctly comment that historically the environments of most blacks have been closer to the barren desert. There are probably no places in the Mojave where corn seeds can yield more than a stubble, whereas in Iowa there are doubtless some fields that produce far richer crops than others. For variation in actualized genetic potential to occur, there must be a high enough average level of expression around which such variation can be manifested. If, as the result of barren soil, all growth is stunted, there is little variety to be observed. In addition, if the variation in growth is potentially of environmental origin, then the environment itself has to be varied. In the Mojave Desert, there is little opportunity for either of these conditions to be met, whereas the farmlands of Iowa provide amply for both. Depleted environmental resources preclude potential from being realized.

Herrnstein and Murray assume that if the environment makes a difference, then attaining middle-class status for blacks should have been enough to erode the racial gap in IQ. Such an assumption ignores both the complexity of real-life environments and the link between those environments and IQ development. Concerning the first, when a black reaches a sufficiently high level of economic advantage and schooling to be labeled "middle-class," this does not necessarily mean that his or her life environment is or has been comparable to that of whites who are labeled "middle-class." Many more adult whites than adult blacks who are labeled "middle-class" were, as children, reared in middle-class environments. Simply contrasting black and white middle-class adults does not take this into account. It is no fairer test of the importance of the environment than if we were to find some lower-class individual who won the lottery and suddenly was categorized as middle-class because of his winnings; no one ought to expect such an individual to suddenly score higher on an IQ test, perform better in school, and so on. Similarly, no one ought to expect blacks who, as adults, attain middle-class status for the first time in their lives to have similar IQs to whites who were reared in middle-class environments their entire lives. There is a far greater tendency for middle-class whites than middle-class blacks to have middle-class ancestral origins.

There is also a pronounced difference in economic resources among middle-class blacks versus middle-class whites: middle-class whites are far wealthier than middle-class blacks (see Smith, 1994, cited in Nisbett, 1995), a result of attending elite schools, inheriting wealth, and so on. Herrnstein and Murray's implicit assumptions are that wealth and schooling among middle-class blacks and whites are comparable and that with each year of schooling and with each dollar of income earned, there is an immediate (i.e., within a single generation) increment in IQ.[4] However, the divergent social histories of blacks and whites challenge this assumption. The diversity of microenvironments and wealth levels of families living in different family structures within the same social class belies this claim.

This brings me to the second part of Herrnstein and Murray's seeming paradox about IQ differences being larger among middle-class blacks and whites than among their lowerclass counterparts. Why should one find this surprising? If the environment is important to the development of IQ, and if the difference in environmental resources is actually greater among the middle class than among the lower class, then we ought to find larger differences among the latter. We know that in very poor settings there are insufficient resources to actualize genetic potential. Consequently, potential differences among individuals will not be realized and they will end up being more similar to each other than they might be under more plentiful conditions. Scarcity of resources dampens all psychological effects—a well-known finding. For example, a now substantial body of research on the effects of family structure on cognitive development (Haveman and Wolfe, 1994; McLanahan and Sandefur, 1994) documents consistently higher levels of intellectual outcomes for children raised in two-parent biological families than for those growing up in single-parent or step-parent families. In a recent analysis, Bronfenbrenner and Ceci (1994) went a step further to demonstrate that this pattern of contrasting outcomes by family structure was significantly more pronounced among youngsters from middle-class homes than from lower-class homes.

In short, these findings provide three answers to Herrnstein and Murray's challenge to explain the finding of greater black-white IQ differences in middle-class than in lower-class families. First, as I have argued, black middle-class families are significantly poorer than white middle-class families, due to less inherited wealth and attendance at less elite schools (see Smith, 1994, cited in Nisbett, 1995). Second, family-structure differences are accentuated in middle-class homes, so that single-parenthood, divorce, and being reared by a never-married mother take a relatively greater toll on middle-class children than on poor children, and relatively more black middle-class children than white middle-class children fall into these categories. And third, comparing black and white middle-class adults ignores the unequal developmental histories that these two groups experienced and, in particular, the far greater likelihood that white middle-class adults were themselves the products of multigenerational middle-class homes. In short, Herrnstein and Murray's conundrum may not be a conundrum.

On Herrnstein and Murray's "General Rule"

The New Interpreters frequently assert that equating the environments of individuals will make them more similar rather than less similar. But is this really true? Let's consider the "general rule" that Herrnstein and Murray lay down, and then apply it to the interpretation of specific findings, and then spell out the implication of these findings for society. Commenting on the efforts of modern societies to equalize opportunity, they write:

> As a general rule, *as environments become more uniform, heritability rises.* When heritability rises, children resemble their parents more, and siblings increasingly resemble each other; in general, family members become more similar to each other and more different from people in other families. It is a central irony of egalitarianism: Uniformity in society makes the members of families more similar to each other and members of different families more different. (Herrnstein and Murray, 1994, p. 106)

This "rule" can be called into question on two counts. First, if what they mean by a "uniform environment" is a situation in which a group of human beings is exposed to similar environmental conditions, then on what basis do the New Interpreters conclude that lower-class families have fewer common experiences than upper-class families? To the extent that uniformity is relevant at all, a more accurate generality might be: "In a uniformly 'good' environment, the heritability would be high; in a uniformly 'bad' environment, it would be low." So restated, the proposition exposes the "central irony," not of egalitarianism, but of the authors' seemingly preordained notions about how the cognitive elite is identified and promoted.

The second challenge to Herrnstein and Murray's environmental uniformity rule comes from empirical data and speaks to their claims that as heritability rises, siblings become more similar to each other. Their formulation, however, suggests a lack of understanding of the heritability coefficient, how it works, and what it does and does not measure. Consider the following hypothetical case: A middle-class family has fraternal twins, one of whom displays an early propensity for mathematics and the other an early interest in music. Being middle class, the parents have the resources to foster each child's potential talents; for example, purchasing a computer for the first child and sending this child to math camp, while enrolling the second child in a school for the performing arts, and providing music lessons. In short, these hypothetical parents have done what real middle-class parents often do: they have used their resources to enable their children to actualize their differing genetic potentials. Because of this, these two hypothetical middle-class children will end up becoming more dissimilar than they would if their environment had lacked the resources to enable them to bring their latent talents to fruition. Unfortunately, the latter outcome is precisely what happens in a poor environment, with the result being that the two siblings end up being less dissimilar than they otherwise might have been had their environments been good enough to actualize their different potentials. We thus observe high levels of heritability concomittant with high levels of dissimilarity, not the reverse. Herrnstein and Murray (1994) have it exactly backwards.[5]

Putting aside empirical realities, what should one expect on purely theoretical grounds? Because monozygotic twins are genetically alike, their shared exposure to a developmentally favorable versus unfavorable environment is likely to affect both twins equally. The only way they can become more similar is through exposure to a more favorable environment that actualizes common genetic potential not yet realized. In the opposite direction, the only way they can become dissimilar is through differential treatment from their environment that serves to bring to fruition genetic potentials in one twin that are not brought to fruition in the other. But dizygotic twins have an additional way of becoming different from each other; because they share only half of their genes, the actualizing power of a favorable environment works to magnify their differences, whereas this differentiating vector would be weaker in a developmentally unfavorable milieu. This means that improvements in the quality of the environment will have a greater impact on increasing dissimilarity between fraternal twins than on increasing similarity between identical twins. In accord with this expectation, Sundet and his colleagues (1988) found that rises in heritability were more a product of increasing differences between fraternal twins than of increased similarity between identical twins (a range of twenty-four percentage points for the former compared with nine points for the latter). Further confirmation comes from a meta-analysis carried out by McCartney, Harris, and Bernieri (1990). They found that changes in heritability coefficients over time were chiefly a function of differences in the degree of dissimilarity between fraternal twins rather than in the extent of similarity between identical twins. Additional evidence bearing on this issue was provided by Plomin and Daniels (1987), who commented that "the research converges on the remarkable conclusion that . . . two children in the same family [are] as different from one another as are pairs of children selected randomly from the population" (p. 1). In short, there is ample empirical and theoretical support for the conclusion that the New Interpreters have misconstrued the meaning of the heritability coefficient and what it measures. Contrary to Herrnstein and Murray's formulation, both theory and research findings indicate that, at least in the case of MZ-DZ contrasts, when heritability rises, DZ twins decreasingly resemble each other, at least when they reside in environments that contain sufficient resources to actualize their potential differences.

IS IT BETTER TO BE BORN RICH OR SMART?

The New Interpreters attempt to complete their genetic meritocracy argument by linking inherited variations in IQ to variations in poverty, adult earnings, and other social indicia of well-being (e.g., criminality, completion of high school, welfare dependency, matriculating in college). Again, I turn to Herrnstein and Murray (1994) for a quotation, though equally explicit arguments can be found in the writings of other New Interpreters:

> Cognitive ability is more important than parental SES in determining poverty . . . In sum, low intelligence means a comparatively high risk of poverty. If a white child of the next generation could be given a choice between being disadvantaged in socioeconomic status or disadvantaged in intelligence, there is no question about the right choice. (p. 135) . . . IQ has a strong effect independent of socioeconomic background. (p. 137) . . . What is the role of socioeconomic background after we take IQ into account? Not much. (p. 139)

Once again, the New Interpreters have oversimplified the complex path linking childhood IQ to adult economic status, omitting from their discussion relevant counterevidence.

As described in Chapter 4, Charles Henderson and I, in response to evidence similar to that cited by other New Interpreters, analyzed a sample of adults from *Project Talent*. Our goal was to replicate and extend the analyses in support of the claim that IQ variation is causally linked to economic variation. We applied extensive statistical and mathematical models to data obtained from a nationally representative sample to assess the impact of adolescent IQ on adult economic success. In the most fully specified statistical model in which parental social status and high school intelligence scores were each entered as covariates, the effect of IQ as a predictor of adult income was totally eliminated. In contrast to the impotence of IQ as a predictor of earnings, socioeconomic background and education were strongly significant positive predictors of adult income. Basically, in our most completely specified statistical model, differences in IQ were not linearly related to variance in adult earnings. This same analysis simultaneously revealed the potency of social and educational variables as important predictors of differences in adult earnings. (The full details of the various mathematical and statistical models we tested can be found in Henderson and Ceci, 1992). So does this mean that the conclusions drawn by New Interpreters such as Herrnstein and Murray are wrong? The answer is more complicated than a simple yes or no. It is possible to construct alternate models in which IQ does play an important causative role, leading from family socioeconomic background to adult economic achievement. An entire school of sociology, called "status attainment," was begun in the 1960s and thrived through the early 1970s on just such models; e.g., Duncan, Featherman, and Duncan, 1972; Waller, 1971. In addition, Henderson's and my results say nothing about the possibility that genetic sources of variance over and above those that "load" on IQ may be predictive of accomplishments (though this becomes a tortured argument, requiring more and more assumptions and less and less parsimony in the models).

Consistent with the argument advanced above, one could ask what difference it would make to the income distribution of Americans if everyone in Herrnstein and Murray's NLSY sample was remunerated in accord with his or her non-IQ characteristics (e.g., if everyone was paid in accord with his or her family social origins, motivation, skill levels, and not in accord with differences in IQs). The answer is that statistically eliminating IQ variation as a source of income differences, and distributing wealth solely according to non-IQ characteristics, would not change the present income distribution of Americans much; in fact, it changes it hardly at all, as shown in a recent analysis by Dickens, Kane, and Schultze (1995).

In contrast, one can ask what difference it would make to the current income distribution of Americans if everyone was remunerated exactly in accord with his or her IQ and not as a result of non-IQ attributes (i.e., eliminating differences in skill, motivation, and family education as bases of income differences, and distributing money solely according to IQ). The answer is that paying workers solely according to their IQs would dramatically reshape the American income distribution: many more people would earn around the average salary than is the case at present because income varies much more as a result of non-IQ differences than as a result of IQ differences. In the words of Dickens and his colleagues: "If all that mattered was (IQ) scores, U.S. society would clearly be very egalitarian. Eliminating differences due to IQ would have little effect on the overall level of inequality" (Dickens et al., 1995, p. 20).

Another way to think about this is to compare the incomes of those who possess the top 10 percent of IQs with those who possess the top 10 percent of wages. The incomes of those with the top 10 percent of IQs in Herrnstein and Murray's sample earn 55 percent more than persons with average-IQs. In contrast, the top 10 percent of wage earners in Herrnstein and Murray's sample earn 200 percent more than the average person! Hence, the proportion of the variation in income that can be explained on the basis of differences in IQ seems to be quite small.

The bottom line from the analyses reported above (Dickens et al., 1995; Henderson and Ceci, 1992) is that the New Interpreters have overlooked evidence that calls into question another prong of their genetic meritocracy syllogism. Contrary to their unqualified assertion that it is better to be born smart than rich, the most recent evidence would seem to suggest the opposite. Earning variation is far more the result of variation in nonintellective characteristics than in differences in IQ. This point is worth bearing in mind when we consider the thrust of the New Interpreters' claim that society is a meritocracy in which the "haves" and "have-nots" are sorted according to cognitive ability. As we saw, the sorting is largely according to noncognitive bases.

One additional point is relevant here: Herrnstein and Murray have provided what some regard as a *tour de force*, reporting no fewer than nine-

teen separate analyses in which they pit social class against IQ in predicting some developmental outcome, such as poverty, crime, or completing
schooling. In all nineteen of these analyses, they find that IQ is a far more
potent predictor than social class. Take, for instance, the relationship
between IQ and entering college vs. the relationship between social class
and entering college. New Interpreters argue that variations in social class
are unimportant in predicting who will attend college if we take into consideration differences in IQ, whereas variations in IQ turn out to be
extremely important in predicting who will attend college even after we
have controlled for social-class differences. In other words, the implication
is that low-IQ individuals have a greatly reduced chance of attending college, no matter what their parents' social class, whereas low social class
individuals have the same chance of attending college as high social class
individuals—if their IQs are comparable. These nineteen analyses lead
Herrnstein and Murray to assert that social class (and, by proxy, all aspects
of the environment that are associated with it) are relatively unimportant
in predicting real-world attainments like the college attendance example
just described: they assert that IQ drives social attainment, irrespective of
the environment. But is this really true?

One thing that readers need to bear in mind is that although statistics
are powerful tools for reducing data and establishing linkages and testing
hypotheses, they are like any other tools—they can be used in a variety of
ways. The statistics used by Herrnstein and Murray are no exception. In
all nineteen of their analyses these authors have done the following: they
fixed the control variable (IQ or social class) at its average or mean value
when they estimated the impact of the other variable (either IQ or social
class) on the outcome in question—in this example "attending college."
Thus, logically, they are asking the following two questions: "How important are differences in IQ in predicting who goes to college *among middle-
class individuals?*"; and "How important are differences in social class in
predicting who goes to college *among average-IQ individuals?*" Note that I
underscored two phrases that most readers will not realize are implicit in
the way Herrnstein and Murray did their statistics. They did not ask how
important IQ is in predicting who goes to college among low versus high
social class individuals, nor did they ask how important the environment
is in predicting who goes to college among low- and high-IQ individuals.
They set values of their control variables (IQ and socioeconomic status) at
their midpoints, which is traditionally the way such analyses are carried
out. If, however, one is interested in assessing the importance of the environment on "attending college" (or any of the other outcomes in these
nineteen analyses), then the place to look for its impact is not at the midpoint of the distribution, but at its tails, particularly at the left-most tail
(i.e., the low end). What do you suppose would happen if Herrnstein and
Murray had asked the following question: What is the likelihood of going

to college among low-IQ individuals who come from high social class environments versus the likelihood of going to college among average- or even high-IQ persons who come from low social class environments? If the environment is as feeble as these authors claim, then there ought to be a larger portion of high-IQ persons from the lower social strata attending college than there are low-IQ persons from the higher social strata. This is not a question that can be answered from the way Herrnstein and Murray did their analyses, but it is a fairer test of the importance of the environment than is fixing the control variables at their mean value. Interestingly, when the statistical models are run along these lines, it appears that poor individuals with average IQs are less likely to matriculate in college within eight years of graduation from high school than are wealthier individuals who possess lower IQs. Viewed this way, the environment would seem to matter a great deal. Once more, a prong in the New Interpreters' argument is tenuous at best.

Is There a Dysgenesis?

The final claim made by the New Interpreters is that dysgenic trends have been occurring in America for some time and that they are leading to a cognitive and social divergence among Americans. In the words of Herrnstein and Murray (1994):

> The effect is dysgenic when a low-IQ group has babies at a younger age than a high-IQ group ... In the United States women of lower intelligence have babies younger than women of higher intelligence. (p. 351) ... The higher fertility rates of women with low IQs have a larger impact on the black population than on the white. The discrepancies are so dramatically large that the probability of further divergence seems substantial. (pp. 353–354) ... Mounting evidence indicates that demographic trends are exerting downward pressure on cognitive ability in the U.S. and that these pressures are strong enough to have social consequences ... blacks and Latinos are experiencing even more severe dysgenic pressures than whites, which could lead to further divergence in future generations ... Putting the pieces together, something worth worrying about is happening to the cognitive capital of the country. (p. 341)

One might wonder what evidence the New Interpreters rely on for these claims. I can think of three relevant questions to ask in this regard: 1) Are rich and poor individuals' IQs diverging over time? 2) Are blacks and whites' IQs diverging? and 3) Are high and low test scorers' IQs diverging? If the New Interpreters are right, we should find positive answers to these three questions. I doubt we will, however. My colleagues and I have already provided fairly convincing evidence that the answers to the first two questions are negative, and I predict that if we were to tackle the third question we might also find a negative answer. As will be seen, there is no compelling evidence for the claim of dysgenesis, and in fact there is good evidence that the test scores of various groups are actually converging.

Between 1973 and 1990 blacks closed half of the thirty-four-point gap that separated them from whites on various verbal, mathematical, and scientific achievement tests. It is beyond the scope of this Epilogue to discuss these findings in detail, but suffice to say that not only have I arrived at this conclusion on the basis of my analyses of the National Assessment of Educational Progress (NAEP) data, but an independent team of researchers from the RAND Corporation also arrived at strikingly similar results (Grissmer, Kirby, Berends, and Williamson, 1994). But best of all, Herrnstein and Murray themselves have arrived at a similar conclusion! On page 290–291 of *The Bell Curve*, they mention three studies that show that the black-white IQ gap seems to be converging, not diverging, though they attempt to moderate this conclusion by noting statistical concerns about one of the three studies. Further, they note that in their own analysis of the NAEP data they, too, found a convergence of math and verbal fluency measures of blacks and whites, especially for the seventeen-year-olds. Although the magnitude of their convergence is less than what both I and the RAND group (Grissmer et al., 1994) found, there is no denying that for all nine tests (science, math, and verbal fluency, for each of three age groups) that Herrnstein and Murray report in the table on page 291 of *The Bell Curve*, there is convergence as opposed to divergence. Herrnstein and Murray acknowledge this result but then proceed to gainsay it. In their words:

> As the table indicates, black progress in narrowing the test score discrepancy with whites has been substantial on all three tests and across all of the age groups. The overall average gap of .92 standard deviation in the 1969–1973 tests had shrunk to .64 standard deviation by 1990. The gap narrowed because black scores rose, not because white scores fell. Altogether, the NAEP provides an encouraging picture. (p. 291) . . . The question that remains is whether black and white test scores will continue to converge. If all that separates blacks from whites are environmental differences and if fertility patterns for different socioeconomic groups are comparable, there is no reason why they shouldn't. The process would be very slow, however . . . reaching equality sometime in the middle of the twenty-first century . . . If black fertility is loaded more heavily than white fertility toward low-IQ segments of the population, then at some point convergence may be expected to stop, and the gap could begin to widen again. (p. 293)

This is a most amazing meandering across data and assumptions! There is no scientific evidence in the direction of a cognitive divergence of test scores between black and white youngsters, yet Herrnstein and Murray, troubled by the earlier onset of childbearing among black teenagers, cling to a dysgenesis hypothesis. Their worry as to whether fertility among blacks is loaded downward is an example of presupposing what ought to have been the the object of investigation. In fact, if these authors were to stick to the available data, the more warranted conclusion would be that all indicators point in the direction of a continued narrowing of the racial gap that had as recently as twenty-five years ago been approximately twice as large as it is at present.[6] Mutatis mutandem, there is, on the face of it, no evidence that the earlier onset of childbearing among blacks has been "downward loaded."

As if these data are not damaging enough to the claim of dysgenesis, additional findings from James Flynn's lab in New Zealand further damage the hypothesis. The dysgenesis hypothesis leads to the expectation of a growing tendency for good genes for IQ to rise to the top of the occupational scale, and for bad genes to fall to the bottom. Specifically, it leads to the expectation that the IQ gap between the children of the upper- and lower-income groups has been diverging over time. Flynn's data come from the Stanford Binet normative sample tested in 1932 and the WISC samples tested in 1948, 1972, and 1989. Flynn shows that the IQ gap between the children at the top third of SES and those at the bottom third of SES has not diverged during this century. In 1932 there was a twelve-point gap between the richest and poorest individuals' IQs, whereas today that gap is only ten points. Although this may not represent much of a convergence, it is certainly not evidence of a divergence.[7] Another prong in the New Interpreters' genetic meritocracy comes up wanting.

A Summary Assessment

Taken as a whole, my analysis of the New Interpreters' work leads me to conclude that the evidence and argument they present are inadequate to sustain their stated conclusions. Furthermore, other research findings, not cited by them, are incompatible with their position that individual differences in IQ reflect variation in a singular, general intellectual capacity that is primarily determined by biological inheritance and is leading to further racial and cognitive divergence. The principal challenge to their thesis comes from two sources. First, research based on alternative multifactorial conceptions of intelligence challenges the New Interpreters' claims about the unity and pervasiveness of general intelligence as a predictor of life-course outcomes. Attempts to dismiss proponents of multifactorial theories of intelligence by labeling them "revisionists" and "radicals" does little to advance the debate and, more important, cannot lessen their empirical and theoretical support. As I have attempted to explain here, this support is far from trivial. Second, compelling evidence from a number of studies demonstrates that heritability varies systematically as a function of the quality of the environment in which a particular population lives. Rises in heritability occur not because two children become more alike but rather because they become less alike, at least when they are reared in environments good enough to bring their potential differences to fruition. Thus genes appear to code not for static pieces of protein, but rather for potential ranges of reaction that can be more or less realized in various settings (Dobzhansky, 1955). This may be why there are pronounced secular trends in heritability; at some points during this century the size of heritability coefficients has been very small in comparison with other times. With a few exceptions, during times of economic scarcity (depressions and wars) heritability coefficients are smaller than they are during times of plenty (Sundet et al., 1988). Heritability is not something static, cast in stone. Just as it has become fashionable to remind readers that the environment is "genetically loaded," meaning that specific genotypes tend to select specific environments, thus resulting in genetic influences that masquerade as environmental ones, it is equally true that "heritability" is environmentally loaded; its magnitude is a function of how good the environment is.

We have seen that the New Interpreters' derivative conclusion that between-group differences in intelligence are the result of genetic selection is untenable on three counts. First, large IQ fluctuations have occurred over very short periods—often within the subjects' own lifetime, providing no evidence of hereditary transmission. For instance, SAT scores plummeted between 1967 and 1990 by an entire standard deviation, too brief a time frame to be the result of genes. Second, research findings not considered by the New Interpreters directly contradict predictions derived from the genetic distribution hypothesis, but are fully consistent with explanations in terms of environmental influence. This does not mean that genetic selection cannot and does not occur; instead it implies that genetic selection cannot be claimed as a primary or sole source of the between-group differences under consideration. Third, and most critical, the New Interpreters' reliance on differences in heritability coefficients as the primary basis for their conclusion violates explicit warnings (including their own), stressed in basic treatises and texts in human genetics, that such measures are not valid for assessing the contribution of genetics to mean differences between groups.[8]

At a broader level, the New Interpreters' conclusions rest on the assumption that the heritability coefficient provides a measure of the genetic contribution to intelligence that is free of environmental influence. In the light of the findings cited in this Epilogue, such an assumption cannot be sustained. Rather, these same findings point to a testable (and already partially tested) general hypothesis that invokes a more interactive conception of the role of genetics and environment in the development of intelligence; namely, high levels of heritability for IQ occur only in environments that provide the resources and conditions that are conducive to intellectual growth. This is the essence of the bioecological theory of intellectual development. It embraces the interaction of biology with ecology. Biology contributes to virtually everything about us, as I argued at the beginning of this book six years ago. But heritability coefficients are not the proper way to assess the importance of biology in the domain of human intellectual development for one reason: they tell us what part of our biological inheritance has already been actualized by the environment, leaving the unactualized component both unknown and unknowable.

Trend lines. Scientists rarely make predictions except under fairly safe conditions (e.g., if the truth cannot be known in their lifetime, they may hazard a prediction because they will not be embarrassed by refutations). I will go out on a limb and make a prediction that no one else has made, and it is one for which I hope to be alive to be held accountable if I am wrong. I predict that in the coming twenty years the racial gap in IQ that has bedeviled ameliorative efforts by social and educational reformers from the New Deal to the Great Society will, in fact, close to within eight points. Remember that the racial gap has been fifteen to sixteen IQ points throughout the twentieth century. I predict that it will settle around eight points by the year 2015. I have reasons for the specificity of this prediction, but a full explanation of them would carry me beyond the scope of this Epilogue.[9] Remember, the IQ scores of blacks today are identical to those of whites in the 1940s; the latter group has gained approximately fifteen points, and blacks have gained at least this many.

PROXIMAL PROCESSES AS THE ENGINES THAT DRIVE INTELLECTUAL DEVELOPMENT

In closing, it is time to ask about the nature of the resources responsible for intellectual growth. Past research on the influence of the environment has ducked this question, preferring instead to contrast global SES differences on IQ, surmising that some aspects subsumed under the SES rubric must be causative but never specifying precisely what they might be. In a recent article, Urie Bronfenbrenner and I (see Bronfenbrenner and Ceci, 1994) proposed specific mechanisms of organism-environment interaction, called *proximal processes*, through which genetic potentials for intelligence are actualized. We described research evidence from a variety of sources demonstrating that proximal processes operate in a variety of settings throughout the life-course (beginning in the family and continuing in child-care settings, peer groups, schools, and work places), and account for more of the variation in intellectual outcome than the environmental contexts (e.g., family structure, SES, culture) in which these proximal processes take place. Proximal processes refer to sustained interactions between a developing orgasm and the persons, symbols, and activities in its immediate environment. To be effective, these processes must become progressively more complex and interactive over time. So, for example, a proximal process for an infant might be a caregiver's effort to focus her attention on an object; with time, this focusing effort becomes more complex, and the developing child learns to focus using less and less parental input. For an older child, proximal processes might entail parental monitoring of homework assignments or even the interactive nature of reading. Proximal processes are associated with increased competence in all domains in which they have been tested. For instance, in one study it was shown that caregivers trained to interact with children using proximal processes produced larger cognitive gains than caregivers who spent the same amount of time with the children but did not use proximal processes.

I mention this work on proximal processes because the theoretical and empirical findings provide additional evidence for the inadequacy of the kinds of models the New Interpreters have employed, and for the indiscriminate way in which their models have been applied and interpreted.

As can be seen in the figure below, proximal processes are hypothesized to account for more variation in IQ and heritability coefficients than the environmental addresses (social class, political ideology, culture) in which these proximal processes operate. So the straightforward prediction is that if we pit proximal processes against the gross measures of the environment such as SES, the former will account for more of the variation in IQ (and heritability) than the latter.

The distance between the platforms in the figure reflects the hypothesized potency of proximal processes, because the distance between the IQs (and heritability coefficients) associated with good versus poor proximal processes is greater than the distance between the IQs and the heritability coefficients associated with good versus poor environmental addresses (here indexed by SES). In bioecological theory, proximal processes are the engines that drive development, steering it toward one destination or another, depending on its level. With high levels of proximal processes, a child actualizes more of her myriad genetic potentials for intellectual behaviors than she otherwise would; with low levels of proximal processes, fewer of her potentials come to fruition. Psychometric models of the heritability of IQ, in the absence of this concept, are needlessly deterministic and ungenerous.

The root problem with arguments of the kind made by the New Interpreters is that they pose the wrong question. To ask whether genetics or the environment influences intelligence more is to assume that the two sets of forces operate independently, whereas, in fact, they influence each other. The main questions, both for science and for society, is how they influence each other, and how they can best work together. This is where we should be putting our research efforts.

THE BIO-ECOLOGICAL MODEL

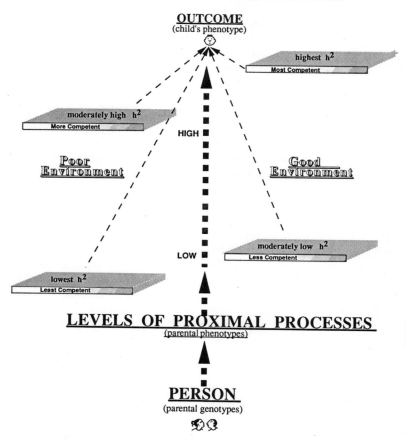

Finally, I return to the least defensible of the New Interpreters' conclusions—their claim that race differences in general intelligence are genetically determined. On the purely scientific side, it is essential to recognize, first and foremost, that heritability is a measure only of that portion of genetic potential that has already been actualized in particular types of environments. At the present time, the extent of unactualized genetic potential for a given population is unknown, and is destined to remain so, pending further advances in molecular genetics and the mapping of the human genome. This means that reported heritabilities apply only to environments that currently exist and have already been studied. Heritability estimates provide no direct information about what the level of actualized potential might be under other environmental conditions that may exist elsewhere or be created in the future either by nature or by design (for a similar argument see Dobzhansky, 1955). In sum, at the present time, no scientific basis exists for claiming a genetically set limit for the level of general intelligence for a particular group of human beings. Furthermore, as pointed out in Chapter 6, the very concept of general intelligence, or g, is fraught with conceptual and empirical difficulties. Horn (1989) has provided a cogent demonstration of the indeterminate nature of g; his argument is roughly analogous to the indeterminate nature of heritability, pointing out that g is a function of the nature and number of tasks being measured, with fluctuating magnitudes and sizable specific factors.

What has emerged from the bioecological analysis is a set of findings that leads to an ironic conclusion. Heritability coefficients turn out to be lower in poorer environments. Given the additive two-factor structure of the heritability formula (i.e., what is not accounted for by genetics must be accounted for by the environment—plus measurement error plus covariation), it necessarily follows that the influence of the environment on differences in IQ among children growing up in straitened circumstances is greater than that for youngsters raised in a more favorable milieu. This in turn implies that efforts to enhance intelligence by improving the environment are likely to be most effective for children living in the most impoverished circumstances, the very group that many of the New Interpreters seem to consider beyond remediation.

The foregoing considerations apply with even greater force to poor black families and their children. It is they and their descendants who, in the New Interpreters' reading of the data, are most surely condemned to a life of genetically predestined mental inferiority, a dysfunctional family, and patterns of behavior destructive to self and society. It is ironic that Bronfenbrenner's and my examination of the same and additional evidence has led us to the opposite conclusion, albeit with less certainty than that exuded by the New Interpreters.

The journey to understanding individual and group differences in intellectual functioning has been a long, winding path, trod by many scholars traveling from different and distant scientific climes. Until the dust settles and a common destination becomes visible, scientists should not be hasty to draw firm conclusions where human destinies are at stake.[10]

NOTES

1. Although heritability coefficients were not reported by Fischbein, I calculated them from the table of intraclass correlations included in her published report (see Fischbein, 1980, p. 58).
2. In addition, because lower-class families in the United States are significantly more disadvantaged than their counterparts in Sweden (Kamerman and Kahn, 1995), the corresponding contrasts in environmentality coefficients for the United

States are likely to be even greater, further eroding the assertions of the New Interpreters about the impotency of the environment to offset genetic differences.

3. Nearly all writers on this topic rely on the eminent statistician John B. Carroll for their support, as he has been the most cogent adherent of the *g* position. Yet Carroll himself has noted that the issue cannot be resolved with existing data, and he acknowledges that the non-*g* depiction I put forward in Chapter 6 may be correct. Describing the example I presented in note 1 of Chapter 6 involving a battery of test scores yielding a general factor not because they are saturated with common variance but because they share adjacent properties that are not wholly shared across all or even most tests, Carroll states:

> It seemed to me that it would be useful to investigate Ceci's proposal by constructing a plasmode in accordance with his hypothesis. Such a plasmode is shown . . . as a hypothetical orthogonal factor matrix . . . Now one problem is immediately apparent: the correlation matrix is only of rank three, and obviously the three factors that could be derived and rotated to simple structure would be correlated, producing a general factor, even though the plasmode is postulated to contain six orthogonal factors . . . I tried to fix this problem by varying the weights in the plasmode . . . Now we know that this is an orthogonal transformation of the plasmode . . . However, applying standard rotation and Schmid-Leiman orthogonalization procedures to the principal factor matrix, we do not recover anything like the plasmode with which we started. We recover only what looks like a 3-factor solution, similar to what was obtained from the 3-factor plasmode. The conclusion I draw is that IF psychological tests reflect a series of "partially overlapping microlevel processes," our standard procedures of factor analysis will not ordinarily reveal this, and the *g* factor we ordinarily derive is an artifact resulting from the operation of these partially overlapping processes. This is a disappointing result if it is valid. I am not sure how to proceed to provide a remedy for the artifactual results. (Carroll, 1993, pp. 16–20)

4. Herrnstein and Murray are on logically treacherous ground here, for the interpretation can go either way. Thus the disjunction between IQ and social class could be explained as a lack of influence of the former on the latter. That is, if a certain level of IQ is thought to be necessary to attain a certain level of socioeconomic status, then the fact that some blacks appear to have attained this status in the absence of IQs as high as their white counterparts could be seen as a refutation of the very premise that a high IQ is causally connected to having a high income. It may be helpful to spell out here the logical problem with this type of reasoning.

If we recast the New Interpreters' assumption (i.e., a certain level of IQ is necessary to attain middle-class status) in terms of "if *p* then *q*" conditional reasoning, we can see the problem more readily. Logically, there exist two classes of hypotheses that are permissible, known as Modus Ponens and Modus Tollens. According to Modus Ponens, if *p* (sufficiently high IQ) is true, then *q* (middle-class status is more likely to be attained) becomes true. Now consider Modus Tollens: *q* is not true, therefore *p* cannot be true. If we accept the New Interpreters' premise that having a high IQ is necessary for attaining a high social status, and if we know individuals who have high IQs, then Modus Ponens leads to the expectation that they will attain high social status, or at least be more likely to do so than lower IQ groups. Additionally, if we know that a group of individuals did not attain high social status, then Modus Tollens leads to the expectation that, as a group, they should not have high IQs. (All of this, of course, assumes that the premise statement is true, but later I will argue that there are good grounds for suspecting that this is not the case either.) However, it is logically impermissible to reason as follows: "Since a group has attained high social status, then the group must have a high IQ." This so-called affirmation of the consequent is illogical because other conditions exist that permit this to occur that are not entailed in the conditional statement if p then q. In other words, we cannot infer that the antecedent is true simply by knowing that the consequent is true. Similarly, we are not permitted by the laws of logic to predict "not q" simply by knowing "not p." This is known as "denying the antecedent." The New Interpreters frequently lapse into treating conditionals as if they were biconditionals. They are not. One can achieve middle-class status in a variety of ways, and the causal role of IQ is unclear in many of them.

5. To extrapolate from the above argument, the magnitude of heritability can increase either because the intraclass correlation for identical twins becomes larger or because the corresponding correlation for fraternal twins becomes smaller. Consider the formula for calculating the heritability coefficient: heritability $= 2(r_{MZ} - r_{DZ})$, where r_{MZ} and r_{DZ} are the intraclass correlation for the IQs of identical and fraternal twins. Thus heritability can go up because MZs' IQs become more alike or because DZs' IQs become less alike. Changes in either term will result in an elevation of heritability. Empirically, it is found that heritability rises because of the latter, and not because of MZs' IQs becoming more alike (Bronfenbrenner and Ceci, 1994).

6. This statement requires explanation. In Herrnstein and Murray's analysis, the racial gap in test scores has been reduced by approximately one-third of its 1972–1978 magnitude. In my own analysis of the same data set used by these authors (the NAEP), the gap has been reduced by approximately one-half, though there are recent signs of a slight widening that, if allowed to continue for another ten years, would erase the narrowing. Grissmer et al. also conclude that the gap has been closed by one-half, and Hauser and Huang (1996) concluded it has closed by approximately one-half. Herrnstein and Murray arrived at their one-third figure because they used an "age-inflated" standard deviation that was inappropriate.

7. In an editorial in the *Wall Street Journal* on December 13, 1994, fifty-two well-known scholars expressed the belief that "there is no persuasive evidence that the IQ bell curves for different racial-ethnic groups are converging. Surveys in some years show gaps in academic achievement have narrowed a bit for some races, ages, school subjects and skill levels, but this picture seems too mixed to reflect a general shift in IQ levels themselves" (p. 26). Although I think that reasonable individuals can disagree about whether the results of a convergence of cognitive scores constitutes "persuasive evidence," there is no evidence that the test scores are diverging, thus no support for the dysgenesis advocated by the New Interpreters.

8. Interestingly, a colleague who is most supportive of the New Interpreters told me that although he agreed that in principle one cannot leap from within-group differences to between-group inferences, there were no animal studies he was aware of in which between-group differences were not genetically based when within-group differences were genetically based.

9. We have become fond of telling ourselves that social problems are intractable, not lending themselves to easy solutions, and that throwing money at them only

causes other problems to supplant them. In reality, however, "throwing money" at the racial gap in achievement test scores during the past three decades does appear to have worked. Programs targeted at minority students and their parents (e.g., Title 1, Title 9, Head Start, desegregation of schools, school breakfast and lunch programs, Affirmative Action, GED programs for school drop-outs, day-care programs for the babies of high school mothers) seem to have been at least partially responsible for the closing of the test score gap described above in the section on racial convergence. The single largest factor in closing this gap (accounting for roughly 25 percent) appears to be the rising educational level of black parents, whose educational progress has outpaced white and Hispanic parents during this same period. It is on the basis of current trends that I predict a further closing of the test score gap. And because I view IQ as conceptually indistinguishable from achievement test scores, it is also the basis for my prediction that the IQ gap will be narrowed. Any wagers?

10. To escape the criticism that the genetic meritocracy is a needlessly ungenerous depiction of ethnic minorities, many New Interpreters enjoin the reader to bear in mind that their assumptions carry no implications for invidious social comparisons, only for large-scale demographic trend analysis. For example:

> It is important to understand that even a difference of 1.2 standard deviations means considerable overlap in the cognitive ability distribution for blacks and whites, a large proportion have IQs that can be matched up. For that matter, millions of blacks have higher IQs than the average white. Tens of thousands have IQs that put them in the top few percentiles of the white distribution. It should be no surprise to see African-Americans functioning at high levels in every intellectually challenging field. This is the distribution to keep in mind whenever thinking about individuals. (Murray and Herrnstein, 1994, p. 28)

Notwithstanding such enjoinders, the New Interpreters' argument would seem to carry undeniable sociopolitical consequences. For example, if one accepts their syllogism that i) IQ scores reflect underlying intellectual potential; ii) the black/white difference on IQ is on the order of 1.1 standard deviations (about sixteen points); and iii) these differences are largely inherited, then whenever a prospective employer requests that the Personnel Office send twenty-four applicants who score one standard deviation above the mean, twenty-three of them will be white. In essence, if Herrnstein and Murray's syllogism is correct, then if an employer currently limits screening to high-IQ persons, one might as well hang a sign that reads: "Blacks need not apply."

REFERENCES

Bronfenbrenner, U., and S. J. Ceci (1994). Nature-nurture reconceptualized in developmental perspective. *Psychological Review, 101,* 568–586.

Carroll, J. B. (1993). Cognitive abilities: A critique. *Proceedings of human cognitive abilities in theory and practice,* 16–20.

Cavalli-Sforza, L.L., and W. F. Bodmer (1971). *The genetics of human populations*. San Francisco: W. H. Freeman.

Ceci, S. J. (1990). The relationship between microlevel and macrolevel processing: Some arguments against reductionism. *Intelligence, 14*, 1–9.

Ceci, S. J., and J. Liker (1986). A day at the races: A study of IQ, expertise, and cognitive complexity. *Journal of Experimental Psychology: General, 115*, 255–266.

Deary, I. J. (1995). Auditory inspection time and intelligence: What is the causal direction? *Developmental Psychology, 31*, 237–250.

——— (1993). Inspection time and WAIS-R IQ subtypes: A confirmatory factor analysis study. *Intelligence, 17*, 223–236.

Detterman, D. K., J. D. Mayer, D. R. Caruso, P. J. Legree, F. A. Conners, and R. Taylor (1992). The assessment of basic cognitive processes in relationship to cognitive deficits. *American Journal of Mental Retardation, 97*, 251–286.

Dickens, W. T., T. J. Kane, and C. L. Schultze (Summer, 1995). Ring true? A closer look at a grim portrait of American society. *The Brookings Review, 13*(3), 18–23.

Dobzhansky, T. (1955). *Evolution, genetics, and man*. New York: Wiley.

Dörner, D., and H. Kreuzig (1983). Problemslosefahigkeit und intelligenz. *Psychologische Rundschaus, 34*, 185–192.

Duncan, G., J. Brooks-Gunn, and P. Klevanov (1994). Economic deprivation in early childhood. *Child Development, 65*, 296–318.

Duncan, O. D., D. Featherman, and B. Duncan (1972). *Socioeconomic background and achievement*. New York: Seminar Press.

Eysenck, H. J. (1982). *A model for intelligence*. New York: Springer-Verlag.

Falconer, D. S. (1989). *An introduction to quantitative genetics*. New York: Wiley.

Fischbein, S. (1980). IQ and social class. *Intelligence, 4*, 51–63.

Fraser, S. (1995). *The Bell Curve Wars*. New York: Basic Books.

Fuller, J. L., and W. Thompson (1976). *Behavior genetics*. New York: Wiley.

Galton, F. (1892). *Hereditary genius*. London: Macmillan.

——— (1883). *Inquiries into human faculty and its development*. London: Macmillan.

Gordon, R. A. (1987). SES versus IQ in the race-delinquency model. *International Journal of Sociology, 7*, 30–96.

Gottfredson, L. (1986). Societal consequences of the *g* factor in employment. *Journal of Vocational Behavior, 29*, 379–410.

Grissmer, D. W., S. N. Kirby, M. Berends, and S. Williamson (1994). *Student achievement and the changing American family*. Santa Monica, Calif.: RAND.

Hauser, R. M., and M. H. Huang (1996). Trends in black-white test score differentials. Paper presented at "Intelligence on the Rise" conference, Emory University, Atlanta, April 12.

Henderson, C. R., and S. J. Ceci (1992). Is it better to be born rich or smart? A bioecological analysis (pp. 705–751). In K. R. Billingsley, H. U. Brown, and E. Derohanes (eds.), *Scientific excellence in supercomputing: The 1990 IBM contest prize papers*. Athens, Ga.: The Baldwin Press, The University of Georgia.

Herrnstein, R., and C. Murray (1994). *The bell curve: Intelligence and class structure in American life*. New York: The Free Press.

Horn, J. (1989). Thinking about human abilities (pp. 645–685). In J. R. Nesselroade (ed.), *Handbook of multivariate psychology*. New York: Academic Press.

Itzkoff, S. W. (1989). *The making of the civilized mind*. New York: Peter Longmans.

———. (1987). *Why humans vary in intelligence*. Ashfield, Mass.: Paideia Publishers.

Jacoby, R., and N. Glauberman (1995). *The Bell Curve Debate*. New York: Random House.

Jensen, A. R. (1985). The nature of black-white differences on various psychometric tests: Spearman's hypothesis. *Behavioral and Brain Sciences, 8*, 193–263.

Johnson-Laird, P. N. (1983). *Mental models: Toward a cognitive science of language, inference, and consciousness*. Cambridge, Mass.: Harvard University Press.

Kammerman, S. B., and A. Kahn (1995). *Starting right: How America neglects its youngest children and what we can do about it.* New York: Oxford University Press.

Keating, D. P., J. A. List, and W. E. Merriman (1985). Cognitive processing and cognitive ability: A multivariate validity investigation. *Intelligence, 9,* 149–170.

Lynn, R. (1991). Race differences in intelligence: A global perspective. *Mankind Quarterly,* 251–277.

McCartney, K., M. J. Harris, and F. Bernieri (1990). Growing up and growing apart: A developmental metanalysis of twin studies. *Psychol. Bull., 107,* 226–237.

Murray, C. (Dec. 2, 1994). The real "bell curve." *Wall Street Journal.*

Murray, C., and R. J. Herrnstein (Oct. 31, 1994). The case for conservative multiculturalism, part 1. *The New Republic.*

Neisser, V., G. Boodoo, T. Bouchard, A. W. Boykin, N. Brody, S. J. Ceci, D. Halpern, J. C. Loehlin, R. Perloff, R. J. Sternberg, and S. Urbina (1996). Intelligence: Knowns and unknowns. *American Psychologist, 51,* 1–25.

Nisbett, R. (1995). Race, IQ, and scientism. In S. Fraser (ed.), *The bell curve wars.* New York. Basic Books, 36–57.

Plomin. R., and D. Daniels (1987). Why are children in the same family so different from one another? *Behavioral and Brain Sciences, 10,* 1–16.

Plomin, R., and R. Rende (1991). Human behavioral genetics. *Review of Psychology, 42,* 161–190.

Rosnow, R. L., A. Skleder, M. Jaeger, and B. Rind (1994). Intelligence and the epistemics of interpersonal acumen. *Intelligence, 19,* 93–116.

Rushton, J. P. (1988). Race differences in behavior: A review and evolutionary analysis. *Personal. and Ind. Differ., 9,* 1009–1024.

—— (1995). Cranial capacity related to sex, rank, and race in a stratified random sample of 6,325 U.S military personnel. *Intelligence.*

—— (1995). *Race, evolution, and behavior: A life history perspective.* New Brunswick, N.J.: Transaction Publishers.

Ryan, A. (Nov. 17, 1994). Apocalypse now? *New York Review of Books,* pp. 7–11.

Scarr-Salapatek, S. (1971). Race, social class, and IQ. *Science, 174,* 1285–1295.

Seligman, D. (1992). *A question of intelligence: The IQ debate in America.* New York: Birch Lane Press.

Smith, J. P. (1994). Racial and ethnic differences in wealth using the HRS. *Journal of Human Resources.*

Sternberg, R. J. (1985). *Beyond IQ: A triarchic framework for intelligence.* New York: Cambridge University Press.

Sternberg, R. J., R. Wagner, W. M. Williams, and J. Horvath (1995). Testing common sense. *American Psychologist, 50,* 912–927.

Streufert, S., and G. Y. Nogami (1989). Cognitive style and complexity: Implications for I/O psychology. In C. Cooper and I. Robertson (eds.), *Internat. Rev. of Industr. and Organizational Psych.* (pp. 93–143). New York: Wiley.

Sundet, J., M. Tambs, P. Magnus, and K. Berg (1988). On the question of secular trends in the heritability of intelligence test scores: A study of Norwegian twins. *Intelligence, 12,* 47–59.

Waller, J. (1971). Differential reproduction: Its relation to IQ test score, education, and occupation. *Social Biology, 18,* 122–136.

Wall Street Journal (Thursday, Dec. 13, 1994). Mainstream science on intelligence, p. A-25.

Endnotes

CHAPTER 1

1. Technically, g is the maximum linear variance among the test scores that can be accounted for independently of rotation, and its magnitude reflects the degree of correlation among the scores. The greater the magnitude of g, the more likely it is that a common source was involved in the different test performances. Some conceive of this common source of test variance as a type of primordial mental energy that "flows" into all of the different intellectual performances. The relationship between other indicants of general intelligence and g are well known; for example, g is correlated with IQ anywhere from .4 to .9, depending on the characteristics of the sample (e.g., age), the IQ test in question, and the battery of tests used to derive g. (Although I have not formally computed a mean value, of the 50 or more studies that I have read that report correlations between IQ and g, an approximate average correlation might be in the .60–.70 range.) Some, like Arthur Jensen (1980), have suggested that a test's g-loading is a reflection of the number or centrality of the neural processes required for its successful completion.

2. As two examples of the influence of the environment on brain growth, I offer a nonhuman and a human study. First, consider the case in which some rats are in small, restrictive cages, while others are allowed to run through large complexes of cages filled with interesting shapes and objects. Even though the two groups of rats were identical prior to the assignment of cages, those free to explore the diverse cages formed over 20% more synapses per neuron in the upper visual cortex than those raised in the restricted cage (Turner and Greenough, 1985). Moreover, when the connective tissue of rats is severed and the rats are provided only monocular experiences, more substantial dendritic fields exist in the occipital cortex of the hemisphere that received the visual input (Chang and Greenough, 1982).

   Next, consider the case of a child born with skin lesions in front of the eardrums. If these lesions are not surgically corrected by age five, there is a good chance that the child will never acquire speech and language properly, even if extensive remedial efforts are made (Goodwin, 1988). Since after surgery there is no longer any physical reason why the child cannot hear, the inference seems to be that the arborization phase of the auditory cortex is over by the age of five. Similar disruptions in brain compartments can be seen in patients whose childhood vision is occluded due to infection and who later have corrective surgery, only to discover that their vision is abnormal. Siegler (1989) suggests that early synaptic overproduction is absolutely critical to certain types of cognitive development, such as the "A-not-B object-permanence" problem in which an object is hidden in a new location. These types of extragenetic or environmental factors are not limitless in their effect on brain growth, but they do exert an influence on developing compartments within a specified range of values.

3. Many psychologists are unaware that the development of the first SAT was commissioned by the College Board Corporation as an "intelligence test to be used in college admissions." Carl Brigham, one of the principal researchers involved in the mass testing of army recruits during World War I, was named as the director of the team developing the first SATs and he proceeded to adapt it directly from the Army Alpha Test of verbal intelligence. Today's SAT still bears a resemblance to that first intelligence test (Owen, 1985), and performance on it is highly correlated with IQ test scores (e.g., Razz, et al., 1983, report a correlation of .81 for a sample of University of Texas undergraduates' SATs and their Cattell IQs).

CHAPTER 2

1. "From nothing, nothing comes."

CHAPTER 3

1. Although most developmental psychologists know of Luria's demonstrations that the construction of logical processes among peasants of Central Asia was linked to their degree of contact with European culture, thus supporting Vygotsky's view of the importance of historically determined cultural factors in cognition, few appreciate Rivers' influence on Vygotsky.

2. The mapping function (over a quarter sine phase) was $x_1, y_1, z_1 =$ random number (0-9), where $x, y, z =$ maxdistance $+ 1.8$ ($0.8 \sin x_1 .10 + 0.6 \sin y_1 \times .10 + 0.4 \sin z_1 \times .10$).

3. The reader may question our use of odds estimation as the criterion for expertise instead of some more obvious index, such as the number of winners picked. However, success at any form of parimutual wagering is dependent on the ability to identify choices (in this case, horses) that the public has undervalued and avoiding those that are overvalued. For example, suppose you are confronted with the possibility of betting on a heavy favorite, say 1:5 ($2.40 return on a $2.00 wager), that an expert regards as overvalued (i.e., the expert has estimated the

"true" odds to be closer to 5:2, or a $7.00 return on a $2.00 wager). Another horse in the race has post-time odds that an expert regards as undervalued, say, 10:1 (a $22.00 return on a $2.00 wager). Let us suppose that in this case the expert has estimated that the latter horse's "true" odds are closer to 4:1 than 10:1. Now, even though a horse whose true odds are 5:2 is likely to beat a horse whose true odds are 4:1 (10 out of 17 times, according to some statistical models), this is not what is important. The expert must determine whether the heavy favorite's advantage is commensurate with the gap between the true and the actual odds. In this example it is wiser to bet against the heavy favorite, even though he is more likely to beat the other horse, because over the long haul you will earn more money by doing so. Given "true" odds of 5:2 for the favorite versus 4:1 for the other horse, the odds of winning a head-to-head race are deterministic according to an independence assumption: The favorite should win 2 out of 7 times (10 out of 35 times), while the other horse should win 1 out of 5 times (7 out of 35 times). Thus, although the favorite ought to win the majority of the races (10 out of 17), at odds of 1:5 it will only return $24.00 over the course of those 10 wins (10 × $2.40) and therefore amount to a $46.00 loss over 35 races (assuming a constant $2 wager each race), while the other horse's 7 wins will yield a profit. So someone who exclusively bets favorites will inevitably lose money over the long haul at these tracks (they win approximately 40% of the time) while someone who picks winners only 10% of the time *could* win money. The moral is that the number of winners one picks is less important than estimation of the "true" odds.

CHAPTER 4

1. Wiseman reports a multiple R of .963 for a combination of social variables (mortality, birth rate, housing density, incidence of tuberculosis, illegitimacy rate, etc.) on IQ. The reason that educational attainment may have been less sensitive to these same social variables is unknown, but one possibility is that the reliability of the IQ score was much higher than the reliability for the attainment measures. Something cannot correlate with anything else higher than the square root of its correlation with itself.

2. Interestingly, the asymmetry between IQ and some components of social class is reflected in statistics such as the following: Among white males, those who score in the top 20% of the population on IQ tests earn about double what those who score in the bottom 20% of the population earn; on the other hand, among white males, those who are classified as the top 20% of wage earners have incomes that are over six times that of men in the bottom 20% of wage earners! This is an enormous gap between the incomes of the top and bottom fifth, and dispersing income according to IQ would not narrow it much. As far as black males are concerned, the relationship is hard to assess due to historical forces that suppressed their earning potential. Philips Cutright assessed the IQs achieved by white and black soldiers in 1952 and again in 1962. He found that whites earned about 43% more than blacks in 1962, but most of this gap was unrelated to IQ

differences; equalizing white and black IQs would only narrow it by about a quarter (Cutright, 1974). (For instance, among blacks and whites with the same IQ, blacks earned almost 32% less!) Similarly, according to Flynn (1987b) when IQ is controlled for between races, the earning and SES gaps are hardly narrowed: When IQ is made identical,

> black Americans are .80 SDs below all whites in income and 1.17 in SES. . . . The bottom 58.4 per cent of white Americans have a mean IQ ten points below all whites, and for them the correlations generate estimates of .20 SDs below average in income and .40 below average in SES. In sum, even if we accept Jensen's conclusions about genes and IQ. . . . 75% of the black deficit in income and 66% of the black deficit in SES cannot be attributed in any simple way to below average genes for intelligence." (p. 235)

CHAPTER 5

1. Take, for example, a phenomenon that was once prevalent in our country, namely, successfully waging a family feud. It is hard to walk away from descriptions of the horrors of these family feuds without questioning the rationality of the players. Rice describes a single harrowing encounter in the infamous feud between the Hatfields and the McCoys:

> Upon hearing that Alifair had been shot, Sarah McCoy, her mother, rushed to the back door. . . . and continued toward her dying daughter. Vance bounded toward her and struck her with the butt of his rifle. For a moment she lay on the cold gound, stunned, groaning, and crying. Finally, she raised herself on her hands and knees and tried to crawl to Alifair. . . . she pleaded with the attackers, "For God's sake let me go to my girl." Then, realizing the situation, she cried, "Oh, she's dead. For the love of God, let me go to her." Sarah put out her hand until she could almost touch the feet of Alifair. Running down the doorsill, where Alifair had fallen, was blood from the girl's wounds. Johnse (Hatfield), who was standing against the outside wall of the kitchen, took his revolver and crushed Sarah's skull with it. She dropped to the ground and lay motionless. (Rice, 1982, pp. 62–63)

Although Alifair and her brother Calvin were both killed on that afternoon, and their mother Sarah and the others were seriously wounded, the attack was only a skirmish in the long war between these two families waged over many generations. Randolph McCoy, who had escaped the attack that afternoon, returned to "make amends" to the Hatfields. And so it went, with both families participating in retaliatory raids against the other that guaranteed subsequent bloodshed. From Rice's account, the Hatfields and the McCoys were not irrational families in other ways (e.g., commerce and family relations). Who is to say that such cycles of violence ought to be labelled irrational, given that both families felt so vehemently that even death to themselves was a price they believed was worth paying for revenge? (It is tempting to draw a parallel between this type of behavior and that which goes on in so-called fundamentalist countries, where acts that Westerners see as irrational continue to find recruits even among the intelligentsia of those countries.)

2. As an example of what I mean, consider a recent *New York Times* article on the relationship between intelligence and myopia (Dean, 1988). It is well known that nearsighted persons do better on IQ tests than those who are not, and they also complete more years of school. A traditional explanation has been that reading itself promotes nearsightedness in persons who are genetically susceptible. However, it was reported that a Danish research team found that the correlation between IQ and myopia was as great as the correlation between years of schooling completed and myopia. This led them to conclude that the close visual exploration of the environment by nearsighted persons during their infancy may foster their later high intelligence as well as their later nearsightedness. Their entire argument depends critically on the assumption that IQ scores index intelligence and that schooling is something quite different. Thus, any variable that influences one ought to be independent of the other. As the researchers noted, "Had reading been the primary cause of the association (between myopia and IQ), one might have expected to find a stronger relation of myopia to educational level than to intelligence test score" (Dean, 1988, p. 3).

3. It may be worth noting that in 1905, when Alfred Binet was commissioned by Parisian school authorities to develop what was to become the first IQ test, he did so by interviewing teachers and asking them to describe the type of questions that children with learning problems found most difficult. He included a large number of these questions on his test, and variants of these questions can still be found in large numbers on most modern IQ tests, in part because the developers of modern tests sought to establish their validity by showing that they yielded comparable results to Binet's (Humphreys, 1984). So it isn't really surprising that IQ tests have predictive validity (i.e., they predict school success with $r \sim .55$), as they were comprised of items that poor learners failed in school. I think that if the tests had been named something other than intelligence tests (e.g., tests to predict school readiness or tests to predict conventional school learning), much of the subsequently passionate controversy about their validity might have been avoided.

That tests made up of items found difficult by children with learning problems can discriminate such children from their normally developing peers is hardly surprising; so can virtually all achievement tests. To call these tests measures of intellectual *aptitude* implies that the ability needed to pass them is something that we have an independent measure of, namely intelligence. The potential circularity here is plain: School failure is both "explained" as a lack of intelligence and is itself the basis for the definition of a lack of intelligence. The extra step in inference here is logically unneccessary, as school failure is the only touchstone. Elsewhere in the book I address the attempts of others to move beyond this circularity by inserting other criteria into the prediction, for instance, the relation between IQ and job success. Such criteria have *prima facie* validity (whatever intelligence is, it should not be antithetical to getting ahead in the world and probably even be positively linked to such an outcome), but their relation to IQ, although statistically reliable, is not nearly as strong as is that of school success. The fact remains that it is difficult to imagine a test of aptitude differing from a test of achievement by the time one has entered the sixth grade. Although early IQ scores may predict later school achievement slightly better than earlier achievement itself predicts later school achievement, IQ scores correlate with achievement scores about as well as the latter correlate with themselves, at least by the sixth grade (Humphreys, 1984).

Finally, in this connection, it is interesting to note that although Binet was occasionally ambiguous in his writings, he was clear in his mind about the nature of what his test of intelligence measured, namely, *judgment*. He did not believe that intelligence or the ability to judge was fixed or innate, nor did he believe that it reflected some quality that was independent of achievement: "Some recent philosophers appear to have given their moral support to the deplorable verdict that the intelligence of the child is a fixed quality. . . .We must protect against this brutal pessimism. . . .A child's mind is like a field for which an expert farmer has advised a change in the method of cultivating, with the result that in place of desert land, we now have a harvest" (quoted in Chase, 1977, pp. 301–302). Yet, when his test was translated into English in 1905 by Goddard and first revised in 1912 by Terman (both had been doctoral students of G. Stanley Hall, a hereditarian at Clark University), they maintained that it was, in fact, measuring innate differences. This contradiction led Marks (1982) to remark; "The remarkable thing about America's most used revision of the Binet test, the Stanford Revision, was that while Binet's tests measured acquired intelligence in France, they measured essentially innate intelligence in America" (p. 18). Prior to its American debut, the ratio of mental age to chronological age that formed the basis for the calculation of the early IQ score (after Stern) had not been used by Binet or others. It caught on only after Terman revised Binet's test in 1916.

4. It is important to bear in mind that I am not claiming that *in principle* one cannot distinguish achievement from aptitude. If someone had an aptitude for learning a foreign language and never went to a high school that taught foreign languages, then clearly his or her aptitude and achievement would be disparate. What I am claiming is that the contents of achievement tests and intellectual aptitude tests, as they are constructed today, are highly similar and inseparable both conceptually and statistically.

5. It is not entirely clear why increased schooling should be associated with higher performance on IQ tests and lower performance on SATs. After all, SATs are often viewed as a surrogate for IQ, since they were developed as a college intelligence test, possess a similar factor structure to that observed with some IQ tests, and have a moderately high correlation with IQ. One resolution of this apparent contradiction comes to mind. It is clear that the population has indeed steadily increased its mean level of education, from under 10 years of schooling in the 1930s when the first Stanford-Binet scales were normed, to approximately 12 years of schooling today. As already argued, this increase in schooling could be directly responsible for some of the improvement in IQ performance during this period. On the other hand, the decline in SAT scores has not coincided with

an increase in schooling, because persons taking the test over this same period of time have always been high school seniors. Thus, while persons today are more likely to complete high school (and take the SATs), those taking the SATs are no more likely to have completed high school than was the case 56 years ago, when the first SAT was administered. (In fact, many students today, including poorly educated minorities, are taking the SATs who probably would not have done so a generation ago, and the mass effect of this could be one of lowering the mean SAT scores.) Furthermore, a number of researchers have commented upon the reduced tendency of high school students today to go beyond the basics in their education: "The problem is not that *primary school children* are not mastering the 'three R's' as well as they used to; if anything, they are doing better. But *teenagers* are not moving as far beyond the 'three R's' as they used to" (Jencks, 1979, p. 33). If one puts together these two arguments (i.e., schooling has not increased among those taking the SATs, and the actual competencies of today's teenagers are not as great as those of teenagers from bygone days who were exposed to similar number of years of schooling), then the decline in SATs alongside an increase in IQ scores begins to make sense. This is admittedly a *post hoc* analysis, but it does have some cogency.

6. In Flynn's (1987b) analysis of Dutch gains in IQ between 1952 and 1972, he reports that only 5% of the increases in IQ were associated with increases in schooling. This raises the possibility that education in the first half of this century differed from what it is like in the second half. For example, perhaps dropping out of high school and going into work-study apprenticeships during the early half of this century has been replaced by completing high school on a nonacademic track (i.e., taking very little academic coursework). There has been an enormous growth of occupational and "alternative" education during the 60s and 70s, and this may have resulted in many students attending school even though little academic work was accomplished, whereas in former times these same students would not have remained in school. This is obviously speculative, and needs careful attention before it can be accepted. As will be seen later in this chapter, schools and schooling are quite complicated and attempts to assess their quality have resisted any facile consensus.

CHAPTER 6

1. Suppose that a battery of three tests (arithmetic, vocabulary, and spatial reasoning) is administered to children and that their scores are factor analyzed and g is extracted. Suppose also, for the sake of illustration, that the arithmetic performance depends on just three microlevel cognitive components a, b, and c. (For example, a could be verbal encoding of the word problems, b could be a spatial mapping skill that is relevant for some of the geometric problems, and c could be some highly specific quantitative skill that is useful for a broad array of arithmetic problems, but not relevant for non-arithmetical problems.) Now, suppose that the two other tests (vocabulary and spatial reasoning) also sample some, but not all, of these same microlevel cognitive components, in addition to some components

that are highly specific to themselves. (For example, maybe vocabulary requires *a* (verbal encoding) as well as *d*, which is the ability to "compare representations" and *e*, which is a highly specific vocabulary skill.) Finally, suppose that spatial reasoning requires *b*, in order to engage in spatial mapping, *d*, in order to compare representations, and *f*, which is a highly specific spatial skill. Then vocabulary and arithmetic performances might be correlated because they share verbal encoding resources (*a*), arithmetic and spatial reasoning might be correlated because they share a mapping resource (*b*), and vocabulary and spatial reasoning might be correlated because they share the need to compare representations (*d*). I have, of course, contrived this example, and it is probably farfetched as far as these three microlevel cognitive components in these specific tasks go. But it does illustrate the point I want to make about partially overlapped sampling of cognitive components: According to this task analysis, *g* could end up being substantial in magnitude without actually representing a single source of processing variance that is common to each of the three tasks! If we replaced this example with a more realistic one, involving dozens of microlevel cognitive components, differentially sampled by various macrolevel tasks, then the possibility that there is a series of partially shared components that give rise to *g* in the absence of a single pervasively shared cognitive resource becomes not so farfetched. Exploratory factor analysis by itself cannot elucidate such possibilities, as any resultant structure (simple or otherwise) can be retrofitted by positing a particular sampling galaxy. So we end up with a technique (factor analysis) that gives rise to a "general" factor that may or may not represent a singular underlying entity, and then we are told that this general factor is correlated with IQ and, finally, that the latter (IQ) is its validation as a truly general factor! As Detterman (1982) points out, this type of reasoning is circular at best. One could just as easily have posited that IQ was itself an amalgam of some of the same microlevel components that were sampled by the three other tests and included it in the battery to begin with. Had this been done, IQ would have been viewed as a marker for the general factor as opposed to its validation!

2. The *a posteriori* significance one attaches to *g* is a function not only of the types of tasks and contexts included in the battery of tests and the type of analysis performed on them (e.g., number, placement, and type of rotated axes), but also of one's *a priori* conceptualization of the nature of intelligence itself. In this sense, a determination of the significance of *g* is impossible because the level of correlation one decrees as necessary to confirm its significance is tied to his or her *a priori* conception of the nature of intelligence. If one's view of intelligence is that it consists of a number of "special purpose" cognitive resources, each of which depends on some distinct neural architecture, then correlated performances across diverse tasks would likely be dismissed as environmental in origin, unless the correlations were much larger than have been reported so far. On the other hand, if one's view of intelligence is that it is substantially unitary in nature, along with several specific abilities, then the presence of even small correlations will be taken as confirmation of the significance of *g*. All of this just goes to further emphasize the circular nature of much of social science reasoning.

3. This strong statement needs some qualification. I am not suggesting here that exceptional talent is a direct result of a biological disposition for some cognitive

skill that underlies the talent. It could also be that there is a biological basis for some personality trait that is instrumental in the talent. And this personality may interact with the environment in ways as to catalyze the skill in question. For example, it is known that Mozart's father had quite a large stake in his young son's musical competence. He set out to prove that he was the best music teacher of his era, something he had already accrued some evidence for because of his teaching Wolfgang's older sister. The temperament of these children had to be especially conducive to the long hours of tutelage to which they were subjected.

CHAPTER 7

1. Nearly 50 years ago, along these lines, Newman et al. (1937) reported the results of the "Chicago Twin Study" in which they concluded that somewhere between 65% and 80% of the differences in IQ were contributed by heredity. Yet, these same investigators reported a correlation of .79 between differences in separated twins' educational levels and differences in their Stanford-Binet IQs as well as a correlation of .91 between the twins' Stanford educational ages and differences in their educational levels. From these values, along with a .51 correlation between the twins' IQ differences and their social differences and a .30 correlation between their IQ differences and their differences in physical appearance, the authors concluded that the differences between the IQs of twins reared apart were 50% attributable to differences in their educations, 10% to social differences, and 12% to the joint influence of educational and social differences. The remainder of the IQ differences were attributed to physical and/or unknown causes.

CHAPTER 8

1. The U.S. Dept. of Labor has been encouraging state personnel agencies to screen job applicants with a psychometrically constructed test known as the General Aptitude Test Battery (GATB). Predictive validity data exists on the GATB for over 500 different types of jobs, indicating moderate levels of predictive validity overall. It is claimed that, at current levels of predictive validity, use of the GATB for all jobs in the U.S. economy will provide a gain of $79.36 billion.
2. Explaining national differences in mental development is an old enterprise, dating back to the ancients. According to Detterman (1982), Juan Huarte, a Spanish physician (1530–1589), wrote a treatise entitled *The Tryal of Wits: Discovering the Great Differences of Wits Among and What Sorts of Learning Best Suits Each Genius.* Huarte maintained that climate influenced differences in national character, e.g., moisture affected memory, dryness influenced understanding, and heat affected the imagination. No doubt, in another four centuries today's theories may look equally quaint to readers.

CHAPTER 9

1. To give you a feel for the complexity and richness of the variations in the transfer literature, consider what is termed a "partially mapped homomorphic transfer" problem. A boat to carry three males and two females across a river can carry at most two people at a time (and always must carry at least one person). But no wife may ever be in the presence of any male, unless her husband is also present. Only one of the pairs is married, that is, the other female is not a wife. So there are no constraints on her consorting with any of the three males. These five individuals can safely cross the river in seven moves, but there are over 60 different seven-move schedules that can be mapped!

2. Suppose that you are given the names of seven months and asked to consider which month differs from the other six. There are numerous ways for them to differ, of course, e.g., in their number of days, number of full weeks, seasons of the year in which they occur, etc. Suppose that you are not allowed to visually inspect a calendar and are required to perform this "oddity task" solely from memory. Do you suppose that it requires abstractness to be able to arrive quickly at the correct answer (assuming that we can identify seven months, among which there is only one defining characteristic that separates six of them from the seventh)? This task is similar to tasks that appear on major IQ tests that ask for underlying similarities, such as *In what way are scissors and a copper pan alike?"* Thus, it is assumed by many that the solution process to this type of IQ problem requires abstractness, even though Howe and Smith (1988) reported that an idiot savant with an IQ in the mentally retarded range could answer it rapidly and accurately without looking at a calendar:

   > Next, he was read aloud a list of seven months falling in particular years, as follows: January, 1971, September, 1972; June, 1973; July, 1974; August, 1975; February, 1974; October, 1976. All of these months except one start on a Friday. However, he was not told this, but was simply asked, 'Which one is the odd one out?' With little hesitation he gave the correct answer (February, 1971), and when asked why he chose that month he said that he did so because, unlike the others, it began on a Monday. Note that the question he was asked provided no hint of the kind of difference that is to be sought. (p. 380)

3. Others before Jensen had postulated that abstract reasoning was primarily the result of physiological processes. For example, Horn (1978) stated that "a relatively large proportion of the reliable variance in fluid intelligence (defined as abstract problem solving) reflects a pattern of physiological influences and a relatively small proportion of this variance reflects acculturation" (p. 247).

References

Abramyan, L. A. (1977). "On the Role of Verbal Instructions in the Directions of Voluntary Movements in Children." *Quarterly Newsletter of the Laboratory for Comparative Human Cognition, 1,* 1–4.

Accredolo, L. P. (1979). "Laboratory versus Home: The Effect of Environment on the Nine-Month Infant's Choice of Spatial Reference System." *Developmental Psychology, 15,* 666–667.

Aebli, H. (1987). "Development as Construction: Nature and Psychological and Social Context of Genetic Constructions." In B. Inhelder, D. de Caprona, and A. Cornu-Wells (eds.), *Piaget Today,* pp. 217–321. Hillsdale, NJ: Erlbaum.

Anastasi, A. (1968). *Psychological Testing.* 3rd Ed. New York: Macmillan.

Angoff, W. H. (1988). "The Nature-Nurture Debate, Aptitudes, and Group Differences." *American Psychologist, 43,* 713–720.

Bakker, D. J. (1972). *Temporal Order in Disturbed Reading.* Rotterdam: Rotterdam University Press.

Baldwin, J. M. (1894). *The Development of the Child and the Race.* New York: MacMillan.

Baltes, P. and Reinert, G. (1969). "Cohort Effects in Cognitive Development in Children as Revealed by Cross-Sectional Sequences." *Developmental Psychology, 1,* 169–177.

Baltes, P. and Willis, S. (1979). "The Critical Importance of Appropriate Methodology in the Study of Aging: The Sample Case of Psychometric Intelligence." In F. Hoffmeister and C. Muller (eds.), *Brain Function in Old Age.* Heidelberg: Springer Verlag.

Baltes, P. and Willis, S. (1982). "Plasticity and Enhancement of Intellectual Functioning in Old Age: Penn State's Adult Development and Enrichment Project (ADEPT)." In F. I. M. Craik, and S. Trehub (eds.), *Aging and Cognitive Processes.* New York: Plenum Press.

Bane, M. J. and Jencks, C. (1977). "Five Myths about Your IQ." In N. J. Block and G. Dworkin (eds.), *The IQ Controversy,* pp. 325–338. London: Quartet Books.

Baron, J. (1985). *Rationality and Intelligence.* New York: Cambridge University Press.

Baumeister, A. (1985). "Age of Acquisition and Meaningfulness as Predictors of Word Availability." *Journal of General Psychology, 112,* 109–112.

Baumrind, D. (1973). "The Development of Instrumental Competence through Socialization." In A. D. Pick (ed.), *Minnesota Symposium on Child Psychology (vol. 7,* pp. 3–46). Minneapolis: University of Minnesota Press.

Baumrind, D. (1967). "Child Care Practices Anteceding Three Patterns of Preschool Behavior." *Genetic Psychology Monographs, 75,* 43–88.

Beal, A. L. (1985). "The Skill of Recognizing Musical Structures." *Memory and Cognition, 13,* 405–412.

Beaumont, J. G. (1983). *Introduction to Neuropsychology.* New York: Guilford Press, Blackwell Scientific Publications.

Bernal, E. (1984). "Bias in Mental Testing: Evidence for an Alternative to the Heredity-Environment Controversy." In C. R. Reynolds, and R. T. Brown (eds.), *Perspectives on Bias in Mental Testing*. New York: Plenum, 189–220.

Berry, J. (1976). *Human Ecology and Cognitive Style*. New York: Wiley and Sons.

Berry, J. (1980). "Cultural Universality of any Theory of Human Intelligence Remains an Open Question." *Behavioral and Brain Sciences, 3*, 584–585.

Berry, J. and Dasen, R. (1973). *Culture and Cognition: Readings in Cross-Cultural Psychology*. London: Methuen and Co.

Berzonsky, M. (1971). "The Role of Familiarity in Children's Explanations of Physical Causality." *Child Development, 42*, 705–715.

Bieri, J. (1966). "Cognitive Complexity and Personality Development." In O. J. Harvey (ed.), *Experience, Structure and Adaptability*. New York: Springer.

Bjorklund, D. F., and Muir, J. (1989). "Knowledge, Mental Effort, and Memory Strategies: Developmental and Individual Differences." In W. Schneider and F. Weinert (eds.), *Interactions among Aptitudes, Strategies, and Knowledge in Cognitive Performance*. New York Springer-Verlag.

Blau, Z. (1981). *Black Children/White Children: Competence, Socialization, and Social Structure*. New York Free Press.

Block, N. J., and Dworkin, G. (1976). *The IQ Controversy*. New York: Pantheon Books.

Bloom, A. D. (1987). *The Closing of the American Mind*. New York: Touchstone Edition, Simon and Schuster.

Boring, E. (1953). *A History of Experimental Psychology*. New York: Appleton-Century-Crofts.

Bortner, M., and Birch, H. G. (1970). "Cognitive Capacity and Cognitive Competence Review." *American Journal of Mental Deficiency, 74*, 735–752.

Borwein, J. M. (1987). "Ramanujan, Modular Equations, and Approximations to π. Ramanujan's Centenary: His Legacy to Mathematics." Symposium presented at the annual meeting of the American Association for the Advancement of Science. Chicago, IL; Feb. 17, Hyatt Regency hotel.

Bosco, J. (1972). "The Visual Information Processing Speed of Lower Middle Class Children." *Child Development, 43*, 1418–1422.

Bossok, M., and Holyoak, K. J. (1989). "Interdomain Transfer between Isomorphic Topics in Algebra and Physics." *Journal of Experimental Psychology: Learning, Memory, and Cognition, 15*, 153–166.

Bouchard, T. J., and Segal, N. (1985). Environment and IQ. In B. Wolman (ed.) *Handbook of Intelligence* (pp. 391–464) . New York: Wiley Interscience.

Bouchard, T. J. (1984). "Twins Reared Together: What They Tell Us about Human Diversity." In S. W. Fox (ed.), *Individuality and Determinism*. New York: Plenum Publishing Corp.

Bouchard, T. J., and McGue, M. (1981). "Familial Studies of Intelligence: A Review." *Science, 212*, 1055–1059.

Bowles, S., and Nelson, V. (1974). "The 'Inheritance of IQ' and the Intergenerational Reproduction of Economic Inequality." *Review of Economics and Statistics, 56*, 39–51.

Boyce, C., and Darlington, R. (1981). "Black Proficiency in Abstract Reasoning." Paper Presented at the Annual Meeting of the Eastern Psychological Association. New York City. March 27.

Bradley, R., Caldwell, B., and Rock, S. (1988). "Home Environment and School Performance: A Ten-Year Follow-Up and Examination of Three Models of Environmental Action." *Child Development, 59*, 852–867.

Bradley, R. H., Caldwell, B. M., Rock, S. L., Ramey, C. T., Barnard, K. E., Gray, C., Gottfried, A. W., Mitchell, S., Hammond, M., Siegel, L. S., and Johnson, D. (1988). "Home Environment and Cognitive Development in the First Three Years of Life: A Collaborative Study Involving Six Sites and Three Ethnic Groups in North America." *Developmental Psychology, 25*, 217–235.

Brand, C. R., and Deary, I. J. (1982). "Intelligence and 'Inspection Time'." In H. J. Eysenck (ed.), *A Model for Intelligence*. New York: Springer Verlag.

Bray, D., and Howard, A. (1983). "The ATandT Longitudinal Study of Managers." In K. W. Schaie (ed.), *Longitudinal Studies of Adult Psychological Development* (pp. 266–312). New York: Guilford Press.

Brody, N. (1985). "The Validity of Tests of Intelligence". In Wolman, B. (ed.) *Handbook of Intelligence* (pp. 353–390). New York: Wiley Interscience.

Bronfenbrenner, U. (1974). "Nature with Nurture: A Re-Interpretation of the Evidence." In A. Montague (ed.), *Race and IQ*. New York: Oxford Univ. Press.

Bronfenbrenner, U. (1989). "Ecological Systems Theory." In R. Vasta (ed.), *Annals of Child Development Research, 6* 185–246.

Brown, A. L. (1975). "The Development of Memory: Knowing, Knowing About Knowing, and Knowing How to Know." In H. Reese (ed.), *Advances in Child Development and Behavior, vol. 10*, New York: Academic Press.

Brown, A. L., and Kane, M. J. (1988) "Preschool Children Can Learn to Transfer: Learning to Learn and Learning by Example." *Cognitive Psychology, 20*, 493–523.

Bruner, J., Olver, R., and Greenfield, P. M. (1966). *Studies in Cognitive Growth*. New York: Wiley and Sons.

Bullock, M. (1981). "Preschoolers' Understanding of Causal Mechanisms." Paper presented at the Biennial Meeting of the Society for Research in Child Development. Boston. April 25.

Burt, C. (1940). *The Factors of the Mind: An Introduction to Factor Analysis in Psychology*. London: University of London Press.

Butcher, H. J. (1970). *Human Intelligence: Its Nature and Assessment*. New York: Harper and Row.

Camp, C., Doherty, K., Moody-Thomas, K., and Denny, N. (1988). "Practical Problem Solving in Adults: A Comparison of Problem Types and Scoring Methods." In J. Sinnott (ed.), *Everyday Problem Solving: Theory and Applications*. New York: Praeger.

Carraher, T. N., Carraher, D. and Schliemann, A. D. (1985). "Mathematics in the Streets and in Schools." *British Journal of Developmental Psychology, 3*, 21–29.

Carroll, J. B. (1982). The Measurement of Intelligence. In R. J. Sternberg (ed.), *Handbook of Human Intelligence* (pp. 29–120). New York: Cambridge University Press.

Carroll, J. B. (1976). "Psychometric Tests as Cognitive Tasks: A New Structure of Intellect." In L. B. Resnick (ed.), *The Nature of Intelligence*. Hillsdale, NJ: Erlbaum.

Carroll, J. B. (1941). "Factorial Representation of Mental Ability and Academic Achievement." *Educational and Psychological Measurement, 3*, 307–332.

Carey, S. (1985). *Conceptual Change in Childhood*. Cambridge: MIT Press.

Case, R. (1985). *Intellectual Development: Birth to Adulthood*. Orlando, FL: Academic Press.

Caspi, A., Elder, G., and Bem, D. (1987). "Moving Against the World: Life Course Patterns in Explosion." *Developmental Psychology, 23*, 308–313.

Cattell, R. B. (1971). *Abilities: Their Structure, Growth, and Action*. Boston: Houghton-Mifflin.

Cattell, R. B. (1963). "Theory of Fluid and Crystallized Intelligence: A Critical Experiment." *Journal of Educational Psychology, 54*, 1–22.

Cattell, R. B. (1950). "The Fate of National Intelligence: A Test of a Thirteen Year Prediction." *Eugenics Review, 42*, 163–148.

Cattell, R. B. (1937).

Ceci, S. J. (1987). "Book Review of Detterman and Sternberg's What is Intelligence?" *American Journal of Mental Deficiency, 92*, 394–396.

Ceci, S. J., Baker, J. G., and Bronfenbrenner, U. (1988). "Prospective Remembering, Temporal Calibration, and Context." In M. M. Gruneberg, P. Morris, and R. Sykes (eds.), *Practical Aspects of Memory: Current Research and Issues* (pp. 360–365). London: Wiley.

Ceci, S. J., Baker, J., and Bronfenbrenner, U. (1987). "The Acquisition of Simple and Complex Algorithms as a Function of Context." Unpublished manuscript. Ithaca, NY: Cornell University.

Ceci, S. J., and Bronfenbrenner, U. (1985). "Don't Forget to Take the Cupcakes out of the Oven: Strategic Time-Monitoring, Prospective Memory, and Context." *Child Development, 56,* 175–190.

Ceci, S. J., Caves, R., and Howe, M. J. A. (1981). "Children's Long-Term Memory for Information Incongruous with Their Knowledge." *British Journal of Psychology, 72,* 443–450.

Ceci, S. J., and Cornelius, S. (1989). "Psychological Perspectives on Intellectual Development." Paper presented at the Biennial Meeting of the Society for Research in Child Development. Kansas City, MO, April 29.

Ceci, S. J., Lea, S. E. G., and Ringstrom, M. D. (1980). "Coding Characteristics of Normal and Learning Disabled 1–year-olds: Modality-Specific Pathways to the Cognitive System." *Journal of Experimental Psychology: Human Learning and Memory, 6,* 785–797.

Ceci, S. J., and Liker, J. (1988). "Stalking the IQ-Expertise Relationship: When the Critics Go Fishing." *Journal of Experimental Psychology: General, 117,* 96–100.

Ceci, S. J., and Liker, J. (1986a). "A Day at the Races: A Study of IQ, Expertise, and Cognitive Complexity." *Journal of Experimental Psychology: General, 115,* 255–266.

Ceci, S. J., and Liker, J. (1986b). "Academic and Non-Academic Intelligence: An Experimental Separation." In R. J. Sternberg and R. K. Wagner (eds.), *Practical Intelligence: Origins of Competence in the Everyday World.* New York: Cambridge University Press.

Ceci, S. J., and Nightingale, N. N. (1990). "Abstract Reasoning and Racial Differences: The Eye of the Beholder." Unpublished Manuscript. Cornell University, Ithaca, New York.

Ceci, S. J., and Tishman, J. (1982). "Developmental Changes in Encoding as a Function of Type of Stimuli." Unpublished Manuscript. Ithaca, New York: Cornell University.

Chang, F., and Greenough, W. T. (1982). "Lateralized effects of monocular training on dendritic branching in adult split-brain rats." *Brain Research, 232,* 283–292.

Charlesworth, W. (1976). "Human Intelligence as Adaptation: An Ethological Approach." In L. Resnick (ed.), *The Nature of Intelligence.* Hillsdale, NJ: Erlbaum.

Charlesworth, W. (1979). "An Ethological Approach to Studying Intelligence." *Human Development, 22,* 212–216.

Chase, A. (1977). *The Legacy of Malthus: The Social Costs of the New Scientific Racism.* New York: Alfred A. Knopf.

Chatterjea, R. G., and Paul, B. (1981). "Ecology, Field Independence, and Geometrical Figure Recognition: A Study in Intelligence Controlled Conditions." *Personality Study and Group Behavior, 1,* 1–10.

Cheng, P. W., Holyoak, K., Nisbett, R. E., and Oliver, L. (1986). "Pragmatic versus Syntactic Approaches to Training Deductive Reasoning." *Cognitive Psychology, 18,* 293–328.

Chi, M. T. H., Hutchinson, J., and Robin, A. (1989). "How Inferences About Novel Domain-Related Concepts Can Be Constrained by Structured Knowledge." *Merrill Palmer Quarterly, 35,* 27–62.

Chi, M. T. H. (1978). "Knowledge Structures and Memory Development." In R. S. Siegler (ed.), *Children's Thinking: What Develops?" Hillsdale, NJ: Erlbaum.*

Chi, M. T. H., and Ceci, S. J. (1987). "Content Knowledge: Its Restructuring with Memory Development." In H. W. Reese and L. Lipsett (eds.), *Advances in Child Development and Behavior, 20,* 91–146.

Chi, M. T. H., and Glaser, R. (1985). "Problem Solving Ability." In R. Sternberg (ed.), *Human Abilities.* New York: Freeman and Company.

Chi, M. T. H., and Glaser, R., and Rees, E. (1982). "Expertise in Problem Solving." In R. J. Sternberg (ed.), *Advances in the Psychology of Human Intelligence, vol. 1.* Hillsdale, NJ: Erlbaum.

Chi, M. T. H., Hutchinson, J., and Robin, A. (1989). "How Inferences about Novel Domain-Related Concepts Can Be Constrained by Structural Knowledge." *Merrill Palmer Quarterly, 35,* 27–62.

Christal, R. E., Tirre, W., and Kyllonen, P. (1984). "Two for the Money: Speed and Level Scores from a Computerized Vocabulary Test." In G. Lee and T. Ulrich (eds.), *Proceedings, Psychology in the Department of Defense, 9th Annual Symposium* (USAFA TR 8–2). Colorado Springs, CO: U.S. Air Force Academy.

Ciborowski, T. (1977). "The Influence of Formal Education on Rule Learning and Attribute Identification in a West African Society." *Journal of Cross-Cultural Psychology, 8,* 17–32.

Clarizio, H. (1979). "In Defense of the IQ Test." *School Psychology Digest, 8,* 79–88.

Clarizio, H. (1982). "Intellectual Assessment of Hispanic Children." *Psychology in the Schools, 19,* 61–71.

Clarke, A. M., and Clarke, A. D. B. (1976). *Early Experience: Myth and Evidence.* London: Open Books.

Cockburn, J., and Smith, P. T. (1988). "Effects of Intelligence on Everyday Memory Tasks." In M. M. Gruneberg, P. Morris, and P. Sykes (eds.), *Practical Aspects of Memory, vol.* 2. London: Wiley.

Cohn, S. J., Carlson, J., and Jensen, A. R. (1985). "Speed of Information Processing in Academically Gifted Youths." *Personality and Individual Differences, 6,* 621–629.

Cole, M. (1976). "A Probe Trial Procedure for the Study of Children's Discrimination Learning and Transfer." *Journal of Experimental Child Psychology, 22,* 499–510.

Cole, M. (1975). "An Ethnographic Psychology of Cognition." In R. W. Brislin, S. Bochner, and W. Lonner (eds.), *Cross-Cultural Perspectives on Learning.* New York: Wiley and Sons.

Cole, M., and D'Andrade, R. (1982). "The Influence of Schooling on Concept Formation: Some Preliminary Conclusions." *Quarterly Newsletter of the Laboratory of Comparative Human Cognition, 4,* 19–26.

Cole, M., Gay, J., Glick, J. A., and Sharp, D. W. (1971). *The Cultural Context of Learning and Thinking.* New York: Basic Books.

Cole, M., and Scribner, S. (1974). *Culture and Thought: A Psychological Introduction.* New York: Wiley.

Cole, M., and Scribner, S. (1977). "Cross-Cultural Studies of Memory and Cognition." In R. V. Kail and J. W. Hagen (eds.), *Perspectives on the Development of Memory and Cognition.* Hillsdale, NJ: Erlbaum.

Cole, M., Sharp, D., and Lave, J. (1976). "The Cognitive Consequences of Education." *Urban Review, 9,* 218–232.

Coles, G. (1978). "The Standard LD Battery." *Harvard Educational Review, 48,* 313–340.

Coltheart, V., and Walsh, P. (1988). "Expert Knowledge and Semantic Memory." In M. M. Gruneberg, P. Morris, and P. Sykes (eds.), *Practical Aspects of Memory, vol.* 2. London: Wiley.

Confrey, J. (1988). *Exploratory Research on Student Understanding of Exponential Functions: A Progress Report to the Faculty of Education, Cornell University Research Office.*

Coon, H., Fulker, D. W., DeFries, J. C., and Carey, G. (in press). "The Home Environment as a Predictor of Cognitive Ability of 7-Year-Old Children in the Colorado Adoption Project." *Behavior Genetics.*

Cornelius, S., Kenny, S., and Caspi, A. (1989). "Academic and Everyday Intelligence in Adulthood: Conceptions of Self and Ability Tests." In J. D. Sinott (ed.), *Everyday Problem Solving: Theory and Application.* New York: Praeger.

Cronbach, L. J. (1970). "Test Validation." In R. L. Thorndike (ed.), *Educational Measurement* (pp. 443–507). Washington, DC: American Council on Education.

Cronbach, L. J. (1975). "Five Decades of Public Controversy Over Mental Testing." *American Psychologist, 30,* 1–14.

Cutright, P. (1974). "The Civilian Earnings of White and Black Draftees and Non-Veterans." *American Sociological Review, 39,* 217–227.

Dasen, P. R. (1973). "Piagetian Research in Central Australia." In G. Kearney, P. deLacey, and G. Davidson (eds.), *The Psychology of Aboriginal Australians.* Sydney: Wiley.

Dawson, J. L. M. (1967). "Cultural and Physiological Influences upon Spatial-Perceptual Processes in West Africa, Part 1." *International Journal of Psychology, 2*, 115–128.

Dean, C. (1988). "Study Links Intelligence and Myopia." *New York Times*. Dec. 20, Section C, p. 3.

DeFries, J., Plomin, R., Vandenberg, S., and Kuse, A. (1981). "Parent-Offspring Resemblance for Cognitive Abilities in the Colorado Adoption Project." *Intelligence, 5*, 245–277.

DeGroot, A. D. (1951). "War and the Intelligence of Youth." *Journal of Abnormal and Social Psychology, 46*, 596–597.

de la Rocha, O. (1985). "The Reorganization of Arithmetic Practice in the Kitchen." *Anthropology and Education Quarterly, 16*, 193–198.

Detterman, D. K. (1982). "Does 'g' Exist?" *Intelligence, 6*, 99–108.

Detterman, D. K. (1986). "Human Intelligence is a Complex System of Separate Processes." In R. J. Sternberg, and D. Detterman, *What Is Intelligence?* (pp. 57–61). Norwood, NJ: Ablex.

Detterman, D. K. (1986). "Basic Cognitive Processes Predict IQ." Paper presented at the 27th annual meeting of the Psychonomic Society, New Orleans, LA, November.

Detterman, D. K. (1987). "What Does Reaction Time Tell Us about Intelligence?" In P. A. Vernon (ed.), *Speed of Information Processing and Intelligence*. Norwood, NJ: Ablex.

Detterman, D. K. (1989). Paper presented at the Annual Meeting of the Psychonomics Society.

Detterman, D. K., and Daniel, M. H. (in press). "Correlations of Mental Tests with Each Other and with Cognitive Variables Are Highest for Low IQ Groups. *Intelligence*.

Deutsche, J. M. (1937). *The Development of Children's Concepts of Causal Relations*. Minneapolis: University of Minnesota Press.

di Sessa, A. A. (1982). "Unlearning Aristotelian Physics: A Study of Knowledge-Based Learning." *Cognitive Science, 6*, 37–75.

Dixon, L., and Johnson, R. (1980). *The Roots of Individuality*. Belmont, CA: Wadsworth Publishers.

Done, D. J., and Miles, T. (1988). "Age of Word Acquisition in Developmental Dyslexics as Determined by Response Latencies in a Picture Naming Task." In M. M. Gruneberg, P. Morris, and P. Sykes (eds.), *Practical Aspects of Memory, vol. 2*. London: Wiley.

Doris, J., and Ceci, S. J. (1988). "Varieties of Mind." *Forum, 68*, 18–22.

Dornbusch, S., Ritter, P., Leiderman, P., Roberts, D., and Fraleigh, M. (1987). "The Relation of Parenting Style to Adolescent School Performance." *Child Development, 58*, 1244–1257.

Dörner, D., and Kreuzig, H. (1983). "Problemlosefahigkeit und Intelligenz." *Psychologische Rundschaus, 34*, 185–192.

Dörner, D. Kreuzig, H., Reither, F., and Staudel, T. (1983). *Lohhausen: Vom Umgang mit Unbestimmtheit und Komplexitat*. Bern: Huber.

Douglas, J. W. B. (1964). *The Home and the School*. London: McGibbon and Kee.

Duncan, O. D., Featherman, D., and Duncan, B. (1972). *Socioeconomic Background and Achievement*. New York: Seminar Press.

Durkheim, E., and Mauss, M. (1901–1902). "Dequelques formes Primitives de Classification." *L'Annee socioloqique, 6*, 1–71.

Economos, J. (1980). "Bias Cuts Deeper than Scores." *Behavioral and Brain Sciences, 3*, 342–343.

Edgerton, R. (1981). "Another Look at Culture and Mental Retardation." In M. J. Begab, H. C. Haywood, and H. L. Garber (eds.), *Psychosocial Influences in Retarded Performance, (Vol. 1*, pp. 309–324). Baltimore: University Park Press.

Edmonds, R. (1986). "Characteristics of Effective Schools." In U. Neisser (ed.), *The School Achievement of Minority Children: New Perspectives*. Hillsdale, NJ: Erlbaum.

Elder, G. (1986). "Military Times and Turning Points in Men's Lives." *Developmental Psychology, 22*, 233–245.

Elder, G., Hastings, T., and Pavalko, E. (1989). "Adult Pathways to Greater Distinction and Disappointment." Paper Presented at the Life-History Research Society Meeting. Montreal, June 14.

Ellis, N., and Hennelly, J. (1980). "Welsh Children's Time-Based Sound Durations and Short Term Memory." *British Journal of Psychology, 71*, 43–51.

Epstein, H. T. (1979). "Brain Growth Spurts During Brain development." In H. Epstein (ed.), *Education and the Brain*. Chicago, IL: University of Chicago Press.

Erlenmeyer-Kimling, L., and Jarvik, L. (1964). "Genetics and Intelligence: A Review." *Science, 142,* 1477–1479.

Evans, J., and Segal, M. (1969). "Learning to Classify by Color and by Function: A Study of Concept Discovery by Ganda Children." *Journal of Social Psychology 77,* 35–52.

Eyferth, K. (1961). "Leistungen Verschiedener Gruppen von Besatzungskinddern in Hamburg-Wechsler Intelligenztest fur Kinder (HAWIK)." *Archives fur die Gesante Psychologie, 113,* 222–241.

Eysenck, H. J. (1988). "The Biological Basis of Intelligence." In S. H. Irvine and J. W. Berry (eds.), *Human Abilities in Cultural Context* (pp. 87–104). New York: Cambridge University Press.

Eysenck, H. J. (1986). "Inspection Time and Intelligence: A Historical Introduction." *Personality and Individual Differences, 7,* 603–607.

Eysenck, H. J. (1982). *A Model for Intelligence.* New York: Springer-Verlag.

Eysenck, H. J. (1979). *The Structure and Measurement of Intelligence.* New York: Springer-Verlag.

Fahrmeier, E. D. (1975). "The Effect of School Attendance on Intellectual Development in Northern Nigeria." *Child Development, 46,* 281–285.

Feldman, D. (1980). *Beyond Universals and Cognitive Development.* Norwood, NJ: Ablex.

Felsten, G., and Wasserman, G. (1980). "Visual Masking: Mechanisms and Theories." *Psychological Bulletin, 88,* 329–353.

Ferguson, G. A. (1956). "On Transfer and the Abilities of Man." *Canadian Journal of Psychology, 8,* 121–131.

Feurstein, R. (1979). "Cognitive Modifiability in Retarded Adolescents." *American Journal of Mental Deficiency, 83,* 539–550.

Fischbein, S. (1980). "IQ and Social Class." *Intelligence, 4,* 51–63.

Fischer, K. (1980). "A Theory of Cognitive Development: The Control of Hierarchies of Skill." *Psychological Review, 87,* 477–531.

Flynn, J. R. (1984). "The Mean IQ of Americans: Massive Gains 1932 to 1978." *Psychological Bulletin, 95,* 29–51.

Flynn, J. R. (1987a). "Race and IQ: Jensen's Case Refuted." In S. Modgil and C. Modgil (eds.), *Arthur Jensen: Consensus and Controversy* (pp. 221–232). Sussex, UK: Falmer Press.

Flynn, J. R. (1987b). "Massive IQ Gains in 14 Nations: What IQ Tests Really Measure." *Psychological Bulletin, 101,* 171–191.

Flynn, J. R. (1988). "The Ontology of Intelligence." In J. Forge (ed.), *Measurement, Realism, and Objectivity* (pp. 1–40). Dordrecht, Netherlands: D. Reidel.

Fobih, D. K. (1979). *The Influence of Different Educational Experiences on Classificatory and Verbal Reasoning Behaviour of Children in Ghana.* Unpublished Doctoral Dissertation. University of Alberta.

Fodor, J. A. (1983). *The Modularity of Mind.* Cambridge, MA: M.I.T. Press.

Ford, M., and Keating, D. P. (1981). "Developmental and Individual Differences in Long-Term Memory Retrieval: Process and Organization." *Child Development, 52,* 234–241.

Forrester, P. J. (1987). "Rogers-Ramanujan Identities and the Hard Hexagon Model." Symposium presented at the Annual Meeting of the American Association for the Advancement of Science. Chicago, IL, Feb. 17, Hyatt Regency Hotel.

Frank, R. (1988). *Passions within Reason: The Strategic Role of Emotions.* New York: Norton.

Frederiksen, N. (1986). "Toward a Broader Conception of Human Intelligence." In R. J. Sternberg and R. Wagner (eds.), *Practical Intelligence: Nature and Origins of Competence in the Everyday World* (pp. 84–116). New York: Cambridge University Press.

Freeman, F. S. (1934). *Individual Differences.* New York: Henry Holt and Co.

Fulker, D. W., DeFries, J. C., and Plomin, R. (in press). "Genetic Influence on General Mental Ability Increases between Infancy and Middle Childhood." *Nature.*

Fuller, B. (1987). "Defining School Quality." In J. Hannaway and M. Lockhead (eds.), *The*

Contribution of the Social Sciences to Educational Policy and Practice: 1965–1985 (pp. 33–69). Berkeley, CA: McCutchan.

Gabennesch, H. (1972). "Authoritarianism as a World View." *American Journal of Sociology, 77,* 857–875.

Galbraith, R. C. (1982). "Sibling Spacing and Intellectual Development: A Closer Look at the Confluence Models." *Developmental Psychology, 18,* 151–173.

Galton, F. (1892). *Hereditary Genius.* London: Macmillan.

Garcia, J. (1981). "The Logic and Limits of Mental Aptitude Testing." *American Psychologist, 36,* 1172–1180.

Gardner, H. (1988). "Creative Lives and Creative Works: A Synthetic Scientific Approach." In R. J. Sternberg (ed.), *The Nature of Creativity.* New York: Cambridge University Press.

Gardner, H. (1987). "Developing the Spectrum of Human Intelligence." *Harvard Educational Review, 57,* 187–193.

Gardner, H. (1983). *Frames of Mind: The Theory of Multiple Intelligences.* New York: Basic Books.

Gay, J., and Cole, M. (1967). *The New Mathematics and an Old Culture.* New York: Holt, Rinehart, and Winston.

Gelman, R. (1979). "Why We Will Continue to Read Piaget." *Genetic Epistemologist, 8,* 1–3.

Geschwind, N. (1979). "Specializations of the Human Brain." *Scientific American, 241,* 158–168.

Gettinger, M. (1984). "Individual Differences in Time Needed for Learning: A Review of Literature. *Educational Psychologist, 19,* 15–29.

Gholson, B., Eymard, L. A., Long, D., Morgan, D., and Leeming, F. (1988). "Problem Solving, Recall, Isomorphic Transfer, and Nonisomorphic Transfer among Third Grade and Fourth Grade Children." *Cognitive Development, 3,* 37–53.

Gilbert, C. D., and Weisel, T. N. (1979). "Intra-Cortical Projections of Functionally Characterized Neurons in the Cat Visual Cortex." *Nature, 280,* 120–125.

Gladwin, H. (1971). *East Is a Big Bird.* Cambridge, MA: Harvard University Press.

Glaser, R. (1984). "Education and Thinking: The Role of Knowledge." *American Psychologist, 39,* 93–104.

Goldsmith, H. (1983). "Genetic Influences on Personality from Infancy to Adulthood." *Child Development, 54,* 331–355.

Goldstein, K., and Scheerer, M. (1941). "Abstract and Concrete Behavior: An Experimental Study with Special Tests." *Psychological Monographs, 53,* 1–151.

Goleman, D. (1989). "An Emerging Theory." *New York Times,* Sec. 12, April 10, 22–24.

Goodwin, P. A. (1988). *Efficacy of the Formal Education Process in Rural Alaska. Vol. 1.* Honolulu: Applied NeuroDynamics Corp.

Gordon, R. A. (1987). "SES versus IQ in the Race-IQ-Delinquency Model." *International Journal of Sociology, 7,* 30–96.

Gordon, R. A. (1984). "Digits Backward and the Mercer-Kamin Law." In C. R. Reynolds and R. T. Brown (eds.), *Perspectives on Bias in Mental Testing* (pp. 357–506). New York: Plenum.

Gordon, R. A. (1980). "Labelling Theory, Mental Retardation, and Public Policy: Larry P. and Other Developments since 1974." In W. R. Gove (ed.), *The Labelling of Deviance: Evaluating a Perspective.* Beverly Hills, CA: Sage.

Gordon, R. A. (1976). "Prevalence: The Rare Datum in Delinquency Measurement and Its Implications for a Theory of Delinquency." In M. Klein (ed.), *The Juvenile Justice System.* Beverly Hills, CA: Sage.

Gottfredson, L. S. (1986). "Societal Consequences of the g Factor in Employment." *Journal of Vocational Behavior, 29,* 379–410.

Gottfried, A. W. (1984). "Issues Concerning the Relationship Between Home Environment and Early Cognitive Development." In A. W. Gottfried (ed.), *Home Environment and Early Cognitive Development.* London: Academic Press.

Goward, L., and Rabbitt, P. (1988). "What Intelligence Tests Don't Measure." In M. M.

Gruneberg, P. Morris, and P. Sykes (eds.), *Practical Aspects of Memory, vol. 2.* London: Wiley.

Greenfield, P. M., and Childs, C. P. (1972). "Weaving, Color Terms, and Pattern Representation: Cultural Influences and Cognitive Development among the Zinacantecos of Southern Mexico." Paper Presented at the Meeting of the International Association of Cross-Cultural Psychology, Hong Kong. September 21.

Greenfield, P. M., Reich, L. C., and Olver, R. R. (1966). "On Culture and Equivalence." In J. S. Bruner, R.R. Olver, and P. M. Greenfield (eds.), *Studies in Cognitive Growth.* New York: Wiley.

Gregory, R. L. (1981). *Mind in Science.* Cambridge, England: Cambridge University Press.

Greulich, W. W. (1957). "A Comparison of the Physical Growth and Development of American-Born and Japanese Children." *American Journal of Physical Anthropology, 15,* 489–515.

Grigsby, O. J. (1932). "An Experimental Study of the Development of Concepts of Relationships in Preschool Children as Evidenced by Their Expressive Ability." *Journal of Experimental Education, 1,* 144–162.

Guilford, J. P. (1967). *The Nature of Human Intelligence.* New York: McGraw-Hill.

Guilford, J. P. (1985). "The Structure-of-Intellect Model." In B. Wolman (ed.). *Handbook of Intelligence.* New York: Wiley Interscience.

Hall, J. (1972). "Verbal Behavior as a Function of Amount of Schooling." *American Journal of Psychology, 85,* 277–289.

Hardy, G. H. (1963). *A Mathematician's Apology.* London: Cambridge University Press.

Harnquist, K. (1968). "Relative Changes in Intelligence from 13 to 18." *Scandinavian Journal of Psychology, 9,* 50–64.

Hartup, W. W. (1983). "Peer Relations." In M. Hetherington (ed.), *Handbook of Child Psychology, Vol. 4. Socialization, Personality, and Social Development.* (pp. 103–196). New York: Wiley.

Hayes., D., and Grether, J. (1982). "The School Year and Vacations: When Do Students Learn?" *Cornell Journal of Social Relations, 17,* 56–71.

Hayes, K. J. (1962). "Genes, Drives, and Intellect." *Psychological Reports, 10,* 299–342.

Hellige, J., and O'Boyle, M. (1989). "Cerebral Hemisphere Asymmetry and Individual Differences in Cognition." *Learning and Individual Differences, 1,* 7–36.

Hendrickson, A. E., and Hendrickson, D. E. (1982). "The Psychophysiology of Intelligence." In H. J Eysenck (ed.), *A Model for Intelligence* (pp. 151–228). New York: Springer-Verlag.

Hendrickson, D. E., and Hendrickson, A. E. (1980). "The Biological Basis of Individual Differences in Intelligence." *Personality and Individual Differences, 1,* 3–33.

Herrmann, D., Grubs, L., Sigmindi, R., and Gruenich, R. (1985). "Awareness of Memory Ability Before and After Memory Experience." *Human Learning, 5,* 91–107.

Herrnstein, R. J. (1973). *IQ in the Meritocracy.* Boston: Little Brown.

Herrnstein, R. J., Nickerson, R. S., DeSanche, M, and Swets, J. A. (1986). "Teaching Thinking Skills." *American Psychologist, 41,* 1279–1289.

Heyns, B. L. (1978). *Summer Learning and the Effects of Schooling.* New York: Academic Press.

Hick, W. E. (1952). "On the Rate of Gain of Information." *Quarterly Journal of Experimental Psychology, 4,* 11–26.

Hilliard, A. (1984). "IQ Testing as the Emperor's New Clothes." In C. R. Reynolds and R. T. Brown (eds.), *Perspectives on Bias in Mental Testing* (pp. 189–220). New York: Plenum.

Hoffman, L. (1985). "The Changing Genetics/Socialization Balance." *Journal of Social Issues, 41,* 127–148.

Hoosain, R., and Salili, F. (1988). "Sound Durations, Digit Span, and Mental Calculations." In M. M. Gruneberg, P. Morris, and P. Sykes (eds.), *Practical Aspects of Memory, vol. 2.* London: Wiley.

Horn, J. (1983). "The Texas Adoption Project: Adopted Children and Their Intellectual Resemblance to Biological and Adoptive Parents." *Child Development, 54,* 268–275.

Horn, J. (1985). "Bias? Indeed!" *Child Development. 56,* 779–780.

Horn, J. L. (1978). "Human Ability Systems." In P. Baltes (ed.), *Life-span Development and Behavior, Vol. 1* (pp. 211–256). New York: Academic Press.

Horn, J. L., and Donaldson, G. (1980). "Cognitive Development in Adulthood." In O. G. Brim, Jr. and J. Kagan (eds.), *Constancy and Change in Human Development*. Cambridge, MA: Harvard University Press.

Howe, M. J. A. (1990). *Origins of Genius: The Psychology of Exceptional Abilities*. Oxford, England: Blackwell.

Howe, M. J. A. (1988). "The Hazards of Using Correlational Evidence as a Means of Identifying the Causes of Individual Ability Differences: A Rejoinder to Sternberg and a Reply to Miles." *British Journal of Psychology, 79,* 1–7.

Howe, M. J. A. (1972). *Understanding School Learning*. New York: Harper and Row.

Howe, M. J. A. (1988). "Intelligence as an Explanation." *British Journal of Psychology, 79,* 1–12.

Howe, M. J. A. (1989). *The Making of Genius and Talents*. London: Basil Blackwell.

Howe, M. J. A., and Smith, J. (1988). "Calendar Calculating in 'Idiot Savants': How Do They Do It?" *British Journal of Psychology, 79,* 371–386.

Huang, I. (1943). "Children's Conceptions of Physical Causality: A Critical Summary." *Journal of Genetic Psychology, 63,* 71–121.

Hudson, W. (1960). "Pictorial Depth Perception in African Groups." *Journal of Social Psychology, 52,* 183–208.

Humphreys, L. (1984). "General Intelligence." In C. R. Reynolds and R. T. Brown (eds.), *Perspectives on Bias in Mental Testing* (pp. 221–248). New York: Plenum.

Humphreys, L. (1962). "The Organization of Human Abilities." *American Psychologist, 17,* 475–483.

Humphreys, L. (1979). "The Construct of General Intelligence." *Intelligence, 3,* 105–120.

Hunt, E. (1984). "The Correlates of Intelligence." In D. Detterman (ed.), *Current Topics in Human Intelligence, Vol. 1,* (pp. 167–178). Norwood, NJ: Ablex.

Hunt, E. (1985). "Verbal Ability." In R. J. Sternberg (ed.), *Human Abilities: An Information Processing Approach*. San Francisco: Freeman.

Hunt, E. (1980). "Intelligence as an Information Processing Concept." *British Journal of Psychology, 71,* 449–474.

Hunt, E., Lunneborg, C., and Lewis, J. (1975). "What Does It Mean To Be High Verbal?" *Cognitive Psychology, 7,* 194–227.

Hunt, J. McV. (1961). *Intelligence and Experience*. New York: Ronald Press.

Hunter, J. (1983). *The Dimensionality of the General Aptitude Test Battery and the Dominance of General Factors Over Specific Factors in the Prediction of Job Performance in the U.S. Employment Service*. (Uses Test Research Report #44). Washington, DC: U.S. Dept. of Labor, Employment, and Training Administration, Division of Counselling and Test Development.

Hunter, J. (1986). "Cognitive Ability, Cognitive Aptitudes, Job Knowledge, and Job Performance." *Journal of Vocational Behavior, 29,* 340–363.

Hunter, J., and Schmidt, F. (1982). "Fitting People to Jobs: The Impact of Personnel Selection on National Productivity." In M. Dunnette and E. Fleishman (eds.), *Human Performance and Productivity, vol. 1* (pp. 233–284). Hillsdale, NJ: Erlbaum.

Hunter, J., Schmidt, F., and Rauschenberger, J. (1984). "Methodological, Statistical, and Ethical Issues in the Study of Bias in Psychological Tests." In C. R. Reynolds, and R. T. Brown (eds.), *Perspectives on Bias in Mental Testing* (pp. 41–97). New York: Plenum.

Husén, T. (1967). "International Study of Achievement in Mathematics." In Husén, T. (ed.), *International Study of Achievement in Mathematics: A Comparison of Twelve Countries*. Stockholm: Almquist and Wiksell. New York: Wiley,

Husén, T. (1951). "The Influence of Schooling Upon IQ." *Theoria, 17,* 61–88.

Inhelder, B., and deCaprona, D. (1987). "Introduction." In B. Inhelder, D. de Caprona, and A. Cornu-Wells (eds.), *Piaget Today* (pp. 1–14). Hillsdale, NJ: Erlbaum.

Irvine, S. H., and Berry, J. W. (1988). *Human Abilities in Cultural Context.* New York: Cambridge University Press.

Irwin, M. H., and McLaughlin, D. H. (1970). "Ability and Preference in Category Sorting by Mano School Children and Adults." *Journal of Social Psychology, 82,* 15–24.

Itzkoff, S. W. (1989). *The Making of the Civilized Mind.* New York: Peter Longmans.

Itzkoff, S. W. (1987). *Why Humans Vary in Intelligence.* Ashfield, MA: Paideia Publishers.

Jackson, A. and Morton, J. (1984). "Facilitation of Auditory Word Recognition." *Memory and Cognition, 12,* 568–574.

Jackson, M. (1980). "Further Evidence for a Relationship Between Memory Access and Reading Ability." *Journal of Verbal Learning and Verbal Behavior, 19,* 683–694.

Jaeger, M. E., and Rosnow, R. L. (1988). "Contextualism and Its Implications for Inquiry." *British Journal of Psychology, 79,* 63–75.

Jaques, E. (1978). *Levels of Abstraction in Logic and Human Action: A Theory of Discontinuity in the Structure of Mathematical Logic, Psychological Behavior, and Social Organization.* London: Heinman Press.

Jarman, R. F. (1980). "Modality-Specific Information Processing and Intellectual Ability." *Intelligence, 4,* 201–217.

Jencks, C. (1979). "What's Behind the Drop in Test Scores?" *Working Papers,* July/August, 29–41.

Jencks, C., Smith, M., Acland, H., Bane, M. J., Cohen, D., Gintis, H., Heyns, B., and Mitchelson, S. (1972). *Inequality: A Reassessment of the Effects of Family and Schooling in America.* New York: Basic Books.

Jencks, C., and Crouse, J. (1982). "Aptitude versus Achievement: Should We Replace the SAT?" In W. B. Schrader (ed.), *New Directions for Testing and Measurement, Guidance, and Program Improvement, No. 13.* San Francisco: Jossey Bass.

Jensen. A. R. (1969). "How Much Can We Boost IQ and Scholastic Achievement?" *Harvard Educational Review,* Winter, 1–123.

Jensen, A. R. (1972). *Genetics and Education.* New York: Harper and Row.

Jensen, A. R. (1979). "*g:* Outmoded Theory or Unconquered Frontier?" *Creative Science Technology, 2,* 16–29.

Jensen, A. R. (1980). *Bias in Mental Testing.* New York: Free Press.

Jensen, A. R. (1981). *Straight Talk about Mental Tests.* New York: The Free Press.

Jensen, A. R. (1982). "Reaction Time and Psychometric *g.*" In H. J. Eysenck (ed.), *A Model for Intelligence.* New York: Springer-Verlag.

Jensen, A. R. (1984). "Test Bias: Concepts and Criticisms." In C. R. Reynolds and R. T. Brown (eds.), *Perspectives on Bias in Mental Testing* (pp. 507–586). New York: Plenum.

Jensen, A. R. (1986). *The "g" beyond Factor Analysis.* Paper Presented at the Buros-Nebraska Symposium on Measurement and Testing: The Influence of Cognitive Psychology on Testing and Measurement. Lincoln: NE, April.

Jensen, A. R. (1987). "Unconfounding Genetic and Nonshared Environmental Effects." *Behavioral and Brain Sciences, 10,* 26–27.

Jensen, A. R. (1977). "An Examination of Culture Bias in the Wonderlic Personnel Test." *Intelligence, 1,* 51–64.

Jensen, A. R. (1968). "Social Class, Race, and Genetics: Implications for Education." *American Educational Research Journal, 5,* 1–42.

Jensen, A. R., and Figueroa, R. (1975). "Forward and Backward Digit Span Interaction with Race and IQ: Predictions from Jensen's Theory." *Journal of Educational Psychology, 67,* 882–893.

Jepson, C., Krantz, D., and Nisbett, R. E. (1983). "Inductive Reasoning: Competence or Skill?" *Behavioral and Brain Sciences, 3,* 494–501.

Jeyifous, S. W. (1985). *Developmental Changes in Conceptual Structure: Cross-Cultural Findings.* Unpublished Doctoral Dissertation. Ithaca, NY: Cornell University.

Johnson, R. C., DeFries, J. C., Ahern, F., and Mi, M. P. (1977). "Resemblances of Collateral Relatives in Cognitive Abilities." *Behavior Genetics, 8,* 102–121.

Johnson, W., Dark, V., and Jacoby, L. (1985). "Perceptual Fluency and Recognition Judgments." *Journal of Experimental Psychology: Learning, Memory, and Cognition, 11*, 3–11.

Johnson-Laird, P. N. (1983). *Mental Models: Toward a Cognitive Science of Language, Inference, and Consciousness.* Cambridge, MA: Harvard University Press.

Jones, A. P., and Butler, M. C. (1980). "Influences of Cognitive Complexity on the Dimensions Underlying Perceptions of the Work Environment." *Motivation and Emotion, 4*, 1–18.

Jones, H. (1955). "Perceived Differences among Twins." *Eugenics Quarterly, 5*, 99–102.

Kagan, J., and Klein, R. (1981). "Cross-Cultural Perspectives on Early Development." In E. M. Hetherington and R. D. Parke (eds.), *Contemporary Readings in Child Psychology.* New York: McGraw-Hill.

Kahneman, D., Slovic, P. and Tversky, A. (1982). *Judgment Under Uncertainty.* New York: Cambridge University Press.

Kamin, L. J. (1974). *The Science and Politics of IQ.* Hillsdale, NJ: Lawrence Erlbaum.

Karier, C., Violas, P., and Spring, J. (1973). *Roots of Crisis.* Chicago: Rand McNally.

Kasprow, W., Schachtman, T., and Miller, R. R. (1987). "The Comparator Hypothesis of Conditioned Response Generation: Manifest Conditioned Excitation and Inhibition as a Function of Relative Excitatory Strengths of CS and Conditioning Context at the Time of Testing." *Journal of Experimental Psychology: Animal Processes, 13*, 395–406.

Kearins, J. (1981). "Visual Spatial Memory in Australian Aboriginal Children of Desert Regions." *Cognitive Psychology, 13*, 434–460.

Keating, D. P., List, J. A., and Merriman, W. E. (1985). "Cognitive Processing and Cognitive Ability: A Multivariate Validity Investigation." *Intelligence, 9*, 149–170.

Keating, D. P. (1984). "The Emperor's New Clothes: The 'New Look' in Intelligence Research." In R. J. Sternberg (ed.), *Advances in the Psychology of Human Intelligence, vol. 2.* Hillsdale, NJ: Erlbaum.

Keil, F. (1985). "On the Structure-Dependent Nature of Stages of Cognitive Development." In S. Strauss and I. Levine (eds.), *Stage and Structure in Children's Development* (pp. 144–163). New York: Ablex.

Keil, F. (1984). "Mechanisms in Cognitive Development and the Structure of Knowledge." In R. J. Sternberg (ed.), *Mechanisms in Cognitive Development* (pp. 81–100). New York: W. H. Freeman.

Keil, F. (1981). "Constraints on Knowledge and Cognitive Development." *Psychological Review, 88*, 197–227.

Kemp, L. (1955). "Environmental and Other Characteristics Determining Attainments in Primary Schools." *British Journal of Educational Psychology, 25*, 67–77.

Keogh, B. K. (1982). "Children's Temperament and Teachers' Decisions." In R. Porter and G. Collins (eds.), *Temperamental Differences in Infants and Young Children* (pp. 269–279). London: Pitman.

Kilbride, P., and Leibowitz, H. (1975). "Factors Affecting the Magnitude of the Ponzo Perspective Illusion among the Baganda." *Perception and Psychophysics, 17*, 543–548.

Klahr, D., and Wallace, J. G. (1976). *Cognitive Development: An Information-Processing View.* Hillsdale, NJ: Erlbaum Associates.

Klayman, J. (1984). "Learning from Feedback in Probabilistic Environments." Unpublished Manuscript. Chicago: University of Chicago Graduate School of Business.

Klein, P. S., and Tzuriel, R. (1986). "Preschoolers' Type of Temperament as Predictor of Potential Difficulties in Cognitive Functioning." *Israel Journal of Psychiatry and Related Sciences, 23*, 49–61.

Kline, P. (1988). "The British Cultural Influence on Ability Testing." In S. H. Irvine and J. W. Berry (eds.), *Human Abilities in Cultural Context* (pp. 187–207). New York: Cambridge University Press.

Klineberg, O. (1935). *Negro Intelligence and Selective Migration.* New York: Harper.

Kohn, M., and Schooler, C. (1978). "Reciprocal Effects of the Substantive Complexity of Work and

Intellectual Flexibility: A Longitudinal Assessment." *American Journal of Sociology, 84,* 24–53.

Kuhn, D. (1983). "On the Dual Executive and Its Significance in the Development of Developmental Psychology." In D. Kuhn and J. Meacham (eds.), *On the Development of Developmental Psychology* (pp. 80–110). New York: Karger.

Kunda, Z., and Nisbett, R. E. (1986). "The Psychometrics of Everyday Life." *Cognitive Psychology, 18,* 195–224.

Kurtz, B. (1989). "Individual Differences in Cognitive and Metacognitive Processing." In W. Schneider and F. Weinert (eds.) *Interactions among Aptitudes, Strategies, and Knowledge in Cognitive Performance.* New York: Springer-Verlag.

Laboratory of Comparative Human Cognition (1983). "Culture and Cognitive Development." In P. Mussen (ed.), *Handbook of Child Psychology, vol. 1,* pp. 295–356.

Laboratory of Comparative Human Cognition (1986). "Contributions of cross-Cultural Research to Education Practice." *American Psychologist, 41,* pp. 1049–1058.

Labov, W. (1970). "The Logic of Non-Standard English." In F. Williams (ed.), *Language and Poverty.* Chicago: Markham.

Lancy, D. F., and Strathern, A. J. (1981). "Making Two's: Pairing as an Alternative to the Taxonomic Mode of Representation." *American Anthropologist, 83,* 773–795.

Lansman, M., Poltrock, S., and Hunt, E. (1983). "Individual Differences in the Ability to Focus and Divide Attention." *Intelligence, 7,* 299–312.

Larson, G., Merritt, C. R., and Williams, S. E. (1988). "Information Processing and Intelligence: Some Implications of Task Complexity." *Intelligence, 12,* 131–147.

Larson, G. (1989). "Cognitive Correlates of General Intelligence: Toward A Process Theory of G." *Intelligence 13,* 5–31.

Larson, G., Saccuzzo, D. P., and Brown, J. (1989). "Motivation: Cause or Confound in Information Processing/Intelligence Correlations?" Unpublished Manuscript.

Laufer, E. (1985). "Domain-Specific Knowledge and Memory Performance in the Work Place." Paper presented at the annual meeting of the Eastern Psychological Association, Boston, March 23.

Lave, J. (1977). "Tailor-Made Experiments and Evaluating the Intellectual Consequences of Apprenticeship Training." *The Quarterly Newsletter of the Institute for Comparative Human Development, 1,* 1–3.

Lave, J., Murtaugh, M., and de la Roche, D. (1984). "The Dialectic of Arithmetic in Grocery Shopping." In B. Rogoff and J. Lave (eds.), *Everyday Cognition: Its Development in Social Context.* Cambridge MA: Harvard University Press.

Lee, E. S. (1951). "Migration: A Philadelphia Test of the Klineberg Hypothesis." *American Sociological Review, 16,* 227–232.

Leshowitz, B. (1989). "It Is Time We Did Something about Scientific Illiteracy." *American Psychologist, 44,* 1159–1160.

Lewis, D. (1976). "Observations on Route-Finding and Spatial Orientation Among the Aboriginal Peoples of the Western Desert Region of Central Australia." *Oceania, 46,* 249–282.

Lewontin, R., Rose, S., and Kamin, L. (1984). *Not in Our Genes.* New York: Pantheon.

Lewontin, R. (1982). *Human Diversity.* New York: Freeman.

Liker, J., and Ceci, S. J. (1987). "The Role of Experience in IQ." *Journal of Experimental Psychology: General, 116,* 304–306.

List, J. A., Keating, D. P., and Merriman, W. E. (1985). "Differences in Memory Retrieval: A Construct Validity Investigation." *Child Development, 56,* 138–151.

Loehlin, J., Lindzey, G., and Spuhler, J. N. (1975). *Race Differences in Intelligence.* San Francisco: W. H. Freeman

Loehlin, J., and Nichols, R. (1976). *Heredity, Environment, and Personality: A Study of 850 Sets of Twins.* Austin: University of Texas Press.

Loevinger, J. (1940). "Intelligence as Related to Socio-Economic Factors." *Thirty-ninth Yearbook of the National Society for the Study of Education, Part 1: Intelligence: Its Nature and Nurture.* Bloomington, IL: Public School Publishing Co.

Loevinger, J. (1943). "On the Proportional Contribution of Differences in Nature and in Nurture to Differences in Intelligence." *Psychological Review, 40*, 725–758.

Logie, R., and Wright, R. (1988). "Specialised Knowledge and Recognition Memory Performance in Residential Burglars." In M. M. Gruneberg, P. Morris, and P. Sykes (eds.), *Practical Aspects of Memory, vol. 2*. London: Wiley.

Longstreth, L. E. (1984). "Jensen's Reaction-Time Investigations of Intelligence: A Critique." *Intelligence, 8*, 139–160.

Longstreth, L. E. (1981). "Revisiting Skeels' Final Study: A Critique." *Developmental Psychology, 17*, 620–625.

Lorch, R. (1986). "Use of Word Readiness for Studying Word Recognition." *Bulletin of the Psychonomic Society, 24*, 11–14.

Lorenz, C. (1987). "The Structure of the Spatial Domain: An Analysis of Individual Differences." Unpublished Doctoral Dissertation. Cornell University.

Lorenz, C., and Neisser, U. (1986) "Ecological and Psychometric Dimensions of Spatial Ability." *Emory Cognition Project, 10* (July).

Lorge, L. L. (1945). "Schooling Makes a Difference." *Teacher's College Record, 46*, 483–492.

Lorsbach, T., and Gray, (1986). "Item Identification Speed and Memory Span Performance in Learning Disabled Children." *Contemporary Educational Psychology, 11*, 68–78.

Lubin, M. P., and Fernandez, J. (1986). "The Relationship between Psychometric Intelligence and Inspection Time." *Personality and Individual Differences, 7*, 653–657.

Luria, A. R. (1976). *Cognitive Development: Its Cultural and Social Foundations*. Cambridge, MA: Harvard University Press.

Lynn, R. (1982). "IQ in Japan and the U.S. Shows a Growing Disparity." *Nature, 297,* 222–223.

Lynn, R. (1978). "Ethnic and Racial Differences in Intelligence: International Comparisons." In R. T. Osborne, C. E. Noble, and N. Weyl (eds.), *Human Variation: The Biopsychology of Age, Race, and Sex*. New York: Academic Press.

Manning, W., and Jackson, R. (1984). "College Entrance Examinations: Objective Selection or Gatekeeping for the Economically Privileged?" In C. R. Reynolds, and R. T. Brown (eds.), *Perspectives on Bias in Mental Testing* (pp. 189–220). New York: Plenum.

Marks, W. (1982). *The Idea of IQ*. Washington, D.C.: University Press of America.

Marini, Z., and Case, R. (1989). "Parallels in the Development of Preschoolers' Knowledge about Their Physical and Social Worlds." *Merrill Palmer Quarterly, 35*, 63–88.

Marr, D. and Sternberg, R. J. (1987). "The Role of Mental Speed in Intelligence: A Triarchic Perspective." In P. A. Vernon (ed.), *Speed of Information Processing and Intelligence* (pp. 271–294). Norwood, NJ: Ablex.

Marshalek, B., Lohman, D., and Snow, R. (1983). "The Complexity Continuum in the Radex and Hierarchical Models of Intelligence." *Intelligence, 7*, 107–127.

Martin, K. A. C., Somogyi, P., and Witteridge, D. (1983). "Physiological and Morphological Properties of Identified Basket Cells in the Cat's Visual Cortex." *Experimental Brain Research, 50*, 136–151.

Marx, M. H. (1976). "Formal Theory." In M. Marx and F. Goodson (eds.), *Theories in Contemporary Psychology. 2d ed*. New York: MacMillan.

Matarazzo, J. D. (1970). *Wechsler's Measurement and Appraisal of Adult Intelligence*. (5th ed.). Baltimore: Williams and Wilkins.

Mayer, S., and Jencks, C. (1989). "Growing Up in Poor Neighborhoods: How Much Does It Matter?" *Science, 243*, 1441–1444.

McAskie, M., and Clarke, A. M. (1976). "Parent-Offspring Resemblances in Intelligence: Theories and Evidence." *British Journal of Psychology, 67*, 243–273.

McCardle, J. J. and Horn, J. L. (In Press). "A Mega Analysis of Aging and the WAIS: Structural Modelling of Intellectual Abilities Organized by Age Trends." *Psychological Bulletin.*

McClearn, G. E., and De Fries, J. C. (1973). *Introduction to Behavioral Genetics*. San Francisco: W. H. Freeman.

McGillicuddy-DeLisi, A. V. (1982). "The Relationship between Parents' Beliefs about Development and Family Constellation, SES, and Parents' Teaching Strategies." In L. Laosa and I. Sigel (eds.), *Families as Learning Environments for Children* (pp. 261–299). New York: Plenum.

McNemar, Q. (1964). "Lost: Our Intelligence? Why?" *American Psychologist, 19*, 871–882.

Means, M., and Voss, J. (1985). "Star Wars: A Developmental Study of Expert and Novice Knowledge Structures." *Memory and Language, 24*, 746–757.

Merton, R. (1968). *Social Theory and Social Structure*. New York: Free Press.

Miller, P. (1983). *Theories of Developmental Psychology*. San Francisco: W. H. Freeman.

Miller, P., and Weiss, M. G. (1982). "Children's and Adult's Knowledge about What Variables Affect Selective Attention." *Child Development, 53*, 543–549.

Miller, R. R., and Schachtman, T. (1985). "The Several Roles of Context at the Time of Retrieval." In P. Balsam, and A. Tomie (eds.), *Context and Learning*, Hillsdale, NJ: Erlbaum.

Miller, R. R., and Schachtman, T. (1985). "Conditioning Context as an Associative Baseline: Implications for Response Generation and the Nature of Conditioned Inhibition." In R. R. Miller and N. Spear (eds.), *Information Processing in Animals: Conditioned Inhibition* (pp. 51–88). Hillsdale, NJ: Erlbaum.

Minton, H. L. (1988). "Charting Life History: Lewis M. Terman's Study of the Gifted." In J. Morawski (ed.), *The Rise of Experimentation in American Psychology* (pp. 138–162). New Haven, CT: Yale University Press.

Mistry, J., and Rogoff, B. (1985). "A Cultural Perspective on the Development of Talent." In F. D. Horowitz and M. O'Brien (eds.), *The Gifted and Talented: Developmental Perspectives*. Washington, DC: American Psychological Association.

Mohr, A., and Lund, F. (1933). "Beauty as Related to Intelligence and Educational Achievement." *Journal of Social Psychology, 4*, 235–239.

Moore, E. G. J. (1986). "Family Socialization and the IQ Test Performance of Traditionally and Transracially Adopted Black Children." *Developmental Psychology, 22*, 317–326.

Morrison, F. (1987). "The 5 to 7 Shift Revisited: A Natural Experiment." Paper presented at the Annual Meeting of the Psychonomic Society, Seattle, November.

Morton, N. (1974). "Analysis of Family Resemblances I." *American Journal of Human Genetics, 26*, 318–330.

Murtaugh, M. (1985). "The Practice of Arithmetic by American Grocery Shoppers." *Anthropology and Education Quarterly*, Fall.

Myambo, K. (1972). "Shape Constancy as Influenced by Culture, Western Education, and Age." *Journal of Cross-Cultural Psychology, 3*, 221–232.

Nairn, A., and Associates. (1980). *The Reign of ETS: The Corporation That Makes Up Minds*. Washington, D.C.: Ralph Nader Publications.

Nass, M. L. (1956). "The Effects of Three Variables on Children's Concepts of Physical Causality." *Journal of Abnormal and Social Psychology, 53*, 191–196.

Neisser, U. (1985). "The Role of Theory in the Ecological Study of Memory." *Journal of Experimental Psychology: General, 114*, 272–276.

Neisser, U. (1976). "General, Academic, and Artificial Intelligence." In L. Resnick (ed.), *The Nature of Intelligence*. Hillsdale, NJ: Erlbaum.

Neisser, U. (1979). "The Concept of Intelligence." *Intelligence, 3*, 217–227.

Newman, H. H., Freeman, F. N., and Holzinger, K. J. (1937). *Twins: A Study of Heredity and Environment*. Chicago: University of Chicago Press.

Nichols, J., Cheung, P. C., Lauer, J., and Patashnick, M. (1989). "Individual Differences in Academic Motivation: Perceived Ability, Goals, Beliefs, and Values." *Learning and Individual Differences, 1*, 63–84.

Nichols, R. (1981). "Origins, Nature, and Determinants of Intellectual Development." In M.

Begab, H. C. Haywood, and H. Garber (eds.), *Psychosocial Determinants of Retarded Performance*, vol. 1. Baltimore: University Park Press.

Nisbett, R., Fong, G., Lehman, D., and Cheng, P. (1988). "Teaching Reasoning." Unpublished Manuscript. University of Michigan, Ann Arbor.

Nisbett, R., Krantz, D., Jepson, C., and Kunda, Z. (1983). "The Use of Statistical Heuristics in Everyday Inductive Reasoning." *Psychological Review, 90*, 339–363.

Nisbett, R., and Ross, L. (1980). *Human Inference: Strategies and Shortcomings of Social Judgment.* Englewood Cliffs, NJ: Prentice Hall.

Noble, C. E. (1952). "An Analysis of Meaning." *Psychological Review, 59*, 421–430.

Olson, D. (1987). "Adolescents' and Adults' Responses to Disconfirming Evidence Alone and in Conjunction with Implicit and Explicit Alternative Accounts." Paper Presented at the Biennial Meeting of the Society for Research in Child Development. Baltimore, MD. April 22.

Owen, D. (1985). *None of the Above: Behind the Myth of Scholastic Aptitude.* Boston: Houghton-Mifflin.

Parker, K. C. H. (1986). "Changes with Age, Year-of-Birth Cohort, Age-by-Year-of-Birth Cohort Interaction, and Standardization of the Wechsler Adult Intelligence Tests." *Human Development, 29*, 209–222.

Pellegrino, J. W., and Glaser, R. (1979). "Cognitive Correlates and Components in the Analysis of Individual Differences." *Intelligence, 3*, 187–214.

Pellegrino, J. W., and Goldman, S. (1983). "Differences in Verbal Spatial Reasoning." In R. Dillon and R. Schmeck (eds.), *Individual Differences in Cognition* (pp. 140–180). New York: Academic Press.

Pellegrino, J. W. (1986). "Dynamic and Static Assessment of Spatial Ability." Paper Presented at the Annual Meeting of the American Educational Research Association. San Francisco, April 26.

Pellegrino, J. W., and Kail, R. V. (1982). "Process Analysis and Spatial Aptitude." In R. J. Sternberg (ed.) *Advances in the Study of Intelligence*, vol. 1. Hillsdale, NJ: Erlbaum.

Phillips, C., Zeki, S., and Barlow, H. (1984). "Localization of Function in the Cerebral Cortex." *Brain, 107*, 328–361.

Piaget, J. (1952). *The Origins of Intelligence in Children.* New York: International University Press.

Piaget, J. (1953). *Logic and Psychology.* Manchester, UK: University of Manchester Press.

Piaget, J. (1954). *The Construction of Reality in the Child.* New York: Basic Books.

Piaget, J. (1955). *The Language and Thought of the Child.* New York: New York American Library.

Piaget, J. (1976). *The Psychology of Intelligence.* Totowa, NJ: Littlefield, Adams.

Pikas, A. (1966). *Abstraction and Concept Formation.* Cambridge, MA: Harvard University Press.

Pincus, K. V. (1985). "Group Embedded Figures Test: Psychometric Data for a Sample of Accountants Compared to Student Norms." *Perceptual and Motor Skills, 60*, 707–712.

Plomin, R. (1989). "Environment and Genes." *American Psychologist, 44*, 105–111.

Plomin, R. (1985). "Behavioral Genetics." In D. Detterman (ed.), *Current Topics in Human Intelligence*, vol. 1. Norwood, NJ: Ablex.

Plomin, R. (1983). "Childhood Temperament." In B. Lahey and A. Kazdin (eds.), Advances in *Clinical Child Psychology* (vol. 6). New York: Plenum Press.

Plomin, R., and DeFries, J. (1985). *Origins of Individual Differences in Infancy: The Colorado Adoption Project.* Orlando, FL: Academic Press.

Plomin, R., DeFries, J., Fulker, D. W. (1988). *Nature and Nurture During Infancy and Early Childhood.* New York: Cambridge University Press.

Posner, M. I., and Mitchel, R. F. (1967). "Chronometric Analysis of Classification." *Psychological Review, 74*, 392–409.

Pratt, M., Kerig, P., Cowan, P., and Cowan, C. P. (1988). "Mothers and Fathers Teaching 3-Year-Olds: Authoritative Parenting and Scaffolding of Young Children's Learning." *Developmental Psychology, 24*, 832–839.

Provine, W. B. (1986). "Genetics and Race." *American Zoologist, 26,* 857–887.

Pylyshyn, Z. (1980). "Computation and Cognition: Issues in the Foundation of Cognitive Science." *Behavioral and Brain Sciences, 3,* 111–169.

Rabbitt, P. (1988). "Does It Last? Is Speed a Basic Factor Determining Individual Differences in Memory?" In M. M. Gruneberg, P. Morris, and P. Sykes (eds.), *Practical Aspects of Memory, vol. 2.* London: Wiley.

Ramey, C. T., Bryant, D. M., and Suarez, T. M. (1987). "Early Intervention: Why, for Whom, How, and at What Cost?" In N. Gunzenhauser (ed.), *Infant Stimulation: For Whom, What Kind, When, and How Much?* Biscayne Bay, FL: Johnson and Johnson Baby Products Co.

Ramey, C. T., and Landesman, S. J. (1989) "Prevention of Intergenerational Intellectual Retardation." Paper presented at the Gatlinburg Meeting, April 12, Gatlinburg, Tennessee.

Ramphal, C. (1962). *A Study of Three Current Problems of Indian Education.* University of Natal. Unpublished Ph.D. Thesis. (Cited in P. Vernon (1969).)

Rao, D., Morton, N., and Yee, S. (1974). "Analysis of Family Resemblances II." *American Journal of Human Genetics, 26,* 331–359.

Rawling, J. P. (1986). *Semantic Priming Effects in Backward Masking.* Unpublished M.A. Thesis. Cornell University.

Razz, N., Willerman, L., Igmundsen, P., and Hanlon, M. (1983). "Aptitude-Related Differences in Auditory Recognition Masking." *Intelligence, 7,* 71–90.

Regian, J., Shute, V., and Pellegrino, J. (1985). "The Modifiability of Spatial Processing Skills." Paper presented at the 26th Meeting of the Psychonomic Society. Boston, MA, November.

Reuning, H. (1988). "Testing Bushmen in the Central Kalahari." In S. H. Irvine and J. W. Berry (eds.), *Human Abilities in Cultural Context* (pp. 453–486). New York: Cambridge University Press.

Rice, O. (1982). *The Hatfields and the McCoys.* Lexington: University of Kentucky Press.

Rice, T., Fulker, D. W., Defries, J. C. and Plomin, R. (1988). "Path Analysis of IQ During Infancy and Early Childhood and the Index of the Home Environment in the Colorado Adoption Project." *Intelligence, 12,* 27–45.

Rice, T., Fulker, D. W., and Defries, J. C. (1986). "Multivariate Path Analysis of Specific Cognitive Abilities in the Colorado Adoption Project." *Behavior Genetics, 16,* 107–125.

Risse, G. L., Rubens, A. B., and Jordan, L. S. (1984). "Disturbances of Long Term Memory in Aphasic Patients: A Comparison of Anterior and Posterior Lesions." *Brain, 107,* 605–617.

Rivers, W. H. R. (1926). *Psychology and Ethnology.* New York: Harcourt Brace.

Rogoff, B. (1981). "Schooling and the Development of Cognitive Skills." In H. Triandis and A. Heron (eds.), *Handbook of Cross-Cultural Psychology, vol. 4* (pp. 233–294). Rockleigh, NJ: Allyn and Bacon.

Rogoff, B. (1981). "Schooling's Influence on Memory Test Performance." *Child Development, 52,* 260–267.

Rogoff, B. (1978). "Spot Observation: An Introduction and Examination." *Quarterly Newsletter of the Institute for Comparative Human Development, 2,* 21–26.

Rogoff, B., and Waddell, K. (1982). "Memory for Information Organized in a Scene by Children from Two Cultures." *Child Development, 53,* 1224–1228.

Rosch, E., Mervis, C. B., Gray, W., Johnson, D., and Boys-Braem, P. (1976). "Basic Objects in Natural Categories." *Cognitive Psychology, 8,* 382–439.

Rose, R. J., Miller, J. Z., Dumont-Driscoll, M. and Evans, M. M. (1979). "Twin Family Studies of Perceptual Speed Ability." *Behavioral Genetics, 9,* 71–86.

Ross, R. T., Begab, M. J., Dondis, E. H., Giampiccolo, J. S., and Meyers, C. E. (1985). *Lives of the Mentally Retarded: A Forty-Year Follow-Up Study.* Stanford, CA: Stanford University Press.

Roth, C. (1983). "Factors Affecting Developmental Changes in the Speed of Processing." *Journal of Experimental Child Psychology, 35,* 509–528.

Royce, J. R. (1979). "The Factor-Gene Basis of Individuality." In J. R. Royce and L. P. Mos (eds.), *Theoretical Advances in Behavior Genetics.* Alphen aan den Rijn: Sitjhoff and Noordhoff.

Rushton, J. (1988). "Race Differences in Behavior: A Review and Evolutionary Analysis." *Personality and Individual Differences, 9,* 1009–1024.

Russell, E. S. (1976). "Genetics Society of America Resolution on Genetics, Race, and Intelligence." *Genetics, 83,* 99–101.

Salthouse, T. (1982). *Adult Cognition: An Experimental Psychology of Aging.* New York: Springer-Verlag.

Sandoval, J. (1982). "The WISC-R Factoral Validity for Minority Groups and Spearman's Hypothesis." *Journal of School Psychology, 20,* 198–204.

Sandoval, J. (1979). "The WISC-R and Internal Evidence Of Test Bias with Minority Groups." *Journal of Consulting and Clinical Psychology, 47,* 919–927.

Saracho, O. N. (1983). "Cultural Differences in the Cognitive Style of Mexican-American Students." *Journal of the Association for the Study of Perception, 18,* 3–10.

Sarason, S. B., and Doris, J. (1979). *Educational Handicap, Public Policy, and Social History.* New York: Free Press.

Sax, G., and Karr, A. (1962). "An Investigation of Response Sets on Altered Parallel Forms." *Educational and Psychological Measurement, 22,* 371–376.

Saxe, G. B. (1988). "Candy Selling and Math Learning." *Educational Researcher, 17,* 14–21.

Scarr, S. (1968). "Environmental Bias in Twin Studies." *Eugenics Quarterly. 15,* 34–40.

Scarr, S. (1982). "Similarities and Differences among Siblings." In M. E. Lamb and B. Sutton-Smith (eds.), *Sibling Relationships.* Hillsdale, NJ: Erlbaum.

Scarr, S. (1981). "Unknowns in the IQ Equation." In Scarr, S. (ed.), *Race, Social Class, and Individual Differences* (pp. 61–74). Hillsdale, NJ: Erlbaum.

Scarr, S., and Kidd, K. K. (1983). "Developmental Behavior Genetics." In P. Mussen (ed.), *Handbook of Child Psychology, 4th ed. Vol. 2: Infancy and Developmental Psychobiology.* New York: Wiley and Sons.

Scarr, S., and McCartney, K. (1983). "How People Make Their Own Environment: A Theory of Genotype-Environment Effects." *Child Development, 54,* 424–435.

Scarr, S., and Weinberg, R. (1976). "IQ Test Performance of Black Children Adopted by White Families." *American Psychologist, 31,* 726–739.

Scarr, S., and Weinberg, R. (1983). "The Minnesota Adoption Studies: Genetic Differences and Malleability." *Child Development, 54,* 260–270.

Schaefer, E. S. (1987). "Parental Modernity and Child Academic Competence: Toward a Theory of Individual and Societal Development." *Early Development and Care, 27,* 373–389.

Schaefer, E. S. and Edgerton, M. (1985). "Parent and Child Correlates of Parental Modernity." In I. E. Sigel (ed.), *Parental Belief Systems* (pp. 287–318). Hillsdale, NJ: Lawrence Erlbaum.

Schafer, E. W. P. (1987). "Neural Adaptability: A Biological Determinant of g Factor Intelligence." *Behavioral and Brain Sciences,10,* 240–241.

Schaie, W. K. (1988). "Variability in Cognitive Functioning in the Elderly: Implications for Social Participation." In A. Woodhead, M. Bender, and R. Leonard (eds.), *Phenotypic Variation in Populations: Relevance to Risk Management* (pp. 191–212). New York: Plenum Press.

Schaie, W. K., and Willis, S. L. (1988). "Generalizability of Age Patterns of Psychometric Intelligence within and across Ability Domains." Paper Presented at National Institute on Aging Workshop: Generalizing from Experience: An Issue of Development. Bethesda, MD, September 23.

Schiff, M., Duyme, M., Dumaret, A. and Tomkiewicz, S. (1982). "How much can We Boost Scholastic Achievement and IQ Scores? A Direct Answer from a French Adoption Study." *Cognition, 12,* 165–196.

Schliemann, A. (1988). "Understanding the Combinatorial System: Development, School Learning, and Everyday Experience." *Quarterly Newsletter of the Laboratory for Comparative Human Cognition, 10,* 3–7.

Schlotterer, G., Moscovitch, M., and Crapper-McLachlan, C. (1983). "Visual Processing Deficits as

Assessed by Spatial Frequency Contrast Sensitivity and Backward Masking in Normal and Alzheimer's Disease." *Brain, 107,* 309–325.

Schmidt, F. L., and Hunter, J. E. (1981). "Employment Testing." *American Psychologist, 36,* 1128–1137.

Schmidt, W. H. O. (1967). "Socio-Economic Status, Schooling, Intelligence, and Scholastic Progress in a Community in Which Education Is Not Yet Compulsory." *Paedogogica Europa, 2,* 275–286.

Schmidt, W. H. O., and Nzimande, A. (1970). "Cultural Differences in Color/Form Preference and in Classificatory Behavior." *Human Development, 13,* 140–148.

Schneider, W., Körkel, J., and Weinert, F. E. (1989). "Expert Knowledge and General Abilities and Text Processing." In W. Schneider and F. Weinert (eds.), *Interactions Among Aptitudes, Strategies, and Knowledge in Cognitive Performance.* New York: Springer-Verlag.

Schooler, C. (1984). "Psychological Effects of Complex Environments During the Life Span." *Intelligence, 8,* 259–281.

Schooler, C. (1989). "Social Structural Effects and Experimental Situations: Mutual Lessons of Cognitive and Social Science." In K. W. Schaie, and C. Schooler (eds.), *Social Structure and Aging: Psychological Processes.* Hillsdale, NJ: Erlbaum.

Scott, W. A., Osgood, D. W., and Peterson, C. (1979). *Cognitive Structure: Theory and Measurement of Individual Differences.* Washington, DC: V. H. Winston and Sons.

Scribner, S. (1986). "Thinking in Action: Some Characteristics of Practical Thought." In R. J. Sternberg and R. K. Wagner (eds.), *Practical Intelligence: Nature and Origins of Competence in the Everyday World.* New York: Cambridge University Press.

Scribner, S. (1975). "Recall of Classical Syllogisms: A Cross-Cultural Investigation of Error on Logical Problems." In R. Falmagne (ed.), *Reasoning: Representation and Process in Children and Adults.* New York: Wiley and Sons.

Scribner, S. (1976). "Situating the Experiment in Cross-Cultural Research." In K. Riegel and J. Meacham (eds.), *The Developing Individual in a Changing World, vol.1.* Chicago: Aldine.

Scribner, S. (1977). "Modes of Thinking and Ways of Speaking: Culture and Logic Reconsidered." In P. Johnson-Laird and P. Wason (eds.), *Thinking.* New York: Cambridge University Press.

Scribner, S. (1984). "Studying Working Intelligence." In B. Rogoff and J. Lave (eds.), *Everyday Cognition: Its Development in Social Context.* Cambridge, MA: Harvard University Press.

Scribner, S., and Cole, M. (1973). "Cognitive Consequences of Formal and Informal Education." *Science, 182,* 553–558.

Searle, J. (1983). "The World Turned Upside Down." *New York Review of Books,* October 27, pp. 74–79.

Sharp, D., Cole, M., and Lave, J. (1979). "Education and Cognitive Development: The Evidence from Experimental Research." *Monographs from the Society for Research in Child Development, 44* (1–2, Serial No. 178). Chicago, IL: University of Chicago Press.

Sherman, M., and Key, C.B. (1932). "The Intelligence of Isolated Mountain Children." *Child Development, 3,* 279–290.

Siegel, L. S. (1984). "Home Environmental Influence on Cognitive Development in Preterm and Full-Term Children During the First Five Years." In A. W. Gottfried (ed.), *Home Environment and Early Cognitive Development* (pp. 19234). Orlando, FL: Academic Press.

Siegler, R. S. (1989). "Mechanisms of Cognitive Development." *Annual Reviews of Psychology, 40,* 353–379.

Siegler, R. S. (1981). "Developmental Sequences Within and Between Concepts." *Monographs of the Society for Research on Child Development, 81* (Serial No. 189). Chicago, IL: University of Chicago Press.

Skodak, M., and Skeels, H. M. (1949). "A Follow-Up Study of One-Hundred Adopted Children." *Journal of Genetic Psychology, 75,* 85–125.

Sorokin, P. (1956). *Fads and Foibles in Modern Sociology.* Chicago, IL: H. Regnery Co.

Spearman, C. (1904). "General Intelligence Objectively Determined and Measured." *American Journal of Psychology, 15,* 206–221.

Stalnaker, R. C., and Stalnaker, J. M. (1946). "Effect on a Candidate's Score of Repeating the SAT of the College Entrance Examination Board." *Educational and Psychological Measurement, 6,* 495–504.

Stanovich, K. E. (1977). "A Note on the Interpretation of Interaction in Comparative Research." *American Journal of Mental Deficiency, 81,* 394–396.

Stanovich, K. E., and Purcell, D. G. (1981). "Comment on Input Capability and Speed of Processing in Mental Retardation by Saccuzzo, Kerr, Marcus, and Brown." *Journal of Abnormal Psychology, 90,* 168–171.

Staszewski, J. (1989). "Exceptional Memory: The Influence of Practice and Knowledge on the Development of Elaborative Encoding Strategies." In W. Schneider, and F. Weinert (eds.). *Interactions Among Aptitudes, Strategies, and Knowledge in Cognitive Performance.* New York: Springer-Verlag.

Sternberg, R. J. (1986). *Intelligence Applied.* Orlando, FL: Harcourt, Brace, and Jovanovitch.

Sternberg, R. J. (1985). *Beyond IQ: A Triarchic Framework for Intelligence.* New York: Cambridge University Press.

Sternberg, R. J. (1981). "Testing and Cognitive Psychology." *American Psychologist, 36,* 1181–1189.

Sternberg, R. J. (1984). "Toward a Triarchic Theory of Human Intelligence." *Behavioral and Brain Sciences, 8,* 269–315.

Sternberg, R. J. (1977). *Intelligence, Information Processing, and Analogical Reasoning: The Componential Analysis of Human Abilities.* Hillsdale, NJ: Erlbaum.

Sternberg, R. J., and Powell, J. S. (1983a). "The Development of Intelligence." In P. Mussen (series editor) and J. Flavell and E. Markman (eds), *Handbook of Child Psychology, vol. 3* (pp. 341–419), New York: Wiley.

Sternberg, R. J., and Powell, J. S. (1983b). "Comprehending Verbal Comprehension." *American Psychologist, 38,* 878–893.

Sternberg, R. J., Conway, B., Ketron, J., and Bernstein, M. (1981). "People's Conceptions of Intelligence." *Journal of Personality and Social Psychology, 41,* 37–55.

Stevensen, H., Parker, T., Wilkinson, A., Bonnevaux, B., and Gonzalez, M. (1978). "Schooling, Environment, and Cognitive Development: A Cross-Cultural Study." *Monographs from the Society for Research in Child Development, 43,* (3, Serial No. 175). Chicago: University of Chicago Press.

Stewart, I. (1987). "Are Mathematicians Logical?" *Nature, 325,* 386–387.

Streufert, S. (1986). "Individual Differences in Risk Taking." *Journal of Applied Social Psychology, 16,* 482–497.

Streufert, S., and Nogami, G. Y. (1989). "Cognitive Style and Complexity: Implications for I/O Psychology." In C. Cooper and I. Robertson (eds.), *International Review of Industrial and Organizational Psychology* (pp. 93–143). New York: Wiley.

Streufert, S., and Steufert, S. C. (1978). *Behavior in the Complex Environment.* Washington, D.C.: V. H. Winston and Sons.

Streufert, S., and Swezey, R. J. (1986). *Complexity, Managers, and Organizations.* New York: Academic.

Super, C. M. (1980). "Cognitive Development: Looking Across at Growing Up." In C. Super and M. Harkness (eds.), *New Directions for Child Development: Anthropological Perspectives on Child Development, 8,* 59–69.

Svendson, D. (1982). "Changes in IQ, Environmental and Individual Factors: A Follow-Up Study of EMR Children." *Journal of Child Psychology and Psychiatry, 23,* 69–79.

Tanner, J. M. (1962). *Growth at Adolescence: With a General Consideration of the Effects of Heredity and Environmental Factors Upon Growth and Maturation from Birth to Maturity.* (2nd Edition). Springfield, IL: C. C. Thomas.

Terman, L. M. (1925). *Genetic Studies of Genius: I Mental and Physical Traits of a Thousand Gifted Children.* Stanford, CA: Stanford University Press.

Terman, L. M., and Oden, M. H. (1959). *Genetic Studies of Genius: 4 The Gifted Group at Midlife.* Stanford, CA: Stanford University Press.

Terman, L. M. (1921). "Contribution to 'Intelligence and Its Measurement' Symposium." *Journal of Educational Psychology, 12,* 127–133.

Terman, L. M. (1922). "Were We Born That Way?" *World's Work, 5,* 655–660.

Tetewsky, S. J. (1988). *An Analysis of Familiarity Effects in Visual Comparison Tasks and Their Implications for Studying Human Intelligence.* Doctoral Dissertation, Yale University, New Haven CT.

Thomson, G. H. (1948). *The Factorial Analysis of Human Ability* (3rd ed.). Boston: Houghton-Mifflin.

Thorndike, E. L. (1913). "Eugenics with Special Reference to Intellect and Character." *Popular Science Monthly, 83,* 127–128.

Thorndike, E. L. (1924). "The Measurement of Intelligence: Present Status." *Psychological Review, 31,* 219–252.

Thorndike, R. (1985). "The Central Role of General Ability in Prediction." *Multivariate Behavior Research, 20,* 241–254.

Thorndike, R. (1975). "Mr.Binet's Test 70 Years Later." *Educational Researcher, 4,* 3–7.

Thurstone, L. L., and Thurstone, J. (1962). *Test of Primary Mental Abilities* (revised edition). Chicago: Chicago Science Research Assoc.

Tuddenham, R. D. (1948). "Soldier Intelligence in World Wars I and II." *American Psychologist, 3,* 54–56.

Turnbull, W. W. (1980). *Test Scores and Family Income: A Response to Charges in the Nader/Nairn Report on ETS.* Princeton, NJ: Educational Testing Service.

Turner, A., and Greenough, W. T. (1985). "Differential Rearing Effects on Rat Visual Cortex Synapses. I: Synaptic and Neuronal Density and Synapses per Neuron." *Brain Research, 329,* 195–203.

Tyler, L. (1965). *The Psychology of Human Differences* (3rd ed.). New York: Appleton-Century-Crofts.

Vaughn, B. E., Block, J. E., and Block, J. (1988). "Parental Agreement on Child-Rearing During Early Childhood and the Psychological Characteristics of Adolescents." *Child Development, 59,* 1020–1033.

Vernon, P. E. (1969). *Intelligence and Cultural Environment.* London: Methuen.

Vernon, P. A., Nador, S., and Kantor, L. (1985). "Reaction Times and Speed of Processing: Their Relationship to Timed and Untimed Measures of Intelligence." *Intelligence, 9,* 357–374.

Vernon, P. A. (1985). "Individual Differences in General Cognitive Ability." In L. C. Hartlage and C. F. Telzrow (eds.), *The Neuropsychology of Individual Differences.* New York: Plenum.

Vernon, P. A. (1986). "The g-Loading of Intelligence Tests and Their Relationship with Reaction Times: A Comment on Ruchella et al." *Intelligence, 10,* 93–100.

Vernon, P. A. (1987). *Speed of Information Processing and Intelligence.* Norwood, NJ: Ablex.

Vygotsky, L. (1962). *Thought and Language.* Cambridge, MA: MIT Press.

Vygotsky, L. (1978). *Mind in Society.* Cambridge, MA: Harvard University Press.

Wachs, T., and Gruen, G. (1982). *Early Experience and Human Development.* New York: Plenum.

Wagner, D. (1984). "Rediscovering Rote: Some Cognitive and Pedagogical Preliminaries." In S. Irvine and J. Berry (eds.), *Human Assessment and Cultural Factors.* New York: Plenum.

Wagner, D. (1978). "Memories of Morocco: The Influence of Age, Schooling, and Environment on Memory." *Cognitive Psychology, 10,* 1–28.

Wagner, D. (1977). "The Ontogeny of the Ponzo Illusion: Effects of Age, Schooling, and Environment." *International Journal of Psychology, 12,* 161–176.

Wagner, R., and Sternberg, R. J. (1985). "Practical Intelligence in Real-World Pursuits: The Role of Tacit Knowledge." *Journal of Personality and Social Psychology, 49,* 436–458.

Wahlsten, Douglas (in press). "Insensitivity of the Analysis of Variance to Heredity-Environment Interaction." *Behavioral and Brain Sciences.*

Wakai, K. (1985). "Cultural Psychology and Education." Unpublished manuscript.

Walker, C. H. (1987). "Relative Importance of Domain Knowledge and Overall Aptitude on Acquisition of Domain-Related Information." *Cognition and Instruction, 4,* 25–42.

Walker, E., and Ceci, S. J. (1985). "Cognitive Performance and Susceptibility to Backward Masking: Data from Clinical Subgroups." Unpublished manuscript. Atlanta GA: Emory University.

Walker, E., and Emory, E. (1985). "Commentary: Interpretive Bias and Behavioral Genetic Research." *Child Development, 56,* 775–778.

Ward, C. S., and Toglia, M. P. (1987). "Effects of Incentive on Young Children's Memory." Paper Presented at the Annual Meeting of the Eastern Psychological Association, April 11, New York City.

Whitely, S. E. (1980). "Latent Trait Models in the Study of Intelligence." *Intelligence, 4,* 97–132.

Whitely, S. E. (1983). "Construct Validity: Construct Representation Versus Nomothetic Span." *Psychological Bulletin, 93,* 179–197.

Whitely, S. E., and Davis, R. (1975). "The Influence of Test Context on Item Difficulty." *Educational and Psychological Measurement, 35,* 51–66. (Cited in Whitely, 1983.)

Whitely, S. E., and Schneider, L. (1981). "Information Structure for Geometric Analogies: A Test Theory Approach." *Educational and Psychological Measurement, 18,* 383–397. (Cited in Whitely, 1983.)

Wilding, J., and Valentine, E. (1988). "Searching for Superior Memories." In M. M. Gruneberg, P. Morris, and P. Sykes (eds.), *Practical Aspects of Memory,* vol. 2. London: Wiley.

Willerman, L. (1979). "Effects of Families on Intellectual Development." *American Psychologist, 34,* 923–929.

Willis, S., and Schaie, K. W. (1986). "Practical Intelligence in Later Adulthood." In R. J. Sternberg and R. Wagner (eds.), *Practical Intelligence: Origins of Competence in the Everyday World.* (pp. 236–270). New York: Cambridge University Press.

Wiseman, S. (1966). "Environmental and Innate Factors and Educational Attainment." In J. Meade and A. S. Parkes (eds.), *Genetic and Environmental Factors in Human Ability* (pp. 64–79). London: Oliver and Boyd.

Witkin, H. A., and Berry, J. (1975). "Psychological Differentiation in Cross-Cultural Perspective." *Journal of Cross-Cultural Psychology, 6,* 322–334.

Woodsworth, R. S. (1941). *Heredity and Environment: A Critical Survey of Recently Published Material on Twins and Foster Children.* New York: Social Science Research Council.

Yarrow, L., Goodwin, M., Manheimer, H., and Milowe, I. (1971). "Infancy Experience and Cognitive and Personality Development at Ten Years." Paper Presented at the Annual Meeting of the Orthopsychiatric Association, Washington, DC, March 20.

Yeates, K., MacPhee, D., Campbell, F., and Ramey, C. (1983). "Maternal IQ and Home Environment as Determinants of Early Childhood Intellectual Competence: A Developmental Analysis." *Developmental Psychology, 19,* 731–739.

Zajonc, R. (1968). "Cognitive Theories in Social Psychology." In G. Lindzey and E. Aronson (eds.), *Handbook of Social Psychology,* vol. 1. Reading, MA: Addison-Wesley.

Zajonc, R., and Bargh, J. (1980). "The Confluence Model: Parameter Estimation for Six Divergent Data Sets on Family Factors and Intelligence." *Intelligence, 4,* 349–361.

Zigler, E. (1988). "The IQ Pendulum." *Readings,* June Issue, 4–9.

Subject Index